Preventing Occupational Disease and Injury

Preventing Occupational Disease and Injury

Edited by

James L. Weeks, ScD
Barry S. Levy, MD, MPH
Gregory R. Wagner, MD

American Public Health Association
1015 Fifteenth Street, NW
Washington, DC 20005

American Public Health Association
1015 Fifteenth Street, NW
Washington, DC 20005-2605

Mohammad N. Akhter, MD, MPH
Executive Director

The opinions expressed in this publication are the editors' and do not necessarily represent the views or official policies of the Association.

ISBN 0-87553-172-5

5M10/91; 5C4/97; 1.5M7/98

Printed in the United States of America

Cover: Dan Winter
Design: Robert Sacheli
Typesetting: Athens Enterprises, Inc., Ijamsville, MD
Set in: Garamond
Printing and Binding: United Book Press, Baltimore, MD

In memory of
Lorin E. Kerr, MD, MSPH
(1909-1991)

Contents

PART 3: Special Topics

Foreword

The mission of public health is to "fulfill society's interest in assuring conditions in which people can be healthy."[1] In the workplace, the mission of occupational health can be viewed similarly. The congressional drafters of the Occupational Safety and Health Act of 1970 specifically embraced the essence of this idea. They declared it the purpose of Congress "to assure so far as possible every working man and woman in the Nation safe and healthful working conditions."[2] Thus, prevention of hazardous workplace conditions is central to the practice of occupational health as a profession.

Despite the separate and somewhat different origins of occupational health, it is clear today that to meet the diverse needs of working Americans, occupational health can and should be considered an integral part of the broader delivery of public health services. In fact, a 1989 survey of 3000 local health departments showed that 23% currently provide occupational health services to their communities.[3] A higher proportion of state health departments offer some programs in occupational health and safety; most emphasize activities in risk assessment designed to identify problems that require intervention, often by some other agency. Some of these health department functions thus complement the risk management functions of state labor departments.

However, limitations in the scope of federal risk management programs and their counterparts in state labor departments render these programs incapable of meeting the many needs of

1. Institute of Medicine. *The Future of Public Health*. Washington, DC: National Academy Press; 1988.
2. Occupational Safety and Health Act. Pub. L. No. 91-596. Sec. (2)(6).
3. National Association of County Health Officials. *National Profile of Local Health Departments*. Washington, DC: NACHO; 1990.

smaller worksites in preventing work-related diseases and injuries. To meet those needs, effective involvement of local and state public health agencies in occupational health is not only desirable but critical. Without a substantial increase in participation by state and local health departments, there is little hope that the needs of small businesses for meaningful preventive services will ever be met.

This book will be particularly valuable to public health professionals working in public health agencies and other settings who need a concise reference for dealing with day-to-day work-related health problems. It has been developed as a practical guide to those who need assistance in an expanding and complex public health field. We feel that this book will become an essential tool for all who seek to provide a "safe and healthful workplace" for workers everywhere.

Edward L. Baker, MD, MPH
J. Donald Millar, MD, DTPH(Lond.)

Preface

Occupational health is a field within public health devoted to the prevention of occupational disease and injury. Public health professionals and others with responsibilities or interest in controlling or eliminating occupational disease and injury need appropriate tools to act effectively. Prevention depends on four fundamental tasks: (1) anticipation of the potential for disease or injury; (2) surveillance: accurate identification, reporting, and recording of occupational disease and injury; (3) analysis of collected data; and (4) control. This book provides information useful for the performance of these tasks.

Part 1 presents an integrated and multidisciplinary approach to prevention, describing anticipation, surveillance, analysis, and control in the workplace. Good public health practice requires a technical understanding of the problem as well as an appreciation of the social, economic, and political context in which the problem occurs. Therefore, we consider not only the technical aspects of disease entities, causes, and prevention measures, but also the social environment of work. Finally, we discuss regulations as well as workers' compensation and the roles of health professionals.

Part 2 provides the basis for occupational disease prevention. A compendium of adverse health outcomes caused in whole or in part by work, part 2 is organized alphabetically and presented in a consistent format: identification, occurrence, causes, pathophysiology, and prevention. Other issues are discussed and further readings listed. Where possible, disease entries are classified by the *International Classification of Diseases,* 9th revision *(ICD-9).* Sentinel Health Events/Occupational (SHE/O) are identified in part 2, and the concept is described in part 3. The author's initials appear at the end of each entry.

Pamela Vossenas located occurrence data for cancer and respiratory disease entries.

Part 3 presents overviews of occupational musculoskeletal and infectious diseases as well as of occupational cancers, and it addresses a number of broad issues that cut across disease entities.

This book is not meant to be used as a guide to the diagnosis and clinical management of occupational disease and injury. A number of texts devoted to that purpose are widely available. Instead, this book speaks to the public health practitioner as well as to the interested and informed lay reader to provide guidance in identifying and preventing occupational disease and injury.

Topics discussed are drawn from those that are documented in the scientific literature. However, many causes of occupational disease and injury are unknown or poorly documented. In the United States, as elsewhere, sufficient knowledge about the magnitude of the problem is lacking. For example, we do not have an accurate count of one of the simplest measures of job safety: the number of workers killed on the job. Measures of the incidence of traumatic injuries are less reliable, and estimates of occupational disease rates are rough approximations at best. Our list of topics, therefore, is inevitably incomplete.

Given the close relationship between occupational and environmental disease and injury, our discussion has relevance to the prevention of diseases caused by nonoccupational hazards. Because exposures are often more intense in the workplace than in the nonworkplace environment, investigation of workers' health may help identify diseases and conditions that can be caused by exposures occurring outside the workplace.

We do not separately address factors in disease and injury that are largely accepted as being matters of personal choice, such as smoking, drug use, alcohol consumption, food intake, and exercise. Such risk factors are discussed in relation to particular outcomes (eg, smoking and chronic lung diseases, alcohol and traumatic injuries). Similarly, we do not title diseases that have both occupational and nonoccupational etiologies as "occupational," because all conditions described in the book are, at times, occupational. Thus, for example, we list asthma, rather than occupational asthma.

The occupational environment is unique for the practice of public health. This uniqueness arises from both the nature of work and the nature of occupational disease and injury. Both are therefore discussed as they relate to public health practice. Government intervention is a regular feature of public health practice and is one important form of intervention for workplace health and safety. Public health agencies at federal, state, and local levels have the power to impose quarantines; close swimming pools; and withdraw contaminated food, consumer items, or drugs from the market to reduce the risk of disease or injury. It is a logical extension of public health practice for government agencies to have similar powers to enter the workplace to prevent occupational disease or injury. This handbook can provide tools for public health consultation and intervention.

In occupational health, as in other fields of public health, the most common dilemmas remain whether, how, and when to intervene. Prudent public health practice requires action to prevent disease and injury, even in the absence of "definitive" data. Delay can result in more harm. Arguments in favor of preventive measures depend on the assessment of the magnitude of exposure and health effects, on the quality of the data on which this assessment has been based, and on the feasibility of implementing controls.

Public health in the workplace holds out a significant promise: Because the workplace is a wholly human invention, virtually everything that happens there can be changed. It is therefore possible to eliminate or to control—to a degree unmatched in other environments—conditions in the workplace that are harmful, thus significantly reducing the incidence of occupational disease and injury. We expect this book to contribute to that effort.

James L. Weeks
Barry S. Levy
Gregory R. Wagner
September 1991

Acknowledgments

Few endeavors in public health, including this book, are conducted by lone individuals. We are grateful to many people who contributed to this book's development. First and foremost, this project would have been impossible without the patient, cheerful, and industrious assistance and support of Jaclyn Alexander, the Director of Publications at the American Public Health Association. Her predecessor, Adrienne Ash, got us off to a good start. Nancy Kaufman, our liaison with the APHA Publications Board, was consistent in enthusiastically providing us sound advice.

The editorial advisory committee—Edward Baker, Charles Levenstein, Barbara Plog, Linda Rudolph, and Joseph Schwerha—collectively and individually helped us to maintain our focus and provided many useful suggestions.

The writing of a significant portion of this book was supported by a grant from the Charles A. Dana Foundation. Stephen Foster at the Dana Foundation made the process painless. Bernard Goldstein at the Robert Wood Johnson Medical School was especially helpful in securing this assistance. Pamela Vossenas, who was employed by APHA, located occurrence data for many entries. Marilyn Butler did much of the word processing; Jane Gold copy edited the manuscript.

Richard Trumka, President, Cecil Roberts, Vice President, and Joseph Main, Administrator of the Occupational Health and Safety Department at the United Mine Workers of America, were generous in allowing time and resources to be devoted to this project.

Manuscripts were reviewed by many people. We thank you: Miriam Alter, Dean Baker, Frederick Bove, James Cone,

Molly Coye, Mark Cullen, Letitia Davis, Raymond Demers, Howard Frumkin, William Halperin, Patricia Hartge, Ellen Hall, Maureen Hatch, Jeff Johnson, Lorin Kerr, Kathleen Kreiss, Joseph LaDou, Philip Landrigan, Richard Levin, Linda Martin, James Merchant, Charles O'Connell, Peter Orris, John Peters, Glenn Pransky, Kathleen Rest, Mark Rosenberg, Scott Schneider, Barbara Silverstein, Gordon Smith, Zena Stein, Eileen Storey, James Taylor, Paul Vinger, Ellen Widess, Richard Youngstrom, and Sheila Zahm.

Finally, we acknowledge the support and encouragement of our wives and children during the six years in which this book was developed.

We also must acknowledge those who, over the years, have contributed to the APHA publication *Control of Communicable Diseases in Man,* which gave us the idea for the format of this book.

JLW
BSL
GRW

Contributors

Miriam J. Alter, MD
Center for Infectious Diseases
Centers for Disease Control
Atlanta, Georgia

Kenneth Arndt, MD
Department of Dermatology
Beth Israel Hospital
Boston, Massachusetts

Robin Baker
Labor Occupational Health Program
University of California
Berkeley, California

Michael Bigby, MD
Department of Dermatology
Beth Israel Hospital
Boston, Massachusetts

Earnest Chick, MD
West Virginia

David Christiani, MD
Occupational Health Program
Harvard School of Public Health
Boston, Massachusetts

Mark Cullen MD
Occupational Health Program
Yale University
New Haven, Connecticut

Stanley Eller, JD
Natural Resources Council of Maine
Augusta, Maine

Robert Feldman, MD
Department of Neurology
Boston University Medical Center
Boston, Massachusetts

Howard Frumkin, MD
Occupational and Environmental Health
Emory University School of Public Health
Atlanta, Georgia

Robert W. Goldmann, MD
St. Luke's Medical Center
Milwaukee, Wisconsin

Bernard D. Goldstein, MD
Department of Environmental and Community Medicine
UMDNJ—Robert Wood Johnson Medical School
Piscataway, New Jersey

Ian R. Greaves, MD
Division of Environmental and Occupational Health
University of Minnesota
Minneapolis, Minnesota

Michael Grey, MD
Department of Community Medicine
University of Connecticut Health Center
Farmington, Connecticut

Ellen M. Hall, PhD
School of Hygiene and Public Health
Johns Hopkins University
Baltimore, Maryland

Deborah R. Henderson, RN
U.S. Food and Drug Administration
Bethesda, Maryland

Austin Henschel, PhD
Cincinnati, Ohio

Jay Himmelstein, MD
Occupational Health Program
University of Massachusetts Medical Center
Worcester, Massachusetts

Michael Hodgson, MD
Department of Community Medicine
University of Connecticut Health Center
Farmington, Connecticut

Howard Hu, MD
Harvard School of Public Health
Cambridge, Massachusetts

Jeffrey V. Johnson, PhD
School of Hygiene and Public Health
Johns Hopkins University
Baltimore, Maryland

James P. Keogh, MD
School of Medicine
University of Maryland
Baltimore, Maryland

Howard Kipen, MD
Department of Environmental and Community Medicine
UMDNJ—Robert Wood Johnson Medical School
Piscataway, New Jersey

Jess F. Kraus, PhD
School of Public Health
University of California at Los Angeles
Los Angeles, California

Kathleen Kreiss, MD
Occupational and Environmental Medicine Division
National Jewish Center for Immunology and
Respiratory Medicine
Denver, Colorado

Barry S. Levy, MD
Program for Environment and Health
Management Sciences for Health
Boston, Massachusetts

J. Davitt MacAteer
Occupational Health Law Center
Shepherdstown, West Virginia

Marian Marbury, ScD
Minnesota Department of Health
Minneapolis, Minnesota

Susan Proctor, MS
Department of Neurology
Boston University Medical Center
Boston, Massachusetts

Laura Punnett, ScD
Department of Work Environment
University of Massachusetts at Lowell
Lowell, Massachusetts

Leon S. Robertson, PhD
Nanlee Associates
New Haven, Connecticut

Zeda F. Rosenberg, ScD
National Institute for Allergy and Infectious Diseases
Bethesda, Maryland

Kenneth D. Rosenman, MD
College of Human Medicine
Michigan State University
East Lansing, Michigan

Annette Mackay Rossignol, ScD
Department of Public Health
Oregon State University
Corvallis, Oregon

Jonathan M. Samet, MD
University of New Mexico School of Medicine
Albuquerque, New Mexico

Paul Shulte, MD
National Institute for Occupational Safety and Health
Cincinnati, Ohio

Dixie E. Snider, Jr., MD, MPH
Center for Prevention Services
Centers for Disease Control
Atlanta, Georgia

Eileen Storey, MD
Department of Community Medicine
University of Connecticut Health Center
Farmington, Connecticut

Alice Suter, PhD
Alice Suter & Associates
Cincinnati, Ohio

Michael J. Thun, MD, MS
American Cancer Society
Atlanta, Georgia

Arthur Upton, MD
New York University Medical Center
New York, New York

Satya Verma, OD
Pennsylvania College of Optometry
Philadelphia, Pennsylvania

Pamela Vossenas, MPH
Laborers' Health and Safety Fund of North America
Washington, DC

Gregory R. Wagner, MD
Division of Respiratory Disease Studies
National Institute for Occupational Safety and Health
Morgantown, West Virginia

James L. Weeks, ScD, CIH
Laborers' Health and Safety Fund of North America
Washington, DC

David H. Wegman, MD
Department of Work Environment
University of Massachusetts at Lowell
Lowell, Massachusetts

Laura Welch, MD
Occupational Medicine Program
George Washington University Medical School
Washington, DC

Ellen Widess, JD
Center for Public Interest Law
University of San Diego
San Francisco, California

Acronyms

ACGIH	American Conference of Governmental Industrial Hygienists
ANSI	American National Standards Institute
ATSDR	Agency for Toxic Substances and Disease Registry (DHHS)
BEI	Biological Exposure Index
BLS	Bureau of Labor Statistics (DOL)
CDC	Centers for Disease Control (USPHS)
dB	decibel
dB(A)	decibel (A scale)
DHHS	Department of Health and Human Services (U.S.)
DOL	Department of Labor (U.S.)
DOT	Department of Transportation (U.S.)
EPA	Environmental Protection Agency (U.S.)
FAA	Federal Aviation Administration (DOT)
FAO	Food and Agricultural Organization (U.N.)
FIFRA	Federal Insecticide, Fungicide, and Rodenticide Act
FRA	Federal Railroad Administration (DOT)
FRC	Federal Railroad Commission (DOT)
ILO	International Labor Organization
MSHA	Mine Safety and Health Administration (DOL)
MSHAct	Mine Safety and Health Act
NCHS	National Center for Health Statistics (DHHS)
NHTSA	National Highway Transportation Safety Administration (DOT)

NIEHS	National Institute of Environmental Health Sciences (DHHS)
NIOSH	National Institute for Occupational Safety and Health (DHHS)
NSC	National Safety Council
OSHA	Occupational Safety and Health Administration (DOL)
OSHAct	Occupational Safety and Health Act
PEL	Permissible Exposure Limit
ppb	parts per billion
ppm	parts per million
REL	Recommended Exposure Limit
SARA	Superfund Amendments and Reauthorization Act
STEL	Short-Term Exposure Limit
TLV	Threshold Limit Value
TWA	Time-Weighted Average
USPHS	U.S. Public Health Service (DHHS)
WHO	World Health Organization (U.N.)

Part 1
Strategies for Prevention

Synopsis

This synopsis summarizes an integrated strategy for the prevention of occupational disease and injury, which is described in more detail in part 1. Like other public health strategies, it is multidisciplinary, needing people with many different skills and interests to work together to achieve common goals. In general, four tasks are pursued—anticipation, surveillance, analysis, and control—each of which is described separately below. This strategy is built around the elementary premise that occupational disease and injury are caused by exposure to hazards on the job. Primary prevention of these outcomes requires controlling exposure. The central public health questions are what occupational exposures need to be controlled, when, and how.

This strategy is illustrated graphically in Figure 1, which shows activities and events in rectangular boxes and analytic tasks in diamonds. The solid lines with arrows represent a logical (though not always chronological) sequence of events. From left to right, this diagram proceeds from exposure to outcome. Thus, industrial hygiene and safety engineering activities concerned with recognition, evaluation, and control of hazardous exposures are on the left; medical activities concerned with diagnosis of conditions and treatment of people are on the right; and epidemiological and toxicological activities requiring consideration of both exposure and outcome are in the center. Activities that are legally regulated, in one way or another, are marked with an asterisk (*).

From top to bottom, the diagram starts with anticipation and proceeds to surveillance, analysis, and control. There are several possible paths to the control activity, indicating the different ways to determine what exposures should be con-

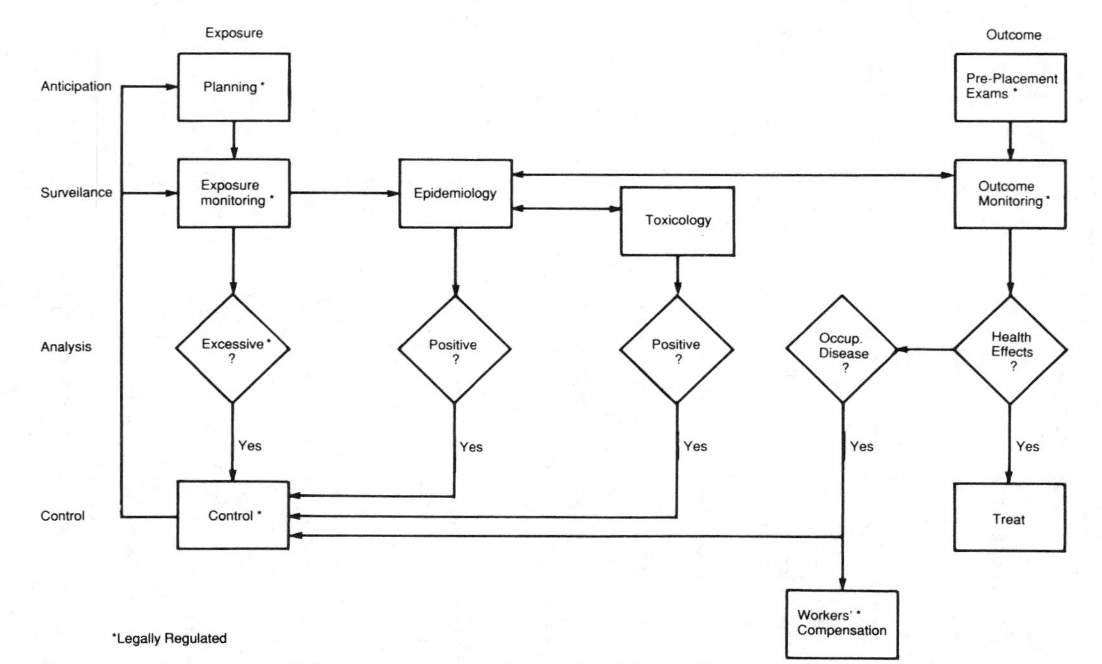

*Legally Regulated

Figure 1.—An integrated strategy for preventing occupational disease and injury

trolled. Education and training are concerned with this entire process of hazard recognition and control.

Chapter 1 of part 1 describes these components of prevention: anticipation of exposures and of outcomes; hazard surveillance and analysis; medical surveillance and analysis; the uses of epidemiology and toxicology; and hazard controls. Chapter 2 describes government regulation, rights to know and sources of information, and worker education and training.

Chapter 1

Anticipation

Whenever changes are made in work, many factors are routinely anticipated, and safety and health hazards should be among them. Safety and health hazards at work can be anticipated before they are manifest, and preventive measures can be implemented before disease or injury occurs. The best time to take preventive action is when new facilities are being built or existing ones modified. Much is known about the hazards of work; this knowledge should be used in planning.

In some instances, workers should be evaluated medically before beginning hazardous work. Medical information can be used in job placement. When exposure to known hazards is anticipated, preexposure health status can be compared with subsequent changes.

Surveillance

General Considerations. Surveillance in the occupational environment includes monitoring both hazardous exposures and health outcomes. It is often desirable to link exposure and outcome data. Surveillance is practiced both at individual workplaces and in settings outside the workplace, such as local, state, or federal government agencies. Information from surveillance is used to detect trends, clusters, associations, and causes of disease and injury and, therefore, to guide prevention activities. Often, the immediate consequence of surveillance is the initiation of more in-depth investigations, but its ultimate purpose is to find disease so that resources for prevention can be allocated rationally.

Hazard Surveillance. Health hazard surveillance consists of both an inventory of hazards and a quantitative assessment of exposure. A hazards inventory should identify each hazard. For purposes of this inventory, there are four classes of hazards: chemical, physical, biological, and psychological. To aid control measures, chemical hazards—liquids, solids, or gases—are classified by the route of absorption. They may be airborne and inhaled, absorbed through the skin or gut, or passed through the placenta. Physical hazards include noise, vibration, radiation, extremes of temperature or barometric pressure, and hazardous motions or postures that may result in cumulative trauma disorders. Biological hazards include infectious agents (bacteria, viruses, parasites, and fungi) and toxins associated with plants or animals. Psychological hazards result from an imbalance of skills and job demands, compounded by lack of personal control or support. Hazards may cause either acute or chronic effects.

Surveillance of hazards that could result in acute traumatic injuries includes assessing sources of energy, vehicles, and circumstances. Sources of energy can be mechanical, electrical, thermal, or chemical. "Vehicles" include stationary or moving machinery; work at heights or depths; and work with electrical circuits, ovens, refrigerators, hot gases, or chemicals. Circumstances include the condition of walking or working surfaces and the necessity of working with or near machinery, power tools, electrical circuits, sources of heat, or acutely toxic chemicals.

Quantitative assessment of exposure consists of measuring the intensity or concentration, the variation in time, and the total duration of exposure, as well as the number of workers exposed. Knowledge of potential effects caused by each hazard should be used to establish priorities for intervention.

If exposure to hazards is excessive, it should be controlled regardless of outcome, and the health of exposed workers should be evaluated. Exposure is considered excessive if it approaches or exceeds established limits, some of which are legally enforceable. Measurement and evaluation of exposure are usually conducted by industrial hygienists, ergonomists, or specialists in injury control. Exposure is also evaluated by non-specialists, such as workers and managers on the job.

Health Effects Surveillance. It is usually physicians, nurses, or other health care providers who evaluate workers' health. Those who can identify occupationally related conditions work in a wide variety of institutional settings—including private offices, hospitals, clinics, health maintenance organizations, chiropractic clinics, poison control centers, and disease or injury registries—and can be allied with virtually any medical specialty or health care profession. To identify a condition as occupationally related, it is essential to acquire an occupational history focused on exposures.

Medical surveillance may be passive, requiring injured or ill people to appear at a medical facility, or it may be active, in which case health professionals recruit and examine workers at risk. Active surveillance is more appropriate for workers at higher risk for disease and injury. It may take many forms, including periodic exams for all workers or for workers exposed to specific hazards, and screening and biological monitoring of selected groups of workers. The precise form of medical monitoring depends on the nature of the effects and occupational exposure—for example, whether effects are acute or chronic, whether screening tests are available, and whether a population of workers is well defined. Often, medical surveillance results in recruitment and examination of similarly exposed workers and in evaluation of their workplaces.

If individuals are sick or injured, if preclinical signs or symptoms are present, or if there are other indicators of adverse effects, there are five different kinds of intervention to initiate: (a) Treat the sick or injured, if appropriate. (b) Obtain an occupational history to aid in diagnosis. (c) Investigate the workplace for possible causes and additional cases. (d) If necessary, control exposure to prevent additional cases. (e) Facilitate workers' compensation if the person became ill or injured by causes " arising . . . out of or in the course of employment," the phrase used in nearly all workers' compensation statutes. In some situations, it may also be necessary to remove the worker from the job if the job presents a continuing and uncorrectable hazard.

Injury surveillance is based on systematic and consistent investigation and recording of incidents resulting in injuries. These records can be used at individual workplaces to manage prevention programs and are available to others outside the workplace who are seeking to prevent workplace injuries.

Analysis

Epidemiological analysis of workers' health is fundamentally useful in this as in other aspects of public health practice. Standard methods and study designs are used, but there are some important differences. In some instances, the specific topics of analysis are unique, such as silicosis or adult lead poisoning. There are also some important logistical issues in studying disease and injury among workers. Exposure must be characterized, and it is important to account for workers being a select and somewhat healthier group than nonworkers, usually by using nonexposed workers as a basis for comparison.

Analysis of occupational hazards is also conducted. This is done in controlled investigations by toxicologists, biochemists, and other scientists, who assess risk by extrapolating results of both animal experiments and in vitro experiments in humans. This analysis may yield information that complements data gained by investigations of humans, or, in some cases, it may be the primary source of information used to establish the need for controls.

Hazard Control

Primary prevention is the preferred means of disease control. In the workplace, this is largely an engineering activity. There is a hierarchy of technical controls derived from a conceptual model that consists of a hazard source, an environment into which the hazard may be released, and the worker. From most to least effective, technical controls include positive engineering that prevents generation of the hazard at its source, environmental controls that are implemented in the work environment, and personal protective devices that individual workers can wear or use. Examples of positive engineering include reducing the use of toxic materials, substituting less toxic materials for more toxic materials, and changing work design to eliminate hazards. Environmental controls include adding ventilation to reduce the concentration of airborne hazards, reducing the duration of exposure, and shielding. Personal protective equipment includes masks, gloves, aprons, and safety devices that workers are required to use. Personal protective devices are less effective than the other two kinds of controls and should be used only as temporary measures or when positive engineering or environmental controls are not feasible.

Chapter 2

Government Regulation

Legal sanctions are necessary but do not sufficiently ensure that the processes described above—anticipation, surveillance, analysis, and control—will be effectively undertaken. Legal requirements are usually carried out by regulatory agencies whose principal jobs are promulgating standards and enforcing them with inspections and penalties for noncompliance. Although most standards limit hazardous exposures, many require employers to conduct medical examinations, monitor exposure, make and maintain records, and report certain incidents. Disputes are adjudicated in administrative, civil, and criminal courts.

Legal sanctions and other means of disease and injury prevention are more likely to be effective when workplaces and workers are organized to control hazards and protect workers' health and safety. Effective organization, which can take many forms, depends on commitments by management and labor unions and on good working relationships between management and labor. Labor-management relationships are governed by collective bargaining contracts and by regulatory and labor relations statutes, which create important rights and responsibilities.

Rights to Know, Sources of Information

Pertinent, accurate, and timely information about the causes of disease and injury is essential to implementing this strategy, as it is for any other public health endeavor. Workers, health professionals, and others have important legal rights to such information, and much of that information has been made available by computerized storage and retrieval systems. But information about workplace hazards is highly specialized and not always accessible because of its economic importance. Such information is often considered proprietary or as trade secrets.

Worker Education and Training

Worker education and training, an essential element of any hazard control program, is most successful when it is part of a comprehensive safety and health program at work. Other elements of such a program are management commitment, systematic workplace analysis, and hazard prevention and control. The objectives of worker education include training workers in basic job skills and preparing workers to take an active role in recognizing and controlling hazards. Training is a continuous process, not an isolated event.

A high-quality training program is conducted through an orderly process. This process consists of assessing needs, gaining support of the target population, defining objectives with the target population, selecting training methods, implementing the training program, and evaluating the program and following up on the training.

Chapter 1: A Public Health Approach to Preventing Occupational Diseases and Injuries

Anticipation

Each year in the United States, between 4400 and 13 000 workers die from acute traumatic occupational injuries. Each year, the same industries—underground mining, construction, agriculture, and logging—report the highest rates of fatal injury, while the construction, manufacturing, and agriculture industries report the highest rates of nonfatal injury. The first and second most common causes of occupational injuries are motor vehicle crashes (on and off the road) and falls from heights. The most common injuries are strains and sprains. Back injuries account for the most lost time. There is, in short, sufficient knowledge to anticipate many fatal and nonfatal occupational injuries.

Similarly, the causes of and control methods for many occupational diseases are well known. Some causes, such as lead poisoning and silicosis, have been known since antiquity. With the development of more sophisticated means of measuring exposure, identifying adverse health effects, and processing large amounts of information, more occupational causes of disease are being identified.

It is thus a rare job that is begun with no knowledge of its potential hazards. The information exists in scientific and technical literature and in the minds of managers, workers, and health and safety specialists. It should be culled from its various sources, integrated, and used to anticipate and control hazards before they become manifest. Occupational diseases and injuries are not an inevitable part of work.

Anticipation can take many forms. These include identifying and controlling hazards in the engineering design of work and workplaces; selecting materials and methods; and training and educating workers, managers, and professional support staff. Other forms include promoting good labor-management relations, forming worksite health and safety committees, and eliminating payment mechanisms that encourage workers to cut corners. Public policy and public visibility support these workplace-based efforts.

Hazard Identification and Control

When decision makers are creating or modifying workplaces, they routinely anticipate many factors, including the price and availability of materials, worker availability and training, work methods, proximity to markets, and transportation. Health and safety hazards should be among these factors. Decision makers should incorporate hazard controls into the planning process so that decisions can be made during a design phase, before commitments are literally cast in concrete.

Preplacement Medical Examinations

Preventive occupational medical services are provided in a variety of settings: industrial medical departments, freestanding urgent care centers, comprehensive primary health care centers, academically affiliated occupational health centers, prepaid and traditional group medical practices, offices of primary care physicians, and elsewhere. A range of personnel—including nurses, technicians, nurse practitioners, physician assistants, and physicians—are involved in providing these services. All these health care professionals have a uniform ethical responsibility to protect the health of the patient. Other considerations, such as source of payment and relationship of the health service provider with his or her employer, should not interfere with this duty.

Many employers require a medical statement of fitness before newly hired workers begin work, despite the questionable value of this practice. Medical personnel should not be involved in the hiring process; however, once an employment determination has been made, preplacement examinations may prove useful to both the employee and the employer.

The primary purpose of the preplacement examination should be to determine the presence of health conditions that are likely to put the worker or co-workers at substantial risk of harm in the workplace. Once this determination is made, it alone should be communicated to the employer. Medical examiners should not be asked to assess who among a group of physically qualified individuals is more qualified.

Knowledge of the workplace is essential to preplacement examinations. Examiners should acquire specific information about any hazardous materials or environments as well as about the physical and psychological demands that will be placed on the worker. Health care providers employed in industrial settings or under contract to employers should visit workplaces for which they examine workers and should supplement firsthand observation with specific information concerning materials and processes used in the enterprise. Exposure standards or guidelines for potentially hazardous materials used in the workplace should be reviewed periodically.

Most people with potential occupational health problems are seen in doctors' offices and emergency rooms. Thus, health care professionals must obtain any information about the work environment through careful questioning of the patient, supplementing it with scientific resources (many of which are referenced in this book).

Medical information about any individual must be considered confidential and should not be communicated to nonmedical personnel, except at the request of the worker or when required by regulation. Such information should be kept in a file separate from any personnel information. Group information used for surveillance purposes, as described below, can and should be collected and analyzed. Results of the analysis should be communicated with personal identifying information removed.

Medical examinations are a necessary part of respiratory protection programs where use of respirators is expected. Although the Occupational Safety and Health Administration (OSHA) rules merely require physician certification that a worker can wear a respirator, these preplacement examinations should be done only after the examiner is knowledgeable about both the characteristics of available respirators and the hazards to which the worker will be exposed. Regulations pertaining to respirator use are available, as are guidelines for performance of medical examinations (Code of Federal Regulations, Ch. 29, Sec. 1910.134, cited as 29 CFR 1910.134).

"Baseline" health data such as audiometry or lung function tests may be of value, depending on the specific hazards potentially found in the workplace and on the program for periodic follow-up to identify and act on changes from baseline. Biological monitoring and medical screening, which are discussed more fully below, should not substitute for exposure control and risk minimization. However, information obtained before placement may be useful in case exposure controls break down.

The preplacement examination also provides an opportunity to discuss work hazards with newly hired workers, review their rights to health protection, and orient them to whatever health services and information resources are available.

Return-to-Work Examinations

Some employers require certification before permitting employees to return to work after an illness or injury. These examinations are similar to preplacement exams in that they require a provider to match what is known about job demands to the worker's physical capacity. The central issue in a return-to-work examination should be whether there has been a temporary or permanent change in health status that would put the worker or co-workers at substantial risk.

At times, a worker is temporarily incapable of returning to his or her usual job but may be able to perform less demanding work until fully recovered. Some employers permit a return to "light duty," but many employers and employee organizations require that a worker be completely capable of performing all job duties before returning to work. The health care provider must be aware of both policies and practices concerning workplace accommodation of impaired workers before certifying that someone is able to return to work. Laws concerning handicap discrimination or workers' compensation may be relevant (*see* Government Regulations below).

"Work-hardening" programs of physical and psychological rehabilitation following absence due to occupational injury or disease have been promoted as a way to reduce compensation costs to employers. Some programs have claimed success in reducing the number of days an injured worker is off work. The most effective programs include comprehensive workstation and job-function analyses and are linked to workplace modification to prevent reinjury. In the absence of such efforts at prevention, these programs should be viewed with caution. It may be difficult to distinguish between positive encouragement to return to work and a coercive effort to discontinue benefits to which an injured worker is entitled.

Return-to-work examinations provide an opportunity to ensure that any workplace conditions that may have caused or contributed to the worker's illness or injury have been corrected and that appropriate protective equipment is in place.

Worksite Health Promotion

Efforts to alter personal health practices, such as tobacco and alcohol use, diet, and physical exercise, aim to improve overall health status and reduce absenteeism. Although proponents justify such health promotion or wellness programs with claims that the programs improve productivity and reduce health care costs, these claims have not been rigorously evaluated. In any case, we do not define health promotion programs as occupational health services; we view them as community health services delivered in the workplace. Because they focus attention and resources on personal health habits rather than on workplace hazards, worksite wellness programs should not be permitted to interfere with the primary prevention of occupational disease and injury.

Surveillance

Surveillance, as defined in the *Dictionary of Epidemiology,* is "the ongoing scrutiny [of the occurrence of disease and injury], generally using methods distinguished by their practicality, uniformity, and frequently their rapidity, rather than by complete accuracy. Its main purpose is to detect changes in trends or distributions in order to initiate investigative or control measures." Surveillance is a fundamental part of public health practice. A surveillance system for occupational disease and injury control should (a) acquire information about hazardous exposures and diseases and injuries (outcomes); (b) analyze this information; and (c) disseminate and interpret it to those who need it. Mere information gathering is not sufficient; the point of surveillance is to prevent disease and injury, not merely to document its occurrence. Thus, a surveillance system must be linked with the capability to investigate further and to intervene to prevent disease or injury.

Surveillance strategies differ for acute and chronic outcomes. For acute conditions, in which the time between exposure and outcome is short, exposures that have caused disease or injury are more easily identified and controlled. For chronic conditions that result from repeated exposure or may be latent for long periods, specific causes are more difficult to identify. Whatever exposures caused disease may have changed by the time the disease becomes clinically apparent. Similarly, if the causes of chronic conditions are present, the outcomes may not occur until later. Moreover, for irreversible conditions, control of chronic hazards is effective at preventing disease or injury only prospectively and cannot correct harm that has already occurred. Therefore, surveillance for some chronic effects should persist even after hazards are controlled.

Identification of hazards that cause chronic effects often requires more knowledge than is available from a single workplace or population of workers. Knowledge of effects from other investigations in addition to manifest disease or injury may establish the need for controls. It is not prudent to wait for the development of chronic disease in any single workplace before reducing exposure that has caused problems elsewhere.

For both acute and chronic conditions, surveillance of exposures and of outcomes can be practiced both inside and outside specific workplaces. At the level of individual workplaces, surveillance involves systematic workplace inspections, measurement and evaluation of exposure, examination of workers, record keeping, and reporting of health effects and exposures. Surveillance at the workplace is thus an essential ingredient for managing a disease and injury prevention program. When practiced in settings apart from individual workplaces—for example, at local, state, or national agencies; hospitals; or disease or injury registries—surveillance can involve acquiring and analyzing data from a wide variety of sources, including employer reports, workers' compensation claims, hospital records, police reports, disease registries, and poison control centers. Regardless of the source of surveillance data, however, most information arises from workplaces, and intervention must eventually focus on workers, employers, and workplaces.

Hazard Surveillance

Workplace Inspections. Inspections may occur under three general circumstances: (a) on a regular basis to identify hazards, (b) immediately after injuries or accidents occur to identify manifest causes, or (c) when someone on the job suspects hazards and then requests an inspection. Inspections may be required by statute or regulation, by some insurance carriers, or by a labor-management contract.

Inspections can be conducted by workers at a job site, health and safety professionals, engineers, or inspectors from outside the workplace, such as regulatory agencies, insurance carriers, parent corporate offices, or union representatives. People on the job have regular and sustained experience with the workplace and are essential witnesses to circumstances surrounding specific accidents. Both employers and employees should be represented to give a balanced view of hazards. Safety engineers bring needed expertise and experience to any inspection. Inspections by people not in daily contact with a job, such as government inspectors, insurance agents, or experts from parent corporations or unions, bring fresh perspective and knowledge of pertinent regulations and guidelines. They may also bring needed incentives in the form of citations, changes in insurance rates, and other penalties or rewards.

Regardless of whether inspections are held because of accidents, they are an important source of data for use in later analysis. Therefore, information should be documented in a consistent form.

Periodic inspections not related to particular incidents should be systematic and custom-made for each workplace or industry. Checklists and a plan to visit every job site are useful. A walk-through survey should follow the flow of work from start to finish, accounting for uses, storage, waste, by-products, disposal of all materials, and maintenance operations. Results should be documented and discussed by a worksite committee. Records of prior inspections, committee meetings and actions, and injuries should be available to monitor performance. The frequency of inspections depends on the degree of hazard. In the high-hazard underground coal mining industry, for example, certain inspections are required by statute prior to each work shift.

Recording of accidents that result in injuries is legally required. Such records may be obtained by OSHA or by workers and their representatives. At a specific workplace, these records are useful surveillance tools. Although they are supposed to be used for both diseases and injuries, these records are inherently more likely to be used for acute conditions such as traumatic injuries and certain acute diseases (eg, dermatitis and acute poisoning) than they are for chronic conditions. Under the Mine Safety and Health Administration (MSHA), records of injuries and accidents are required to be reported (not merely recorded) and are available from MSHA.

Hazards Inventory. A hazards inventory is useful for estimating the potential for exposure. For purposes of an inventory and to facilitate evaluating, designing, and implementing controls, hazards can be cross-classified by whether they present potential for acute or chronic outcome and by type of hazard—that is, chemical, physical, biological, or psychological.

Chemical hazards are classified as solids, liquids, or gases that may enter the body by inhalation, ingestion, or absorption through the skin and that may enter a fetus through the placenta. Harmful effects depend on the nature of substances, their concentration in the workplace, and the duration of exposure.

Airborne hazards—aerosols (dusts, mists, and fumes), gases, and vapors—are very common. They are usually inhaled, but some aerosols may be absorbed through the skin; ingested after having been cleared from the lungs; or ingested in contaminated food, drink, or smoking materials. Hazards that become airborne are usually controlled by ventilation and other means described in more detail below.

Aerosols also should be characterized by particle size. Smaller particles (less than 5 to 10 micrometers in diameter) are more likely to be deposited in the regions of the lung where gas exchange occurs (the terminal bronchioles and alveoli) and thus are called respirable particles. Clearance of particles deposited in the respirable region is relatively slow. Respirable particles, therefore, are more likely than larger particles to be retained and absorbed, and thus to injure the lung or cause systemic disease. Both respirable and larger particles may also be deposited in the lung's airways, from which clearance is swifter. Repetitive deposition of aerosols in the airways may result in chronic airways irritation.

The relative solubility of airborne hazards also affects the site of deposition and injury or absorption in the lung. Aerosols, gases, or vapors that are highly soluble in water are more likely to be absorbed in the moist lining of the upper airways than those that are less soluble.

Some airborne materials may also be fire or explosion hazards. These include flammable gases; mists and vapors of flammable liquids; aerosols of organic dusts, such as coal, grain, and sugar; and some metal dusts, such as aluminum and magnesium.

Physical hazards include noise, ionizing and nonionizing radiation, extremes of temperature and barometric pressure, hazardous motions or postures that can result in cumulative trauma disorders (ergonomic hazards), and circumstances that can result in acute traumatic injury. These hazards are described in more detail in part 2 (*see* Hearing Loss, Noise-Induced; Radiation, Ionizing and Nonionizing; Hyperbaric Injury; Injuries, Fatal and Nonfatal; and entries on musculoskeletal disorders, such as Carpal Tunnel Syndrome, Tendinitis, Peripheral Nerve Entrapment Syndromes, and Low Back Pain Syndrome).

Biological hazards include infectious microorganisms and plant or animal toxins. These are described in entries on human immunodeficiency virus (HIV) infection; tuberculosis; silicosis; hepatitis A, B, and non-A, non-B; and contact dermatitis (irritant and allergic) in part 2.

Psychological hazards result from a complex interplay of job demands, skills, decision-making latitude, personal control of work, and social interactions. Stress and collective stress disorder are also discussed in part 2.

Almost every product used in commerce in the U.S. has a Material Safety Data Sheet (MSDS) that is supposed to describe its acute and chronic health effects, physical properties, flammability, and relevant emergency procedures, among other things. A health professional not on the job may obtain MSDSs from a worker's employer. Federal regulations (29 CFR 1910.1200) and some state and local statutes provide specific rights of access to this information for health professionals, workers, and their representatives. The quality of the information on these sheets varies considerably. Some information may be designated a trade secret. Under OSHA's hazard communication standard, health professionals with a specific need to know may obtain trade secret information directly from the manufacturer.

Industry-specific information about occupational health hazards has been collected by the National Institute for Occupational Safety and Health (NIOSH) in the National Occupational Exposure Survey for both general industry and mining. This survey is an inventory of toxic materials used by employers; quantitative assessments of exposure are not made.

Exposure Measurements. Quantitative assessment of exposure is necessary for more exact hazard assessment. Measurement of exposure is appropriate whenever hazards are suspected or reasonably predictable based on a hazards inventory combined with an assessment of work practices. The purposes of measuring exposure include (a) identifying hazards in order to implement controls, (b) evaluating controls, (c) determining compliance with standards, and (d) assessing exposure for epidemiologic research.

Measurement of exposure should take account of variables inherent in any occupational setting. The major variables include work practices, level and intensity of production, use of industrial hygiene controls, and accidents. Like any scientific measurement, exposure measurements vary by the analytical methods. A sampling strategy should be able to account for major variables by an appropriate selection of the type, number, and timing of samples. Any written report should describe these features.

Measurement methods depend on the particular hazard and circumstances under which exposure occurs. Airborne hazards are typically measured by analyzing a sample of air. An air sample may consider exposure either at a point in time (a "grab sample") or averaged over an appropriate time period (a time-weighted average, or TWA), such as a work shift.

Physical hazards, such as noise, heat, ionizing and non-ionizing radiation, and barometric pressure, are measured by specialized methods unique to each hazard. Methods are discussed in part 2.

Work-related skin or eye problems or other acute irritant reactions typically result from work practices involving direct contact of a harmful substance with the skin, eyes, or mucosal membranes. In such cases, the specific work practices need to be assessed. Some skin problems result from systemic poisoning, and some toxic substances may be absorbed directly through the skin.

Exposure measurement for causes of some infectious occupational diseases requires standard methods of infectious disease control (*see* Benenson AS, *Control of Communicable Diseases in Man,* 15th ed., APHA, 1990). Although exposures to some biological hazards are most common in the health care industry, they may also occur in agriculture, among emergency response workers, and in other industries and occupations. Indoor air quality of offices and schools may also be affected by bioaerosols (*see* Building-Related Illness).

Monitoring and measuring some occupational hazards are required by law (*see* Table 1). Employers under OSHA's jurisdiction are required to maintain exposure records and make them available to OSHA or to workers or their representatives. When analyzed collectively, surveillance data can be used to identify potential hazards and, in some instances, to estimate exposure for individuals or populations over the period these regulations have been in effect. Exposure monitoring data for the mining industry are available from MSHA.

Table 1.—Selected Exposure Monitoring Regulations

Hazard	Federal Regulation
General Industry:	
acrylonitrile	29 CFR 1910.1045
asbestos	29 CFR 1910.1001
arsenic	29 CFR 1910.1018
benzene	29 CFR 1910.1028
coke oven emissions	29 CFR 1920.1029
ethylene oxide	29 CFR 1910.1047
formaldehyde	29 CFR 1910.1048
lead	29 CFR 1910.1025
vinyl chloride	29 CFR 1910.1017
Mining:	
In coal mines:	
respirable dust and quartz	30 CFR 70.201–220 30 CFR 71.201–220
In non-coal mines:	30 CFR 56.5002

Standards and Exposure Limits. Measurements are often evaluated by comparison with exposure limits and standards. Some health standards are legal requirements; others are not. Legal standards are set by federal, state, or local governmental regulatory agencies. If exposure exceeds a legal standard, the employer is then legally obligated to reduce it. If exposure exceeds a standard that is not a legal requirement, a public health need may require reduction of exposure. The principal federal government agencies that set standards pertaining to occupational hazards are OSHA and MSHA, both in the Department of Labor. Both OSHA and MSHA refer to legally enforceable exposure limits for toxic substances or other hazards as permissible exposure limits (PELs). There are limits on TWA exposure, short-term exposure limits (STELs), ceilings (C), and, for some hazards, conditions that might result in skin absorption. There are PELs for about 600 chemical substances commonly found at workplaces.

As described below (*see* Government Regulation), there are limits in the scope and completeness of standards. Therefore, there are inevitably instances when health effects are associated with work with little or no evidence that any exposure limits have been violated. In these situations, it may be necessary to investigate working conditions in detail to identify the cause, or causes, of workers' ill health. State or local public health agencies may conduct small- or large-scale epidemiological investigations as described below. Both the OSHAct and the MSHAct provide for health hazard evaluations (HHEs), which are usually focused investigations of health effects and occupational exposures. These investigations are conducted by NIOSH and can be requested by employers or by workers or their representatives. The names of requestors may be kept confidential. To request an HHE, call 1-800-35NIOSH.

Medical Surveillance

The purpose of medical surveillance is to promote prevention by identifying the distribution and trends in the occurrence of disease and injury in populations. Outcome measures may range from asymptomatic signs and symptoms to well-defined diseases. Most often, medical surveillance results in secondary rather than primary prevention, because it can identify only individuals already affected by occupational exposures. When combined with hazards surveillance, however, analysis of medical surveillance data can complement primary prevention efforts by also identifying hazards. Surveillance is distinguished from screening (described below) by its concern with a target population; screening is more concerned with individuals.

Medical surveillance may be conducted at specific workplaces or in community settings. Employer medical departments usually have easy access to workers and to workplaces and thus are well situated to detect acute conditions, monitor active workers' health regularly, link medical with exposure surveillance, and implement programs for the early detection and prevention of occupational conditions. However, conditions caused by multiple exposures at different workplaces and chronic conditions that may not appear until after retirement are harder to detect by workplace-based surveillance programs. Moreover, workplace-based medical departments serve only a small minority of the work force, usually those situated in large or exceptionally high-risk workplaces.

Therefore, community-based surveillance efforts are also appropriate. Such efforts may be designed and implemented by government agencies, hospitals, or clinics, or they may be based on networks of health care providers. For example, the SENSOR (Sentinel Event Notification System for Occupational Risk) program developed by NIOSH is a small-scale model of disease surveillance targeted on selected disease outcomes with well-defined clinical features. This program depends on reports of cases from a select group of health care providers. A state agency collects and analyzes individual reports, and preventive measures are taken at worksites in relation to individual workers and their co-workers.

Medical surveillance may be active, in which case populations of workers are selected, recruited, and examined; or it may be passive, requiring affected individuals to appear at a medical facility. Passive surveillance usually detects only symptomatic disease and cannot be relied on to uncover conditions earlier in their natural history. It also requires that health professionals be able to recognize the effects of occupational exposures in individuals in clinical settings. However, because the effects of many occupational exposures resemble effects of nonoccupational exposure, passive surveillance cannot be relied on to detect many occupation-related diseases without also assessing occupational exposure.

Active surveillance for occupation-related conditions requires assessment of exposure prior to conducting surveillance in order to define and select a population. Workers selected for active surveillance are usually at high risk for disease or injury. Selection criteria include assessment of current or past exposure based on measurements (if possible), employment history, or similar parameters. Because health effects depend on the identity of hazards, assessment of exposure is also required prior to selecting medical procedures. Thus, for example, workers exposed to lead, silica, or noise should receive regular laboratory tests for blood lead levels, chest x-rays, or audiometric examinations, respectively. For a limited number of diseases discussed in part 2, conventions for case definition that could be used for surveillance are included or referenced. Certain conditions are considered sentinel health events/occupational (SHE/O) (*see* part 3), and are indicated as such in the individual entries in part 2.

The occupational contribution to illness often cannot be recognized when individuals are considered in isolation from similarly exposed workers. Thus, results of surveillance should be analyzed in populations classified by exposure. Basic epidemiological methods are used to analyze such data, so that when aggregate findings are linked with assessment of occupational exposure, the results can be used to identify and evaluate hazards.

Intervention is an essential part of surveillance. If surveillance reveals the presence of SHE/O, identifies sick or injured people, or identifies people with preclinical signs or symptoms or other effects, there are four different kinds of intervention to consider: (a) assure access to appropriate treatment, if possible; (b) investigate the workplace for possible additional cases and causes; (c) control causative or excessive exposure; and (d) arrange for workers' compensation, depending on state statutory requirements.

Medical Removal. If there is a continuing and uncorrectable hazard, it may also be necessary to remove the worker from the job, either temporarily or permanently. This is a more controversial issue. Temporary or permanent removal of the worker could cause serious income and other losses and should be considered only when there is no other option. Even if removal is appropriate, it may not be practical for a worker to find a suitable job with the same employer. As a general rule, it is better to control workplace hazards than to subject the worker to unnecessary dislocation.

For medical removal protection to be effective, pay and employment benefits must be maintained for workers moved from high-risk to lower-risk work. Job retraining and income maintenance are critical. The issues involved in medical removal programs are complex, involving not only health concerns, but also a web of interrelationships between a worker, the employer, co-workers, union contract provisions, and the workplace. Statutes and regulations include OSHA and MSHA regulations, labor relations case law, and prohibitions against discrimination against handicapped workers. Clinicians unfamiliar with these issues must exercise caution in recommending medical removal where there is no explicit assurance of protection for the worker. Otherwise, jobs may be jeopardized unwittingly. Nevertheless, for some conditions such as asthma or heart disease, medical removal may be the only effective way to preserve the health of the affected worker.

OSHA or MSHA regulations provide for medical removal protection only for workers with elevated blood lead levels (29 CFR 1910.1025) and for underground coal miners with chest x-rays that are positive for coal workers' pneumoconiosis (30 CFR 90). For workers with elevated blood lead levels, removal may be temporary because the condition is reversible. Underground coal miners with irreversible pneumoconiosis may choose to transfer at any time to a job with reduced exposure to dust (*see* Lead Poisoning, Coal Workers' Pneumoconiosis).

Biological Monitoring. Biological monitoring is sampling and analysis of biological material (eg, blood, urine, or exhaled air) to estimate exposure to a hazardous substance. For example, blood lead levels are routinely monitored in exposed people to determine the amount of lead absorbed. For some substances, guidelines have been established for additional workplace investigation, removal from exposure, and possible treatment. Within the regulatory sphere, results of specific assays, like other standards, represent compromises based on considerations of feasibility as well as on scientific information.

Biological monitoring is a form of bioassay using exposed workers rather than laboratory animals. Data from such examinations may be used to establish levels of exposure in formal epidemiological investigations and may, at times, be an important complement to industrial hygiene. Nevertheless, biological monitoring should never be a substitute for effective environmental monitoring of toxic exposures.

Screening. The purpose of medical screening is early detection of disease or conditions for which treatment can successfully affect morbidity or mortality. Screening is a form of secondary prevention. Guidelines for screening have been developed for the nonoccupational setting, and, more recently, screening has been used and promoted for preventing occupational disease. Although similar in some ways, there are some important differences in the purposes and practices of screening in the occupational setting compared with that in the nonoccupational setting. Guidelines for nonoccupational screening are as follows:

Disease characteristics:

- causes significant morbidity/mortality
- can be identified early
- responds to acceptable, available, and effective intervention/treatment
- is prevalent in the targeted population

Test characteristics:

- is acceptable to those at risk for disease
- has acceptable sensitivity, specificity, and predictive value
- is available at reasonable cost

In the occupational setting, these principles may be modified because of special opportunities and responsibilities. Because a population with a well-defined but rare exposure may exist in the workplace, screening tests that are too costly to use elsewhere may be appropriate. Access to such a population may also justify departure from the principles appropriate for more general screening programs.

Other guidelines for screening for occupational disease are the same as those for screening in the community setting. Screening programs should not be established without adequate justification. Each test or examination should be able to stand on its own merits. The adverse consequences of misclassifying an individual in a screening program should be considered carefully. Screening tests that yield positive results should trigger valid confirmatory examinations to which the worker must have access. Confidentiality of test results must be maintained so that problems (eg, those concerning employment or insurability) do not arise as a result of false-positive screening test results that are not confirmed by additional examination.

The purpose of screening for occupational disease also can be broader. In the community setting, the purpose is case finding that leads to secondary prevention; in the occupational setting, the purpose can include hazard finding that would lead to primary prevention. If a screened population is carefully selected and recruited, if participation is high, and if exposures are characterized, screening results can be analyzed collectively to identify occupational causes.

An important issue in the use of screening tests is whether individuals are correctly classified as positive or negative. *Sensitivity, specificity,* and *predictive value* are standard terms used to evaluate screening tests, and it is important to understand their meaning. The sensitivity of a screening test is a measure of its ability to classify correctly people who would be diagnosed with the condition; it is the proportion of people with a condition who test positive (*see* Table 2). Specificity is the analogous term for people whose tests yield negative results; it is the

Table 2.—Evaluating a Screening Test

		True Diagnosis +	True Diagnosis −	
Screening results	+	a	b	predictive value (+) = a/(a+b)
	−	c	d	predictive value (−) = d/(c+d)

true positives = a	sensitivity = a/(a+c)
false positives = b	specificity = d/(b+d)
false negatives = c	prevalence rate = (a+c)/(a+b+c+d)
true negatives = d	

proportion of people who do not have the condition who test negative.

In practice, the utility of a screening test depends on the prevalence rate of the condition in the screened population. If this rate is low, even screening tests with high sensitivity and specificity may produce a large number of false positives. For example, if the sensitivity and specificity of a screening test are both 95% and the prevalence rate of a targeted condition is 1%, more than 90% of those who screen positive would be false positives. This effect is measured by the predictive value of a test: the proportion of people who screen positive or negative who are correctly classified (*see* Table 2).

Unfortunately, for many tests, data do not yet exist to permit determination of these specific values, and there is no consensus concerning appropriate values before implementing screening programs. In some instances, the importance of the condition and the consequences of failing to identify it must be weighed against a test's presumed predictive value.

Data Sources

Of the several data sources for information on occupational diseases or injury, none provides a comprehensive account, and their utility varies by outcome. Sources include direct medical surveys, hospital discharge summaries, physician reports of illnesses, laboratory reports, workers' compensation case files, employer reporting, death certificates, disease registries, medical examiner reports, reports to poison control centers, and cases of some infectious diseases reported to the CDC.

Occupational Diseases. Data sources for surveillance of occupational diseases are undeveloped in the U.S. Employers are required to keep records of occupational injuries and illnesses and to make these available to OSHA and to workers or their representatives. However, employer reporting is more sensitive to acute conditions with single causes than to chronic and multifactor conditions. The Bureau of Labor Statistics (BLS) summarizes these employer reports annually by surveying a sample of employers from which industry-specific incidence rates are estimated. In 1989, for example, BLS reported 190 000 cases of occupational illnesses (up from 125 000 in 1984). However, estimates from other sources place the number at more than twice this.

Because employer reports are required by OSHA regulations, these numbers wax and wane with enforcement policies. Over the past decade, for example, there have been large and well-publicized increases in OSHA penalties for failure to record repetitive trauma disorders. During the same time, from 1984 to 1987, the annual estimated incidence rate of these disorders increased from 5.1 to 10.0 cases per 10 000 workers. OSHA and MSHA also require medical surveillance for workers exposed to a limited number of hazards, including asbestos, cotton dust, lead, and coal mine dust. Records generated by these efforts are required to be made available to OSHA or MSHA, to individual workers, or to labor unions that represent the workers. NIOSH maintains results of medical surveillance of coal miners.

Workers' compensation systems are a poor source of data for chronic occupational conditions. Records for acute conditions, such as many dermatologic disorders, or for conditions with unambiguous occupational causes are more reliable. This disparity arises because it is difficult to satisfy statutory requirements that a chronic condition arose from work. As a result, many workers are discouraged from filing, and cases are not counted.

Poison control centers also have data on acute poisonings, some of which are occupational. Some infectious diseases that may result from occupational exposure (eg, hepatitis B virus infection, legionella pneumonia, tuberculosis, and human immunodeficiency virus infection) are reportable to the CDC, which is a source of surveillance information for these conditions.

Occupational Injuries. Sources of data for injuries can include employer reports required by regulatory agencies (OSHA, MSHA) or workers' compensation plans, death certificates for traumatic fatalities, reports of investigations of fatalities by state or federal agencies, hospital emergency department reports or discharge summaries, police reports of automobile crashes or assaults, Coast Guard reports of boating accidents, or FAA reports of airplane crashes.

Under OSHA regulations (29 CFR 1904), employers with 11 or more employees are required to maintain records on individual injuries and to enter them in a log in a standard form (OSHA 200). For reporting purposes, an "injury" is carefully defined. These records must be made available to workers, their unions, and OSHA, if requested. Because injury data required by OSHA do not include hours worked, it is impossible to use them to calculate injury rates. Under MSHA regulations (30 CFR 50), however, mine operators are required to report (not merely record) all injuries and hours worked for each mine. Thus, it is possible to calculate mine-specific rates, which are useful for focusing intervention efforts. These data may be obtained from MSHA.

As mentioned above, the Bureau of Labor Statistics conducts an annual survey of a sample of employers' reported injuries and illnesses (as recorded on the OSHA 101 form) and hours worked. It is used to estimate annual injury incidence rates classified by four-digit Standard Industrial Classification (SIC) codes. This data system is the most comprehensive surveillance system in the U.S. and provides a uniform basis for measuring trends and for intra-industry comparison of injury rates. However, according to a panel convened by the National Research Council, the OSHA/BLS system is currently inadequate for providing OSHA with data necessary for effective disease and injury prevention. Because the system is based on a sample, establishment-specific rates are not available. However, regional data without the identity of specific employers are available from BLS.

The National Center for Health Statistics (NCHS) also collects data by interviewing individuals in households in its Health Interview Survey. This information is used to estimate industry- and occupation-specific injury rates. These data are available from NCHS. BLS and NCHS injury incidence rates are compared under Injuries, Nonfatal, in part 2.

Workers' compensation case records are managed by each state according to its workers' compensation plan. Statutory requirements for eligibility differ from state to state. The primary purpose of workers' compensation is to compensate workers whose injury or illness arose out of and in the course of employment. It is designed neither for surveillance nor for prevention. Nevertheless, data from state compensation systems can be used for surveillance, provided limitations are recognized. To systematize data, BLS has created a supplementary data system (SDS) that acquires and summarizes workers' compensation data from about half the states. The National Center for Compensation Insurance in Boca Raton, Florida, obtains similar data from participating states.

Occupational Fatalities. Fatalities (and serious nonfatal injuries resulting in loss of consciousness or the hospitalization of five or more workers) occurring among employers under OSHA's jurisdiction are required to be reported to and investigated by OSHA. OSHA has records of most such fatalities, but many are not reported or investigated. Fatalities among miners are similarly required to be reported to and investigated by MSHA.

Fatalities may also be counted by examining the death certificates maintained by each state that indicate death due to external causes. Based on death certificates, NIOSH has compiled data on traumatic occupational deaths in the National Traumatic Occupational Fatality data set, which includes state, occupation, age, race, and sex-specific census data.

NIOSH also investigates fatalities under its Fatal Accident Circumstances and Epidemiology (FACE) program. This program is designed to provide systematic and detailed investigations intended to identify causes rather than to give a comprehensive accounting. It is currently limited to investigating fatalities from falls, confined spaces, and electrocutions.

State or local government agencies also record fatalities. For instance, medical examiners in some states are either required or authorized to investigate traumatic fatalities, including occupational fatalities, often for forensic purposes. Police reports can be used to identify fatalities from motor vehicle crashes or assaults that may have occurred in the course of work.

Transportation fatalities are recorded by agencies in the U.S. Department of Transportation. The Coast Guard records boating-related injuries or fatalities; the FAA records air-transport accidents, near-misses, and fatalities; the FRA records railroad fatalities; and the NHTSA compiles and analyzes traffic fatalities based on local police reports. Work-relatedness is not always noted but can usually be deduced from incident narratives or by other means.

No single source has been shown to provide a complete count of fatal occupational injuries in the United States.

Work Force Demographics. To calculate rates of occurrence of occupational diseases or injuries, it is necessary to acquire information about the population of workers within which cases occur. Sources can include BLS surveys, employer records, union membership lists, or census data. Because these data sources are not equivalent, it is essential to use rates based on the same denominator when making comparisons.

BLS collects information about the labor force for other purposes; such information can be used as denominators. The Employment and Wages Program (ES-202), a quarterly collection of information on employment and wages by SIC codes from state employment security departments, is a virtual census of nonagricultural employees. Detailed data for each state are available from BLS. These data can be used to calculate state- or industry-specific injury or disease rates. BLS also manages the Occupational Employment Statistics Survey to obtain periodic information by occupation collected by state agencies from a sample of employers. Data about occupation and industry can then be cross-classified to estimate the distribution of jobs within industries. In its 10-year census, the U.S. Bureau of the Census collects information about occupation, age, sex, and race, making it possible to calculate rates adjusted by these variables for counties, states, or the nation. Depending on the data source, rates can be calculated based on the population of workers in the nation or by industry, employer, or job. The more specific the data source, the more useful it can be for identifying control measures.

Calculation of disease or injury rates need not be limited to information about the population or potential victims of injury but also can include information about hazards. Such information is often gathered in transportation-related injuries (eg, the age, size, and number of vehicles). It can also be gathered for injuries in other settings. In crane accidents, for example, information could include an accounting of the number of different types and ages of crane in use in the construction industry. Sources for such information vary but might include industry trade groups and equipment manufacturers.

Analysis

Systematic analysis is required to answer the key question of whether to intervene to control hazards. As illustrated in Figure 1, there are four conditions that require control: (a) if exposure is excessive, (b) if epidemiological or (c) toxicological analysis demonstrates a positive relationship between exposure and outcome, or (d) if there are manifest occupational diseases. We have already discussed (a) and (d) above. Below, we describe selected issues in epidemiological and toxicological investigations and discuss control methods.

Epidemiological and toxicological studies are occasionally introduced in litigation and are subject to legal scrutiny when regulatory agencies set standards and exposure limits. Consequently, it is important to be familiar with legal evaluation of epidemiological and other forms of scientific analysis. These are discussed briefly at the conclusion of this section. (In chapter 2, we discuss government regulations, rights to know and sources of information, and worker education as important factors affecting our ability to control hazards.)

Epidemiology

Epidemiological investigations of occupational disease and injury are used for decision making in much the same way as they are in other aspects of public health practice. The purposes of occupational epidemiology are (a) to identify and assess causes of disease and injury and, thereby, (b) to identify opportunities for prevention, (c) to evaluate or determine exposure limits, and (d) to evaluate control measures.

The methods and problems in occupational epidemiology are the same as those in other public health endeavors, but there are some important differences. Basic issues in epidemiology common to public health in general and to occupational disease and injury prevention in particular include designing studies, measuring disease or injury occurrence rates, controlling bias and confounding, assessing statistical precision and validity, measuring effects, and assessing causes. These issues are discussed extensively in texts in the field, and readers are urged to consult them. Issues in occupational epidemiology discussed here include (a) characteristics of working populations, (b) measurement and classification of worker exposures, and (c) measurements in occupational injury epidemiology. We also discuss criteria for concluding that exposure(s) "caused" disease.

Characteristics of Working Populations. For an epidemiological investigation of a population of workers to be conducted, two practical conditions must be met. First, a population must be available to study, and second, exposure must be characterized. For conclusions to be scientifically valid, the population has to be large enough for stable estimates of disease occurrence to be made (ie, large enough to yield satisfactory statistical power), and it must be accessible.

Because of practical limits on the availability of sufficiently large populations for study, the occupational hazards that epidemiologists study are often those affecting workers in larger establishments or in industries with low turnover and who are accessible. Epidemiologists are less likely to investigate workplaces that are small and widely dispersed and where labor turnover is high, because such places present significant logistical problems. Such workplaces are often hazardous and may employ racial or national minorities.

Active working populations are also select groups. Participation in the labor force is restricted for some people who are handicapped or ill; other people are selected out when they become ill. These selection processes result in the so-called healthy worker effect, in which workers as a group are somewhat healthier than nonworkers. Therefore, it is often necessary to select comparison populations also composed of workers and to find former or absent workers and determine their health status.

Occasionally, clusters of disease are investigated in occupational groups. Whether such diseases are caused by occupational exposures can be determined if the disease is SHE/O, is very rare (such as mesothelioma or hepatic angiosarcoma), or almost always arises from occupational exposure (such as pneumoconiosis or adult lead poisoning). Often, however, occupational causality may not be confirmed, because the cluster had not been identified in a rigorous manner. Even with a completely random distribution of disease or injury, clusters will appear by chance, and so, it is possible that a cluster being investigated is a chance event. Following identification and investigation of a cluster, then, it may be necessary to validate observations with a rigorously designed investigation. The CDC has recently published guidelines for investigating clusters.

Exposure Measurement and Classification. Because a fundamental purpose of occupational epidemiology is to identify possible occupational causes of disease or injury, it is important to characterize these causes. This is done by assessing exposure. Exposure refers to the presence of a hazard in the workplace and is not equivalent to "dose." Dose refers to the concentration of a biologically active toxin at its target organ. Theoretically, dose can be estimated from exposure if sufficient information is known about absorption, clearance, and metabolism. In most instances, however, such information is not known, so we refer only to exposure.

Exposure assessment is based on measuring the *concentration* or intensity of a hazard (employing standard industrial hygiene methods discussed above) and the *time* course of exposure. Ideally, exposure assessment includes quantitative measurements of concentration for each individual in a study over a relevant period. The period can be very short—such as the time surrounding an incident resulting in acute effects—or much longer—such as an individual's working lifetime.

For example, in evaluating the causes of carbon monoxide poisoning, it would be useful to estimate the concentration of carbon monoxide (in parts per million, ppm) to which workers have been exposed for the hours immediately prior to manifest cases. Because cigarette smoking is a source of carbon monoxide, the smoking status (ie, smokers versus nonsmokers) of individuals should be considered also. But to assess effects of exposure to crystalline silica, which is primarily a chronic hazard, it would be necessary to estimate the concentration of exposure (in milligrams per cubic meter [mg/m^3] or a surrogate) for many years.

However, complete information for measuring concentration or time of exposure is rarely available. Thus, exposure often must be inferred by using surrogate measures. Some measures of concentration, in approximate rank order of the quality of information, include personal exposure measurements, area measurements, quantity of a substance used on an individual's job, presence of a substance on a job or in a particular department or division of a workplace, employment in a particular workplace, employment in an industry, and residence in a census tract with a large concentration of workers with the potential for a common exposure. Measures of the time of exposure may include, in rank order, time in particular phases of work (eg, for a painter, time spent spraying, using a brush, mixing, or climbing a scaffold); duration of employment at a job, in a department or division, at a workplace, or in an industry; and any previous employment at a job or industry.

Measuring exposure for chronic health hazards is a prominent problem. Obviously, one cannot measure today exposure that occurred in years past. But exposure records may be held by an employer, the employer's compensation insurance carrier, a government regulatory agency, or others. In the absence of such records, a detailed work history is useful, including tenure in specific jobs and a detailed description of job responsibilities, substances used, by-products encountered, and accidents that occurred. The production and industrial hygiene history of a workplace—what was produced when and how, with what ingredients, and under what industrial hygiene controls—is useful for establishing the potential for exposure. Changes in production methods or materials should be dated; if there have been no changes, past exposure may be inferred from present exposure. Documenting a work history to estimate exposure to health hazards can be done by interviewing workers or their spouses, fellow workers, or supervisors. Industrial hygienists can estimate exposure based on a work history. Time is also a factor if there is a period of latency in the natural history of the disease outcome. For a particular exposure to cause such a disease, exposure must precede outcome; consequently, when such a disease is being studied, it is important to document when exposure occurred as well as how long it lasted.

Once exposure has been estimated, individuals in a study can be classified. In the absence of quantitative classification by a continuous scale (eg, gram-hour/m^3), ordinal (eg, high, medium, and low) and dichotomous (exposed versus not exposed) scales are often used. Workers can also be classified by occupation or industry. These scales are listed in their approximate rank order by the quality of information (ie, the most informative scale is continuous, followed by ordinal, dichotomous, and then classification by occupation and industry). The potential problem with using poorer-quality measures for exposure is misclassification of study subjects and the consequent introduction of bias in assessing results.

Workers are commonly exposed to multiple occupational and nonoccupational hazards during their working lives and afterward. However, assessment of the interactions of multiple exposures is complex, and analytic methods are undeveloped with a few exceptions. Effects may be synergistic (as with the risk of lung cancer associated with cigarette smoking and asbestos exposure) or additive (as with the risk of chronic lung disease associated with smoking and coal mine dust exposure).

When subjects have been classified by both health status and exposures, incidence or prevalence rates are calculated and analyzed in relation to possible causes. Standard epidemiological methods are used.

Measurements in Occupational Injury Epidemiology. Epidemiological analysis of occupational injuries uses variations of the standard measures of incidence, prevalence, and duration. The usual injury incidence rate is the number of injuries per 200 000 person-hours worked, which is equivalent to the number of injuries per 100 full-time workers per year (40 hours/week x 50 weeks/year x 100 workers = 200 000 worker-hours per year). Incidence rates based on census data or on employer or union rosters may also be reported as the number of injuries per 100 workers, but these rates are not equivalent to rates based on hours worked. Another common measure used historically in the U.S. is the number of injuries per million hours worked.

The conventional injury severity rate is the number of lost workdays per 200 000 person-hours per year. Average severity is the average number of disability days or days lost per injury that resulted in lost workdays. Severity measured by lost workdays is a function not only of an individual's physical disability but also of job demands. Clinically identical foot injuries, for example, may require an attorney to miss only a day's work but could require a construction worker to be out for a month. Therefore, this measure of severity combines both clinical and economic effects of injuries.

In some circumstances, injured workers may return to work prior to full recovery and engage in a modified work assignment or restricted work activity. Thus, restricted workdays and rates are also used as measures of injury severity.

Injury severity may also be evaluated clinically. The most widely used standardized clinical measure of injury severity is the Abbreviated Injury Scale (AIS). This scale was developed to estimate the likelihood of death or permanent disability following severe injuries. A simpler dichotomous measure of clinical severity is whether the injury required hospitalization.

These three measures—incidence rate, severity rate (by lost workdays), and average severity—are analogous to the standard epidemiological measures of incidence rate, prevalence rate, and average duration and have the same arithmetic relation: $Sr = Ir \times Sav$ (severity rate = incidence rate x average severity), just as $P = I \times D$ (prevalence rate = incidence rate x average duration).

Evaluating Epidemiological Studies. In evaluating epidemiological studies, it is important to assess questions of both validity and causality. Validity is an assessment of whether the study measures what it claims to measure. It depends on controlling bias and reducing the effects of naturally occurring random variation. Methods for controlling bias and measuring random variation are treated extensively in textbooks on epidemiology and statistics.

Bias may arise from either confounding or misclassification of workers by exposure or outcome, and it may be random or systematic. Random misclassification bias results in a "no association" conclusion, and systematic misclassification may result in either positive or negative effect. Confounding bias occurs when an extraneous variable is associated with both exposure and outcome. Problems of misclassification can be controlled by fastidious attention to assessment of exposure and outcome. Confounding can be controlled either in the design of a study—for example, by matching study subjects on potential confounding variables—or it can be controlled in the analysis of data—for example, by stratification of data by the confounder.

Random variation or chance is measured as statistical power, or as P value (the probability that an estimated association is a chance occurrence) or with confidence intervals when reporting results. By convention, associations are considered statistically significant if the P value is less than 0.05. It is more informative, however, to report either precise P values (eg, $P = 0.048$ rather than $P < 0.05$) or confidence intervals. Whether an association is statistically significant depends on both the magnitude of random variation and the number of independent observations. Significance is easier to achieve when variation is small or observations are numerous. Thus, it is possible for meaningful results not to be statistically significant just because the number of observations is small or random variation is large.

More fundamentally, inferences based primarily on statistical reasoning assume rigorous experimental designs with appropriate attention to randomization, number of observations, and other matters. Often, however, epidemiology is *not* experimental. This disparity has provoked considerable debate within the scientific community, resulting in a de-emphasis of strict statistical reasoning.

Solving problems of bias and chance does not, by itself, directly address the question of whether an outcome is, in fact, *caused* by a particular exposure or condition. Causal associations may be obscured by bias or may not be statistically significant. Likewise, some associations may be significant but unimportant. To address the question of whether an exposure caused a particular outcome, the following issues need to be considered: (a) the magnitude of risk (eg, the size of the rate ratio), (b) whether the association is consistent with other studies, (c) whether there is an exposure-response relationship (ie, is there an increase in the risk with increasing exposure?), (d) whether the association is biologically or otherwise plausible, (e) whether disease or injury followed exposure, and (f) whether control of exposure results in reduction of risk.

In practice, it is rare for any single study to meet all these criteria. Therefore, a conclusion that a demonstrated association is a cause-and-effect relationship inevitably requires careful judgment, about which reasonable people may differ.

Toxicology

It is also useful to evaluate hazards using other sciences, specifically toxicology and biochemistry. The toxic effects of many exposures are known primarily by their effects on animals or through biochemical investigations such as in vitro investigations using bacteria or other cells. Information from such investigations can help evaluate biological plausibility and disease mechanisms. Dose-response relationships can also be evaluated. Risk to humans can be estimated by considering differences in physiology and by comparing the biochemical or toxicological potency of some substances with that of other substances whose effects on humans are better understood.

The clear advantage of toxicological and biochemical investigations is that they are experimental. Dose (as distinguished from exposure) and bias can be controlled, and outcome can be assessed to a degree unmatched in epidemiological investigations. The disadvantage is that investigators must extrapolate from one species to another and often from high to low dose, or they must make inferences from a biochemical reaction in cells to effects in whole organisms. In making extrapolations, investigators must compare humans with other animals concerning disease mechanisms and routes of exposure (eg, whether substance was inhaled, ingested, or injected). To aid in assessing risk, NIOSH and regulatory agencies such as EPA and OSHA have developed methods of extrapolation.

Evidence and Reasoning in Public Health

The need to *prove* causality is typical of the stringency required in all sciences. But there is an inescapable dilemma for public health. The study that meets all tests of validity and causality is the exception. Many inferences must be made based only on theoretical models. Convincing proofs could take many years, and during this time, the health or safety of workers, their families, or their communities could be needlessly harmed. Given that scientific information inevitably is incomplete, it is better to err in favor of protecting workers' health than to risk controllable disease or injury by failing to act.

The basis for taking preventive actions is the best available evidence—the same legal requirement embodied in both the OSHAct and the MSHAct. In occupational health policy, as in all matters of public policy, decisions must be made with less than perfect information. We should not allow the illusory promise of scientific certainty and the always-appropriate need for more research to prevent action that otherwise seems prudent. The purpose of research is to facilitate prevention, not to impede it; the absence of proof is not proof of absence.

Control

Once it is established that intervention is required—because exposure is excessive, because epidemiological or toxicological analysis is positive, or because there are manifest cases of occupational disease or injury—it is necessary to design specific hazard controls. In the occupational environment, primary prevention by controlling exposure to the causes of disease or injury is largely a problem of engineering, which uses basic methods of industrial hygiene and safety engineering. This discussion describes for public health practitioners the essentials of industrial hygiene control methods as methods of primary prevention. Issues are treated in more detail in textbooks in the field, to which readers are referred.

Hierarchy of Controls

To devise controls of health hazards, it is useful to conceive of hazards being generated by a source, being released into an environment, and coming into contact with workers. Controls can focus attention on each of these elements: the source, the environment, or the worker. The hierarchy of controls, from the most to the least effective, is to use (a) positive engineering at the source, (b) environmental control of hazards, and (c) personal protective devices. Administrative controls may also be used. These controls are discussed below in relation to types of hazards. Worker training and education is discussed separately.

Positive Engineering. Positive engineering requires a consideration of all health and safety hazards in the design and organization of work. It is self-evident that elimination of hazards at their source is inherently superior to adding on controls. For example, less hazardous ingredients, processes, or machines can be used instead of those that are more hazardous, and practices that expose workers to hazards can be eliminated or redesigned. Neurotoxic organic paint solvents, for example, may be replaced with water. Noise from saw blades can be reduced by redesigning the arrangement of teeth on the saw. Cadmium and lead can be removed from solder. Farming can be practiced with less reliance on chemical pesticides or fertilizers. Given the appropriate opportunity, effective solutions can be developed with industrial hygienists, engineers, and workers who are familiar with a particular job.

Similar considerations apply to control of traumatic injury hazards. For example, vehicles can be designed with a low center of gravity to prevent rollover and with rollover structures to protect occupants. Working at heights can be prohibited during inclement weather. Hoists can be used to lift patients from hospital beds.

Positive engineering is implemented best and most easily when new jobs or processes are being developed and before commitments are made that might have to be undone, sometimes at significant cost. This fact is the basis for our discussion of anticipation as a key process in preventing occupational disease and injury. However, positive engineering may also be used to revise a job, process, or industry. The opportunity to devise positive engineering controls is not lost if the need becomes apparent only after hazards appear. Revisions in jobs are routine in industry, and positive engineering to eliminate hazards should take its place alongside other types of revision and redesign.

Positive engineering also includes designing jobs to reduce or eliminate exposure. For example, exposure to respirable coal dust in underground mines can be reduced by keeping workers upwind of dust sources. Exposure to volatile organic solvents during cleaning operations can be reduced by not spraying or heating the solvent, given that both actions promote evaporation. Workers should never enter confined spaces without first testing the atmosphere for contamination or donning personal protective equipment and having the same equipment available for a rescue worker.

Environmental Controls and Personal Protective Devices. In considering environmental controls, hazards are classified as airborne (eg, dusts, vapors, gases, mists, and fumes), as physical (eg, noise, cold or heat, electromagnetic radiation, and hyperbaric environments), and as skin contact hazards. General principles for controls are discussed below for each class of hazard. Controls for specific biological hazards and for stress are discussed under separate entries in part 2.

To control exposure to chemical hazards, it is also useful to consider the route of absorption into the body: by inhalation, skin contact, or ingestion. Risk of inhalation is a function of the volatility of the substance, its temperature, the volume released into the work environment, and the total surface area of the substance that is exposed to the air. Absorption through the skin usually results from direct contact with liquids, although some vapors may be absorbed through the skin also. Risk of ingestion or inhalation exists if food, drink, or smoking materials become contaminated or if workers place contaminated fingers or implements into their mouths.

Airborne chemical hazards are common and can usually be controlled by ventilation. Airborne hazards that can be controlled by ventilation include not only toxic substances but also oxygen deficiency, air temperature, and humidity.

In general, there are two types of ventilation: local exhaust and dilution. Local exhaust ventilation removes contaminated air from as close to its source as possible, taking it from the worker's breathing zone, cleaning it by means appropriate to the hazard (eg, dust particles may be removed by a filter, organic vapors with an absorbent material), and releasing it outside the workplace. Only under certain circumstances should "cleaned" air be recirculated into the workplace. Local exhaust ventilation should enclose the source as completely as possible, consistent with the need to gain access to a process for production or maintenance. In designing local exhaust ventilation systems, it is common practice conceptually to enclose the source of the hazard completely and provide access only as necessary. This allows fewer opportunities for hazards to escape into the worker's breathing zone and reduces the need to remove and treat large quantities of air. There are standard designs of local exhaust systems for a wide variety of industrial processes, so practitioners need not redesign systems from first principles for each installation.

Dilution ventilation is less efficient for controlling hazards from particular sources, because it simply circulates fresh air into a workplace. It is used for controlling less toxic hazards and for situations in which the source of a hazard is not limited to discrete locations. Ventilation in underground mines, where the workplace and the airway are the same, is a special type of dilution ventilation.

Respirators are the third form of protection from airborne hazards. NIOSH and MSHA regulate the testing and certification of respirators (30 CFR 11), and OSHA and MSHA regulate their use at the worksite (29 CFR 1910.134; 30 CFR 56.5005, 57.5005, 70D). There are two circumstances under which respirators may be used: (a) as temporary measures (ie, during emergencies or while other controls are being implemented), or (b) when engineering controls are not feasible.

Effective protection by respirators requires systematic evaluation of possible respirators, of the hazards and demands of the work environment, and of the people who might wear the respirators. Potential wearers need to be educated about the specific hazards where they might work, trained in respirator use, and evaluated medically before using respirators and periodically thereafter.

There are two general classes of respirators: air-cleaning (often called air-purifying) and air-supplied systems. Air-cleaning respirators use a filter or absorbent material in a cannister to remove hazards from inhaled air. They may be a half mask (covering the nose and mouth), full-face mask, or gas mask. Some are single use or disposable. Usually, workers must pull air through the filter or cannister, but powered air-cleaning respirators, with a portable battery-powered fan, may pull air through a cleaner and provide it to the mask. Air-cleaning filters or cannisters have limited and highly variable service life, depending on their capabilities, the concentration of hazards in the environment, and the breathing patterns of the worker.

Air-cleaning respirators increase both the resistance of breathing and the physiological dead space in the respiratory tree, which may prevent some people from wearing them.

Supplied-air respirators provide clean air from a source independent of the work environment—either a tank that the worker wears (a self-contained breathing apparatus) or an air line connected to a tank or other air supply. Supplied-air respirators may have either positive or negative pressure inside the mask. The lifetime of a self-contained breathing apparatus depends on the amount of air it contains and the oxygen needs of the wearer. Air line systems do not, by themselves, limit the amount of time the worker may be in the hazardous environment (unless air is supplied by a tank), but they do limit mobility. The worker may not go beyond the length of the air line and usually must return by the same route as he or she entered. The self-contained breathing apparatus permits greater mobility but is heavier and more cumbersome.

Respirators should be properly stored, inspected, and maintained. Some have a limited storage life. Following use, filters or cannisters should be replaced, air tanks refilled, and the face mask and other apparatus inspected and repaired, if necessary. Face masks should be cleaned, sanitized, and stored in a clean and safe place.

Before respirators are selected, however, the hazards themselves must be fully evaluated. Criteria for respirator use are specific for each airborne hazard. The following hazard attributes affect the selection of a respirator and therefore need to be evaluated: form (aerosol, gas, vapor, or mixture), concentration, effects (acute and chronic health effects, risk of explosion or fire), routes of absorption, and warning properties in the event of respirator malfunction or leakage. If the hazard can be immediately dangerous to life or health (IDLH), or if its concentration significantly exceeds exposure limits, if it is carcinogenic, or if there is insufficient concentration of oxygen, supplied-air respirators are usually needed. If the hazard can reach flammable or explosive limits, the respirator must be "permissible"—that is, not capable of generating a spark that could ignite a flammable or explosive mixture. If the hazard is an eye irritant, eye protection may also be needed. If the hazard is an aerosol, filters are appropriate; if it is a gas or vapor, cannisters with appropriate absorbent material are required; and if it is a mixture, both filters and absorbent cannisters should be used. If the hazard can be absorbed through the intact skin, additional protection is needed.

A hazard's warning properties—that is, its odor or irritant threshold at concentrations below exposure limits—are sometimes useful to alert the wearer that the respirator is not working properly. Warning properties, however, are highly variable between individuals and so are not fully reliable indicators of respirator leakage. For warning properties to be relied on, the threshold of detectability would have to be determined for each respirator wearer.

Work practices that can be impaired by the use of the respirator should also be evaluated, just as specific work demands also need to be evaluated to ensure that use of a respirator is compatible with the job. If the workplace is a confined space, additional precautions are needed to provide for emergency escape and rescue. Communication among workers who have to wear respirators should be organized before work is begun in a hazardous environment. The physical demands of work should also be evaluated to ensure that a respirator would not excessively impair breathing.

The person who will wear a respirator should be educated about the specific hazards in the work environment where respirator use is intended. The worker should know the acute and chronic health effects, including consequences of respirator failure. The worker should also know what alternative hazard controls are intended or why they are not being used. Training in the proper use of the respirator is needed, as is fit testing to ensure a proper fit. People with facial hair or with facial scars or deformities may not obtain an airtight seal with a face mask. Eyeglasses may similarly prevent a good seal. Contact lenses should not be worn without eye protection if the hazard is an aerosol or eye irritant. Potential wearers should be evaluated medically for any physiological impairment or psychological factor that may be aggravated by use of a respirator.

Because respirators are often uncomfortable, they might not be used properly, if at all. Even when used properly, they often do not perform as predicted. The protection factor of a respirator—the ratio of hazard concentration outside to inside the mask—is usually determined in a laboratory setting either by the respirator manufacturer or by NIOSH. It represents the theoretical upper limit of protection that can be obtained with a respirator. Actual protection achieved in use, however, is often much lower than the theoretical value. As a general rule, respirators and other personal protective devices are not as effective as either positive engineering or environmental controls. Respirators appear to be convenient, but an effective respirator program requires more thought and effort than may seem apparent.

Contact chemical hazards usually result from specific work practices or accidental spills. Engineering control of contact hazards is usually accomplished by changes in materials or in those work practices that result in exposure. Specific causes, such as chemical ingredients or contaminants, need to be identified. Skin contact also may be controlled with personal protective gear, but practical considerations may limit the utility of such measures. More specific recommendations are found in entries on dermatologic conditions in part 2.

Physical Hazards

Physical hazards include processes that transfer harmful amounts of energy—mechanical, electromagnetic radiation, or noise—to workers. As a general rule, risk of harm is increased by higher amounts of energy released, proximity to the source, and increased duration of exposure. There is risk of both acute and chronic injury resulting from overexposure to physical hazards.

Noise, ionizing and nonionizing radiation, lasers, and radiant heat are all forms of radiant energy. The first step in developing controls usually involves analyzing the energy source, the path over which the energy radiates, and the location of workers. Risk of injury from radiant energy can ordinarily be controlled by a combination of implementing engineering controls at the source, reducing the time of exposure, increasing the distance between source and worker, and shielding the worker from the source. Because harmful effects are proportional to the duration and intensity of exposure, reducing exposure time may be an effective means of control, but it often requires careful monitoring. Increasing the distance between the worker and the source reduces the risk of harm. Control of noise is discussed in relation to noise-induced hearing loss in part 2.

Ionizing radiation is electromagnetic radiation with sufficient energy to produce ions in living tissue and thereby cause damage. The principal forms are x-rays; alpha, beta, and gamma radiation; and neutron particles. Evaluation and control of ionizing radiation is a complex field, and readers are advised to consult references listed in the entry on the adverse effects of ionizing radiation in part 2.

In considering controls of exposure to ionizing radiation, it is useful to distinguish between radiant energy from a fixed source (eg, an x-ray machine) and airborne radioactive isotopes. Ionizing radiation from a fixed source can be controlled by applying the principles above: engineering controls and reduced time of exposure, increased distance from the source, and use of shielding. Radioactive isotopes, on the other hand, may be inhaled or ingested, or may contaminate the skin and become internal emitters. Either they may lodge at the site of deposition (lung or gastrointestinal tract) and damage tissue locally, or they may become absorbed and localize in other organs, depending on the chemical nature of the particle. Radioactive iodine and radium, for example, lodge in the thyroid and bone, respectively. This type of radiation is best controlled by containment (eg, glove boxes) and local exhaust ventilation.

Nonionizing radiation is electromagnetic radiation that lacks sufficient energy to produce ions. Theoretically, it includes the entire range of the electromagnetic spectrum except for the high-energy portion: above about 10 nam (nanometers) wavelength. Practically, however, we restrict attention to ultraviolet (UV) and microwave radiation and treat lasers and radiant heat (infrared radiation) separately (see Nonionizing Radiation and Heat-Related Disorders in part 2).

Control of exposure to UV and microwave radiation can generally be accomplished by considering engineering controls, time, distance, and shielding, as above. Common sources of UV radiation include the sun, a welder's arc or other electrical discharge, and UV lights. Common sources of microwave radiation include radar and microwave ovens.

Lasers are a special type of electromagnetic radiation in which the energy is monochromatic and synchronized and the path is highly focused. Risk of acute injury is directly proportional to the power of the laser and the probability of intercepting the laser beam. Although the eye is the most vulnerable site of injury, high-power laser beams can also burn tissue. Some high-power beams used in welding operations also generate airborne toxic hazards such as ozone in addition to metal fumes and decomposition products common in welding emissions. ACGIH and ANSI have developed standards for protecting workers from lasers, but government regulatory agencies have not.

Exposure to radiant heat can be controlled like exposure to any other radiant energy source. Sources of radiant heat include any hot object such as the sun, furnaces, or engines.

Heat stress can arise from several sources: air temperature, humidity, air velocity, radiant heat, and metabolic heat from strenuous physical work. To control heat stress, each source must be evaluated and control measures applied. Air conditioning is an obvious way to reduce air temperature and humidity. Worker exposure can also be controlled administratively—for example, by doing "hot" work at night or during the winter. Increasing air velocity over a worker is useful, but only if ambient air temperature is less than 96°F; above this temperature, increasing air velocity adds to heat stress. Exposure to radiant heat is commonly reduced with shielding such as partitions or reflective clothing. Metabolic heat is produced by physical exercise, so to reduce this source, an employer must reduce the physical demands of work. This can be done by substituting mechanical for physical energy, by slowing the pace of work, or by lightening loads or otherwise redesigning a physically demanding job. Because sweating is a principal physiological response to heat stress, it is essential to provide ample supplies of potable water to prevent dehydration. It is rarely, if ever, appropriate to replenish lost electrolytes by adding salt. If workers' electrolytes are lost because of heat, alternative controls are needed.

Vibration is mechanical energy transmitted directly to a worker's body, often by vibrating tools or machines that the worker is operating. Depending on its frequency and on where on the body it is transmitted, vibration can affect limbs, organs, or the whole body. It is best controlled by eliminating the source—for example, by using impact as opposed to pneumatic hammers or by isolating the worker from the source. Low-frequency whole-body vibration from moving vehicles can be controlled by ensuring proper vehicle suspension, maintaining smooth roadways, and providing adequate seats (see Hand-Arm Vibration Syndrome and Low-Back Pain Syndrome in part 2).

Injury Prevention

Injury prevention is the systematic analysis of injury risk and development and the application of control measures for the primary and secondary prevention of injuries. For an injury to occur, four necessary and sufficient causes must exist: (a) energy in sufficient quantities must be accumulated; (b) the energy must have an agent or vehicle; (c) the energy must be released; and (d) the energy must be transmitted to a person, exceeding the body's ability to absorb it. To prevent injury, it is sufficient to control any *one* of these conditions.

Energy can exist in five forms: mechanical, electrical, chemical, heat, and radiation. These forms can produce acute or chronic injury. For example, mechanical energy can cause acute injury from a fall or chronic injury from certain repetitive motions. Electrical energy can cause electrical burns, nervous disorders, or death by electrocution. Chemical energy can cause acute toxic reactions, and heat energy can cause burns, heat stress, and systemic disorders. The effects of radiant energy are discussed above. The acute lack of energy, such as from extreme cold, also can cause injury.

Control of injury hazards includes any intervention that would eliminate or reduce the energy source, eliminate or modify the vehicle, interrupt the path of transmission or reduce the rate at which energy is absorbed. If energy is not accumulated, does not have a vehicle or agent, is not released, or is not absorbed too rapidly, an injury will not occur. If the rate at which energy is absorbed is reduced, injury may not be eliminated but its clinical severity may be reduced.

Similar to control measures for preventing occupational diseases, there is a hierarchy of controls to prevent injury. Passive controls that are intrinsic in job or machine design or that are automatic are preferred to active controls that require a worker's active or continuous participation and cooperation. Engineering controls are preferred to personal protective devices. For example, air bags that work automatically in the event of frontal vehicle crashes are more effective at preventing injury than are seat belts that require occupant participation. (In spite of extensive education and legal penalties, only about 30% of all vehicle occupants wear seat belts.)

This approach to injury prevention is preferred to one that relies on a classification of occupational "accidents" into those caused by "unsafe acts" or "unsafe conditions." Because its principal use is to attribute blame and allocate liability, the unsafe-acts-versus-unsafe-conditions approach is limited in its ability to suggest control measures. It also obscures the fact that nearly everything that happens at work, short of natural disasters, is the result of human acts, unsafe or otherwise. Controlling physical agents through positive engineering is more effective than controlling human behavior.

Although *accidents* is common terminology, we use it sparingly. It is a descriptive term that implies randomness and inevitability and thus suggests that little can be done to identify or control the causes of injuries. It also lacks precision, because it embraces a wide spectrum of unexpected incidents including manifest injuries, near-misses ("near hits" might be more appropriate), and incidents that result in property damage but no personal injury. Such events are important in their own right and may be useful for evaluating hazards. Some effective injury control strategies do little to prevent accidents and are more concerned with preventing injury. Cabs and canopies on construction or mining equipment, for example, do not prevent objects from falling.

Workers, supervisors, and safety professionals must all be engaged to generate and implement successful intervention strategies. People at work have the most immediate knowledge of work and the most vested interest in preventing injury. Therefore, their participation is essential, even in the absence of that of professionals.

Given the considerable variation in jobs, industries, and causes of injury, controls must be specific for each hazard. Targeted controls focused on high-risk jobs tend to be more effective than nonspecific programs such as posters and the use of incentive systems. Some incentive systems also result in underreporting of injuries rather than in actual reductions in injury rates. Examples of successful interventions illustrate the point.

Example 1.—A high injury rate had been identified epidemiologically at a small-parts manufacturing plant. Many data sources were used, including workers' compensation claims, records kept for OSHA, and insurance reports. Strains and sprains, contusions, and lacerations to hands and backs were the most common types of injury and were associated with lifting, pushing, or pulling boxes of parts and miscellaneous metal objects.

Changes were made, including a decrease in the size and weight of boxes and the installation of mechanical lifts and conveyors. Workers received more training, and the workplace was inspected regularly. The following year's injury rate dropped to a third of the prior year's rate, from 41.8 to 13.7 per 100 workers.

In this example, problems were defined epidemiologically and effective engineering solutions developed and implemented. The energy source and vehicle (heavy boxes) and the method of its transmission (manual handling) were defined; primary prevention of injuries was achieved by reducing the magnitude of energy accumulated and transmitted. This demonstrates the application of the general hierarchy of controls: engineering changes were implemented that did not depend on active participation by workers.

Example 2.—In another setting, epidemiology and engineering were combined to identify injuries that were preventable by remote-controlled equipment (power makeup equipment, or PME) used to connect and disconnect segments of drill pipe on offshore drilling rigs. A comparison of preventable injury rates among drill crews that used and did not use PME, as well as of rates before and after PME was installed, demonstrated both its effectiveness and its cost savings. Because this equipment required considerable capital investment, the cost analysis was important.

Example 3.—Analysis of surveillance data can also be used to evaluate public policy. In the coal mining industry, for example, there was a significant decline in the fatality rate for both surface and underground mines following passage and implementation of the federal Coal Mine Health and Safety Act of 1969.

Example 4.—Another example demonstrates the applicability of combining epidemiological with engineering analysis to identify causes of crashes of trucks on interstate highways. A case-control study of truck crashes found that large double-trailer trucks were two to three times overinvolved in crashes, regardless of driver age, hours of driving, cargo weight, or type of fleet. This finding prompted the following formulation:

(weight x speed) + instability = hazardous trucks

and identified the trucks posing the higher risk. The control strategies involved developing public policies to limit the number of vehicles associated with the highest risk.

A similar result was found in analysis of fatal rollover crashes involving multipurpose vehicles. The risk of fatal rollover was directly correlated with a rise in the vehicle's center of gravity.

These findings suggest that an effective way to reduce the risk of vehicle crashes (the leading cause of occupational fatalities) is by changing the configuration of vehicles. As above, problems were clarified epidemiologically, suggesting engineering solutions.

Example 5.—Following transition from piece-rate to hourly wages, the injury frequency and severity rates in the Swedish forestry industry declined by 30%. The decline appeared not to be a random fluctuation and was not associated with confounding factors such as safety activities or reductions in productivity. The shift from piece-rates to hourly wages was brought about by a 10-month strike of forestry cutters. This example suggests that payment schedules may provide incentives to take risks that contribute to injuries.

Further Reading

American Conference of Governmental Industrial Hygienists. *Industrial Ventilation: A Manual of Recommended Practice.* 20th ed. Cincinnati, Ohio: ACGIH; 1989.

American Conference of Governmental Industrial Hygienists. *Threshold Limit Values and Biological Exposure Indices.* Cincinnati, Ohio: ACGIH; published annually.

American Industrial Hygiene Association. *Respiratory Protection Monograph.* Akron, Ohio: AIHA; 1985.

Baker EL, Honchar PA, Fine LJ. Surveillance in occupational illness and injury: concepts and content. *Am J Public Health.* 1989;79(Suppl):9-11.

Baker EL, Melius JM, Millar JD. Surveillance of occupational illness and injury in the United States: current perspectives and future directions. *J Public Health Policy.* 1988;9:198-221.

Baker EL. Sentinel event notification system for occupational risks (SENSOR): the concept. *Am J Public Health.* 1989;79(Suppl):18-20.

Baker SP, O'Neill B, Haddon W Jr, Long WB.The injury severity score: a method for describing patients with multiple injuries and evaluating emergency care. *J Trauma.* 1974;14(3):187-196.

Baselt RC. *Biological Monitoring Methods for Industrial Chemicals.* 2nd ed. Boca Raton, Fla: CRC Press; 1989.

Bell CA, Stout NE, Bender TR, Conroy CS, Crouse WE, Myers JR. Fatal occupational injuries in the United States, 1980 through 1985. *JAMA.* 1990;263:3047-3050.

Benenson AS. *Control of Communicable Diseases in Man.* 15th ed. Washington, DC: American Public Health Association; 1990.

Blanc PD, Rempel D, Maizlish N, Hiatt P, Olson KR. Occupational illness: case detection by poison control surveillance. *Ann Intern Med.* 1989;111:238-244.

Bollinger N, Schutz RH. *NIOSH Guide to Industrial Respiratory Protection.* Washington, DC: U.S. Public Health Service, National Institute for Occupational Safety and Health; 1987. DHHS (NIOSH) publication 87-116.

Burgess WA. *Recognition of Health Hazards in Industry: A Review of Materials and Processes.* New York, NY: John Wiley; 1981.

Checkoway H, Dement JM, Fowler DP, Harris RL Jr, Lamm SH, Smith TJ. Industrial hygiene involvement in occupational epidemiology. *Am Ind Hyg Assoc J.* 1987;48:515-523.

Checkoway H, Pearce NE, Crawford-Brown DJ. *Research Methods in Occupational Epidemiology.* New York, NY: Oxford University Press; 1989.

First M. Engineering control of occupational health hazards. *Am Ind Hyg Assoc J.* 1983;44(9):621-626.

Freund E, Seligman PJ, Chorba TL, Safford SK, Drachman JG, Hull HF. Mandatory reporting of occupational diseases by clinicians. *MMWR.* 1990;39(RR-9):19-28.

Froines J, Wegman D, Eisen E. Hazard surveillance in occupational disease. *Am J Public Health.* 1989;79(Suppl):26-31.

Graham JD, Green LC, Roberts MJ. *In Search of Safety: Chemicals and Cancer Risk.* Cambridge, Mass: Harvard University Press; 1989.

Greenland S. Randomization, chance, and causal inference. *Epidemiology.* 1990;1(6):421-429.

Guidelines for investigating clusters of health events. *MMWR.* 1990;39(RR-11).

Haddon W Jr. The changing approach to the epidemiology, prevention, and amelioration of trauma: the transition to approaches etiologically rather than descriptively based. *Am J Public Health.* 1968;58(8):1431-1438.

Haddon W Jr. A logical framework for categorizing highway safety phenomena and activity. *J Trauma.* 1973;12:193-207.

Hanrahan LP, Moll MB. Injury surveillance. *Am J Public Health.* 1989;79(Suppl):38-45.

Hill AB. *A Short Textbook of Medical Statistics.* London, England: Hodder & Stoughton; 1984.

Kneip TJ, Crable JV. *Methods for Biological Monitoring.* Washington, DC: American Public Health Association; 1988.

Kriebel D. Occupational injuries: factors associated with frequency and severity. *Int Arch Occup Environ Health.* 1982;50(3):209-218.

Last JM. *Public Health and Human Ecology.* East Norwalk, Conn: Appleton & Lange; 1987.

Last JM, ed. *A Dictionary of Epidemiology.* 2nd ed. Oxford, England: Oxford University Press; 1988.

MacKenzie EJ, Shapiro S, Smith RT, Siegel JH, Moody M, Pitt A. Factors influencing return to work following hospitalization for traumatic injury. *Am J Public Health.* 1987;77(3):329-334.

Mohr DL, Clemmer DJ. Evaluation of an occupational injury intervention in the petroleum drilling industry. *Accid Anal Prev.* 1989;21:263-271.

Monson RR. *Occupational Epidemiology.* 2nd ed. Boca Raton, Fla: CRC Press; 1989.

Morrison AS. *Screening in Chronic Disease.* New York, NY: Oxford University Press; 1985.

Plog BA, ed. *Fundamentals of Industrial Hygiene.* 3rd ed. Chicago, Ill: National Safety Council; 1988.

Pollack ES, Keimig DG, eds. *Counting Injuries and Illnesses in the Workplace: Proposals for a Better System.* Report of the Panel on Occupational Safety and Health Statistics, National Research Council. Washington, DC: National Academy Press; 1987.

Proctor NH, Hughes JP, Fischman ML, eds. *Chemical Hazards of the Workplace.* 2nd ed. Philadelphia, Pa: JB Lippincott; 1988.

Robertson LS. *Injuries: Causes, Control Strategies, and Public Policy.* Lexington, Mass: Lexington Books; 1983.

Robertson LS. (Weight x speed) + instability = hazardous trucks. *Am J Public Health.* 1988;78(5);486-487. Editorial.

Rothstein MA. *Medical Screening of Workers.* Washington, DC: Bureau of National Affairs; 1984.

Sax NI, Lewis RL. *Dangerous Properties of Industrial Materials.* 3 vols. New York, NY: Van Nostrand Reinhold; 1989.

Shapiro J. *Radiation Protection: A Guide for Scientists and Physicians.* 2nd ed. Cambridge, Mass: Harvard University Press; 1990.

Smith GS, Kraus JF. Alcohol and residential, recreational, and occupational injuries: a review of the epidemiologic evidence. *Annu Rev Public Health.* 1988;9:99-121.

Stein HS, Jones IS. Crash involvement of large trucks by configuration: a case-control study. *Am J Public Health.* 1988;78(5):491-498.

Sundstrom-Frisk C. Behavioral control through piece-rate wages. *J Occup Accid.* 1984;6:49-59.

Trent RB. Locations of fatal work injuries in the United States: 1980-1985. *J Occup Med.* 1989;31:674-676.

U.S. Congress, Report of the House Committee on Government Operations. *Occupational Illness Data Collection: Fragmented, Unreliable, and Seventy Years Behind Communicable Disease Surveillance,* 98th Cong, 2nd Sess (1984).

U.S. Department of Labor, Bureau of Labor Statistics. *Annual Survey of Occupational Injury and Illnesses,* 1989. Washington, DC: U.S. Government Printing Office; 1990.

Waller J. *Injury Control: A Guide to the Causes and Prevention of Trauma.* Lexington, Mass: Lexington Books; 1985:467-471.

Weeks JL, Fox MB. Fatality rates and regulatory policies in underground bituminous coal mining: United States, 1959-1981. *Am J Public Health.* 1983;73:1278-1280.

Work-related injuries and illnesses in an automotove parts manufacturing company: Chicago. *MMWR.* June 16, 1989;413-416.

World Health Organization. *Epidemiology of Work-Related Diseases and Accidents.* Geneva, Switzerland: World Health Organization; 1989. Technical Report Series 777.

Chapter 2: Social and Public Policy

Government Regulation

Governmental regulation is necessary to any strategy for preventing occupational disease and injury. Therefore, we discuss the legal structures surrounding health and safety problems at work. This discussion is specific for the United States; however, because the social structure of work and the resulting problems of disease and injury control are sufficiently similar in many societies, this discussion has a wider application than may first seem apparent.

There are three main types of relevant law and, therefore, of government intervention: (a) occupational safety and health regulation intended to prevent disease and injury, (b) statutes governing labor-management relations, and (c) methods of compensating workers who are injured or ill because of work.

Occupational Safety and Health Regulation

The principal regulatory agencies in the U.S. are the Occupational Safety and Health Administration (OSHA), which governs employees in the private sector, and the Mine Safety and Health Administration (MSHA), which governs miners. Other regulatory agencies also have occupational health and safety responsibilities. For example, the Environmental Protection Agency (EPA) and the Nuclear Regulatory Commission (NRC) regulate users of pesticides and radioactive materials, respectively, with the authority to issue licenses.

The National Highway Transportation Safety Administration, the Federal Aviation Administration, the Federal Railroad Administration, and the U.S. Coast Guard (all in the Department of Transportation) have responsibility for transportation safety—a major cause of work-related fatalities and injuries—on the highways, airways, railroads, and waterways, respectively. These are simultaneously workplaces as well as places of transportation and recreation for the general public. Therefore, these agencies concern themselves with both occupational and nonoccupational hazards. They govern items such as vehicular and aircraft integrity, limitations on the hours of work, and licensure of airline pilots and water transport captains.

Occupational Safety and Health Administration (OSHA)

OSHA was established in 1971 in the Department of Labor to "assure safe and healthful working conditions for working men and women" (Sec 2(b)). It is responsible for setting health and safety standards, and for enforcing compliance with these standards by inspecting workplaces and issuing citations for noncompliance. The statute creating OSHA also created the Occupational Safety and Health Review Commission (OSHRC), an independent administrative law agency, to adjudicate citations.

OSHA has jurisdiction over employers in the private sector, excluding railroads, offshore facilities, merchant marine and dock employers, mines, and highway traffic. States can implement their own OSHA programs in place of the federal program, provided the state program is "at least as effective" (Sec 18) as the federal program. Federal employees are supposed to be covered by plans implemented by employing agencies.

Under OSHA, employers have a responsibility to know the standards that apply to their workplace, to control hazards, to inform and train workers about these hazards, and to ensure that workers have and use protective equipment. Employers also have a "general duty" to provide a safe and healthful place to work (Sec 5). Employers with 10 or more employees must keep a log of workers' job-related injuries and illnesses (OSHA Form 200) and, when requested, make that information available to OSHA and to workers and their representatives. Public health professionals may obtain this information as an aid in hazard control efforts. Employers under OSHA's jurisdiction are also required to report all workplace fatalities and serious injuries (ie, those resulting in hospitalizations of five or more workers) to OSHA. However, this requirement is often disregarded, resulting in substantial underreporting of occupational fatalities and serious injuries.

Mine Safety and Health Administration (MSHA)

MSHA was first established in the Department of the Interior in 1969 as the Mining Enforcement Safety Administration, with jurisdiction limited to coal mines. The statute was amended in 1977, the name was changed to MSHA, and jurisdiction was extended to all underground and surface mines and attached facilities in the U.S. MSHA is similar in structure and function to OSHA, except that its enforcement powers are significantly stronger. For example, MSHA is required to inspect each underground mine four times every year and each surface mine twice every year. Citations and penalties for noncompliance are mandatory, and MSHA inspectors may shut down all or part of mines they consider to present imminent dangers. OSHA has none of these enforcement powers; it may shut down an operation only with a court order.

Regulatory Functions. Standards Setting. Both the Occupational Safety and Health Act (OSHAct) and the Mine Safety and Health Act (MSHAct) require OSHA and MSHA to adopt legally enforceable standards that (a) are based on the "best available scientific evidence," which will ensure that "no worker will suffer illness if exposed throughout his working life," and (b) are technically "feasible" (OSHAct, Sec 6; MSHAct, Sec 101). With these requirements, developing standards often results in compromises between conflicting interests.

Health standards are supposed to include not only limits on the concentration and duration of exposure but also requirements for work practices, medical monitoring, and personal protective devices. In addition, standards should impose a duty on employers to inform workers about hazards with warning labels and signs, education, and training. Most health standards, however, consist only of PELs.

Exposure to multiple hazards on the job is common, but the health effects of exposure to mixtures are poorly understood. However, both OSHA and MSHA have adopted a formula from the American Conference of Governmental Industrial Hygienists (ACGIH) for exposures to mixtures of substances that affect the same target organ. A violation exists if

$$\Sigma\ [C(i)/L(i)] > 1.0,$$

where $C(i)$ equals the concentration of the ith ingredient and $L(i)$ equals its PEL. Common mixtures include engine exhaust, blends of solvents, and fumes and gases from welding operations. This formula assumes that the effects of each ingredient in a mixture are additive.

Safety standards cover almost as wide a variety of topics as there are industrial processes. With few exceptions, these standards were adopted from existing safety codes developed by professional organizations or by consensus standards organizations. These organizations include the American National Standards Institute (ANSI), the National Fire Protection Association, the National Electrical Code, the American Society for Testing Materials, the American Welding Society, and the American Society of Mechanical Engineers.

For either OSHA or MSHA to issue a citation for noncompliance with a standard, the violation must be observed or overexposure must be documented. Evidence of excess morbidity or mortality without hazards on a job having actually been witnessed is sufficient, but not common, reason for these agencies to act. In some circumstances, OSHA can issue a citation under its general duty clause (Sec. 105c), which is intended to authorize action when a hazard exists but there is no applicable standard.

NIOSH, part of the Centers for Disease Control (CDC) in the U.S. Public Health Service, provides both OSHA and MSHA with recommended exposure limits (RELs). Since its creation in

1970, NIOSH has recommended limits for about 600 substances in the form of criteria documents and current intelligence bulletins, as well as in other ways. These RELs, which are listed in part 3, are often more stringent than PELs adopted by either OSHA or MSHA.

The American Conference of Governmental Industrial Hygienists, a professional association of industrial hygienists who are or have been affiliated with government agencies, has established threshold limit values (TLVs) since 1939. These TLVs are set to protect "nearly all workers" and need not be based on the "best available evidence." The volunteer TLV committee of ACGIH publishes an annual TLV list covering chemical and physical hazards. Some hazards with TLVs are not covered by OSHA or MSHA regulations (eg, limits on exposure to heat, lasers, and microwave radiation). ACGIH also publishes biological exposure indices (BEIs) for estimating exposure to a limited number of hazards by using medical monitoring of exposed workers. Threshold limit values have been adopted by some state agencies and by some other countries and have formed the basis for OSHA's standards. Although it has been a pioneer in setting standards, the ACGIH does not have the resources to evaluate hazards itself, nor is it publicly accountable as are NIOSH, OSHA, and MSHA. Consequently, the TLVs are not as well documented as RELs or current PELs, and the issue of their reliability has stimulated vigorous debate within the scientific community.

Additionally, the American Industrial Hygiene Association (AIHA), the largest association of professional industrial hygienists in the U.S., sets workplace environment exposure limits (WEELs) for about 50 substances. AIHA also accredits laboratories for analyzing workplace exposures.

The American National Standards Institute (ANSI), a consensus standards-setting organization, publishes some standards pertinent to occupational health. These include standards for respiratory protection devices, industrial ventilation, and lasers, as well as quality standards for compressed gas. ANSI assembles a panel of experts and requires that they reach consensus in order to publish a standard. Many ANSI standards are incorporated by reference in OSHA and MSHA regulations. ANSI also has developed criteria for classifying and recording injuries.

Standards set by OSHA and MSHA cover the most common hazards to which workers are exposed, but they are far from comprehensive. Approximately 60 000 chemicals are used in the developed industrial societies, and about 3000 new chemicals are introduced each year. Permissible exposure limits have been adopted for only about 600 of these agents, and there are complete standards, requiring exposure and medical monitoring, and worker education for far fewer. The absence of standards results from knowledge of hazards lagging behind industrial innovation combined with a lengthy and adversarial standards-setting process.

Inspections and Citations. OSHA and MSHA are authorized to enter unannounced and inspect without delay any workplace under their jurisdiction. During OSHA's early years, a few employers challenged its authority to enter and inspect, necessitating a court order. Recently, such challenges have virtually disappeared.

The OSHAct and MSHAct provide that employees or their representatives (ie, unions) may request an inspection and that OSHA (or MSHA) is obligated to respond. Although non-employees with knowledge of workplace hazards (eg, public health professionals) do not have this same right under the federal program (as they do in some states), they can pass information to the agency and request an inspection. However, if a health professional or other person not directly connected with a workplace acquires information from a worker concerning a possible workplace health or safety hazard, that person should never call OSHA or MSHA without the knowledge and consent of the worker involved. Doing so may have adverse consequences to the worker's status.

When requesting an inspection, it is important to be as specific as possible to have an effective inspection. What is the hazard, which workers are affected, how many are affected, for how long, and in what part of the employer's operation are all important questions to answer.

With about 650 inspectors and more than 5 million workplaces under its jurisdiction, OSHA must set priorities for inspection. The priorities are, in this order, (a) imminent dangers, (b) fatalities or catastrophes, (c) employee requests, and (d) general schedule or random selection. (In contrast, MSHA's 1300 inspectors are required to inspect every one of 16 000 mines either twice or four times each year.)

Imminent dangers are conditions in which it is reasonably certain that death or serious bodily harm will occur immediately or before the danger can be eliminated through normal enforcement procedures. Imminent dangers include not only acute hazards but also excessive exposure to serious chronic health hazards such as carcinogens. Formal legal definitions have evolved for such conditions under each act. Under the OSHAct, imminent danger is the only situation in which workers have limited legal protection to refuse to work. It is also the only situation in which inspectors may seek a temporary restraining order to shut down a workplace for a limited time (5 days). Even then, the agency must obtain a court order from the local federal court for permission to close a workplace. In contrast, an MSHA inspector has the authority to shut down all or part of a mine if he or she believes that an imminent danger exists. If a mine is shut down, workers are entitled to receive compensation for limited periods as if they were still employed. If OSHA shuts down a worksite, workers are not guaranteed compensation and ordinarily will not receive it.

Fatalities must be reported by all employers. OSHA inspects to assess violations of standards or of the employer's general duty and recommends measures to prevent recurrences. Annually, OSHA investigates only about 1500 fatalities out of an estimated 8000 that occur under its jurisdiction.

Inspection requests can be made by employers or by individual employees or their representatives. These requests must be in writing and may be made anonymously, if requested. Frequently, an inspection will be limited to the items listed in the complaint. Therefore, a person who is considering a request to inspect should decide carefully what should be inspected. If a so-called wall-to-wall inspection is needed, it should be specifically requested and justified.

General schedule inspections are made on facilities chosen more or less at random. Each field office of OSHA may target certain industries or workplaces based on their known use of hazardous materials, their injury rates, or their history of citations.

Informed employee and employer participation is essential to a thorough inspection because it provides the inspector—usually a stranger to the workplace—with different information, perspectives, and interests. Employees have the right to participate in opening and closing conferences when the inspector arrives at and leaves the workplace. They also have the right to "walk around" with the inspector to point out hazards or abnormal conditions, as well as the right to confer privately with the inspector, regardless of whether they are on the walk-around inspection. Employers are not required to pay employees for this time and usually do not.

If violations of standards or of the employer's general duty are found, the inspector may issue a citation, including an assessment of seriousness and a deadline for correcting the hazard, and may impose a fine. The citation, describing only the hazard and deadline for its correction, is sent to concerned parties and must be conspicuously posted at the workplace until the hazard is corrected.

Adjudication. An employer who disagrees with the citation may appeal to an administrative law judge of OSHRC. (The Mine Safety and Health Review Commission [MSHRC] is the analogous agency under the MSHAct.) Appeals are the first stage of adjudication of any contested claim.

As with inspections, employee participation is essential to fair and balanced adjudication. However, employees may only appeal citation deadlines. They are precluded from challenging the size of the penalty, the seriousness of the citation, or the inspector's failure to issue a citation. Within these limits, employees may participate in all aspects of appeal procedures, but to do so, they must request party status from OSHRC. The legal dispute is between the employer and the government; employees are considered third parties, even though they have much at stake in the proceedings.

Criminal penalties against employers are also possible under OSHA and MSHA. These provisions are rarely used by the federal agencies; most often, criminal charges (eg, for homicide in case of workplace fatalities) are prosecuted under state and local statutes.

OSHA and MSHA also provide for consultative inspections when requested by employers. Such inspections are provided free and carry no risk of citation. They can inform the employer of violations and suggest methods of abatement.

Workers' Rights. Protection from Discrimination. Both the OSHAct and MSHAct stipulate numerous workers' rights, which the agencies are required to protect (Sec 11(c), OSHAct; Sec 105(c), MSHAct). Specific rights include the right to request an inspection, to talk with an inspector, to testify about working conditions, and to participate in OSHA enforcement and adjudication activity. Although these protections exist, the letter and the spirit of the law often do not apply precisely to events on the job.

Right to Refuse Unsafe Work. Under the OSHAct and the MSHAct, workers have limited rights to refuse unsafe work. However, neither of these statutes clearly protects such rights. Under OSHAct and MSHAct, a worker has to show (often after the fact) that an imminent danger existed, that the worker had unsuccessfully requested the employer to eliminate the danger, that there was insufficient time to control the danger by a normal OSHA inspection, that the worker had a "reasonable belief" of death or serious injury, and that there was no other choice. If the worker had talked to others, offered to do other work, and called OSHA, his or her case would be stronger than if these actions had not been taken.

Right to Information. Employees, their representatives, and health professionals also have the right to know the hazards of materials with which they work (29 CFR 1910.1200); to have access to medical records, including records of exposure to toxic substances (29 CFR 1910.20); and to have access to the employer's log of injuries and illnesses (29 CFR 1904.7). Records must be maintained by employers for 30 years and be transferred to new owners if the business is sold.

Information about hazardous materials is required to be summarized in material safety data sheets (MSDSs) provided to each employer by a manufacturer or supplier. These data sheets contain information in a standard format covering the material's name, formula, physical and chemical properties, and health and safety hazards; first-aid procedures in the event of overexposure; and a phone number for emergencies. Employers are required to train workers in the recognition and use of hazardous materials and to label materials in English and other languages of employees. Workers in labor unions also have an independent right to information about hazardous materials for collective bargaining. This right is described in more detail below.

Health Professionals' Right to Information. Employers may restrict access to information if a trade secret is involved. However, health professionals are excluded from this restriction (29 CFR 1910.1200(i)(3)). A health professional who needs to obtain trade secret information must make a written request to the manufacturer of the toxic substance, detailing the medical or occupational health need for the information. In this context, "health professional" includes not only physicians and nurses but also industrial hygienists, toxicologists, and epidemiologists. Needs could include assessing hazards; conducting environmental sampling, medical surveillance, or a medical exam as a prerequisite to work assignment; providing medical treatment; selecting personal protective equipment; or designing environmental controls. The request must describe why disclosure of the information is essential. It must also describe the procedure used to keep the information confidential, and it must include an agreement not to use the information for any but the specified health need or to disclose it except to OSHA.

Citizens may also obtain information from OSHA under the Freedom of Information Act (5 USC 522). Such information can include citations, notes, results of OSHA inspections, test results, photographs, copies of correspondence, and notices of imminent dangers.

Finally, many states and local jurisdictions have statutes that provide for release of information about toxic substances.

Efficacy of Regulation. The ultimate purpose of regulation is to reduce the risk of occupational disease and injury, but the efficacy of such regulation is difficult to measure. The few systematic investigations of this issue show that fatalities in high hazard coal mines were reduced following the 1969 Coal Mine Act, that injuries in individual workplaces decline following inspections, and that workplaces with labor unions receive generally better inspection and follow-up services than nonunion counterparts. When standards are closely related to risk of injury, there is a modest effect in reducing the risk of certain types of injury.

The number of OSHA inspectors is small; consequently, there is only a remote likelihood that a workplace will be inspected. The size of fines is also small ($35 per violation in 1988) in relation to an employer's gross revenues, providing little economic incentive to comply with regulations. Moreover, many of the most dangerous workplaces are small and difficult to find, and they employ minority and transient workers who are often ill-informed of their rights, culturally isolated, and lacking in union representation or other resources to pursue remedies or to recognize and control hazards. Finally, since OSHA's beginning, agency policies have changed significantly with the political orientation of succeeding administrations. Thus, whether a regulatory agency is effective depends on a complex set of factors.

Alternatives to Regulation. The current regulatory framework is not comprehensive or consistent in its coverage of the work force. Public sector employees, agricultural workers, offshore workers, railroad workers, and workers in small workplaces are not provided the same degree of protection as other workers.

Moreover, current health standards are weak in many respects. Along with their limitations, as noted above, most current standards are "horizontal," covering a wide diversity of processes and conditions found throughout industry and commerce. Vertical, industry-specific standards similar to MSHA's for the mining industry might be useful for certain other high-hazard industries, such as logging, agriculture, and construction, each of which has very high fatal injury rates.

Furthermore, standards are not constructed in a way that encourages the many forms of positive engineering, such as reduction in use of toxic substances. Requirements are placed directly on employers but only indirectly on manufacturers of machinery or designers of industrial processes. There are, however, some exceptions: Manufacturers of respirators must conform to NIOSH standards for efficacy, and manufacturers of mining equipment must conform to MSHA standards for safety of operation in explosive environments.

Enforcement measures could include mandatory inspections of certain industries and required formation of workplace health and safety committees, including employees, with powers to inspect and issue or recommend citations. The number of inspectors and the magnitude of fines could be significantly enlarged. Criminal penalties for especially egregious offenses should be considered more often.

Given limited resources, a much more thorough means of setting enforcement priorities is needed. This would require more extensive gathering, auditing, and use of data concerning the incidence of occupational injuries and illnesses to identify problems.

Labor-Management Relations

Laws governing labor-management relations also are important in preventing occupational injury and illness, especially for workers organized to represent themselves in unions. Union workers often have procedures in their contracts pertaining to health and safety problems. These procedures frequently create health and safety committees and provide for resolution of grievances, many of which pertain to health and safety. Unions also often designate a person to assist injured workers in getting compensation for occupational injuries and illnesses.

But existing labor law is concerned more with procedural than with substantive rights of labor and management. The original purpose of the National Labor Relations Act was labor peace, not health and safety. While safety and health on the job are mandatory subjects of bargaining, labor law is relatively indifferent to the specifics of any agreement. Moreover, public health expertise is typically not common among those divisions of management, labor unions, or government agencies responsible for bargaining or for enforcing labor laws. Therefore, achievements in disease and injury control through collective bargaining often depend on economic factors, bargaining skills, and other issues subject to bargaining (such as wages, pensions, and job security) rather than on public health objectives.

Although bargaining as such may be inherently limited as a means of disease and injury control, workers in labor unions are usually better situated to achieve these objectives than are nonunion workers. Unions provide an important independent infrastructure for participating on joint health and safety committees, participating in policy making, knowing about and exercising rights on the job, educating workers, and managing disagreements with employers. Therefore, it is important for health professionals to become familiar with labor unions and the legal and social environment within which unions function.

Unions provide an orderly means for workers to participate in occupational disease and injury control.

Labor-management health and safety committees could play a substantial role in injury and disease control. A report prepared for the U.S. Department of Labor said it well:

> To be substantially empowered, committees would need to be represented by top-level management and well-placed union representatives, with adequate training and expertise in the field. Management representatives would need to have the visible backing of corporate leadership. The union would have the backing, not only of its shop stewards and the local, but of an active union safety and health committee as well. Both would have the benefit of mutually agreed-upon expert consultants. Committee members would have full access to company records and data, not just on health and safety but on planning, finance, and new technology as well. They would meet regularly, on paid working time, and keep formal minutes that would be shared with the work force. They would regularly inspect the facilities and would be involved in monitoring programs and accident investigations. They would have control of a budget, of hiring and firing company safety and health personnel, and of health and safety training. They would have the authority to stop the use of any immediately dangerous equipment or process. And, they would operate in an environment of mutual respect and trust.

Workers' Compensation

Workers' compensation systems pay for lost wages, medical costs, and rehabilitation for persons who become ill or injured by causes "arising out of or in the course of employment"—the phrase adopted in most statutes. Most plans are administered by states; they vary significantly in scope of coverage and magnitude of benefits. The amount of benefits is fixed by schedule and is often low. Workers' compensation has not yet been shown to deter disease or injury or otherwise stimulate effective primary prevention of occupational injury or illness. The burden of proof to show that an injury or illness is occupational is usually on the worker. This is a much different burden than was anticipated.

Workers' compensation was first established as a result of a trade-off between employers and workers made early in the 20th century. At the time, workers injured on the job relied on common-law remedies to sue employers for damages. Awards were unpredictable and occasionally large. Under workers' compensation, workers gave up the right to sue in return for a no-fault system. Employers gave up certain common-law defenses in exchange for a fixed and predictable schedule of awards. Employers could no longer claim that the worker assumed liability for risk of injury by agreeing to employment (the assumption of risk defense); they could no longer claim that injury resulted from the action of another employee (the fellow servant defense); and they could no longer claim that the worker was to blame for his or her own injury (the contributory negligence defense). Workers' compensation thus became the exclusive remedy for workers to obtain compensation from employers for work-related injuries.

Workers' compensation is inherently more responsive to traumatic occupational injury and acute occupational diseases than it is to chronic conditions. There are many reasons for this. The connection between cause and effect is easier to demonstrate for acute conditions than for chronic conditions. Establishing causality between chronic occupational exposure and disease is difficult because the extended time between exposure and disease makes documentation of exposure difficult; moreover, other occupational and nonoccupational exposures may have caused or contributed to disease. In addition, many health professionals are not well-informed about work, occupational exposures, or disease, nor are they willing expert witnesses in litigation. Consequently, it is often difficult to reach the legal threshold of "more likely than not" to prove causation, and this enables many employers to challenge occupational illness claims on the basis that they are not work related. Some jurisdictions have responded to these problems by defining specific compensable diseases or by liberally interpreting the rules of evidence required to show causality.

To determine eligibility for compensation, a distinction is usually (and sometimes artificially) drawn between impairment and disability. Impairment pertains to quantified (if possible) assessment of functional or physiological limitations. Disability pertains to the worker's ability to perform his or her usual work and therefore involves an assessment not only of the impairment but also of the worker's usual job demands.

Because of limitations in workers' compensation plans, some injured or ill workers have bypassed the plans and sued third parties, such as producers or suppliers of toxic materials or manufacturers of hazardous equipment. In a personal injury lawsuit, the burden of proof is different from that in a no-fault workers' compensation claim; the injured or ill persons must prove by a preponderance of evidence that their condition is the result of someone else having negligently exposed them to the toxic substance or harmful device in such a way as to have caused or contributed to the injury or illness. This is often accomplished by showing that the product was defective or abnormally dangerous or that the manufacturer had an obligation to warn of the hazards.

The advantages for injured persons of third-party suits are that the size of an award is potentially much larger and the complainant may find it easier to convince a jury than to convince a panel of experts. The disadvantage is that the prospects for winning are substantially less than they are under workers' compensation.

Rights to Know and Sources of Information

Information about workplace and environmental hazards is essential for effective prevention. The following is a guide to information sources and regulations about record keeping and availability. We include a summary of regulations concerning hazard communication, access to records, and record keeping, in addition to lists of standard textbooks; journals in the field; and descriptions of computerized storage, search, and retrieval services. References to specific diseases or organ systems are included as part of entries in part 2.

Regulations

Hazard Communication. Under OSHA regulations (29 CFR 1910.1200), all manufacturing employers under OSHA's jurisdiction are required to inform employees about potential occupational health and safety hazards. Pesticides, food additives, alcoholic beverages, and consumer products are not covered under these regulations because hazard communication for these substances is provided for under the Federal Insecticide, Fungicide, and Rodenticide Act (FIFRA) (7 USC 136); the Federal Food, Drug, and Cosmetic Act (21 USC 301); the Federal Alcohol Administration Act (27 USC 201); and the Consumer Product Safety Act (15 USC 1261), respectively.

The OSHA Hazard Communication Regulations require employers to provide their employees, the employees' representatives, and health professionals with material safety data sheets (MSDSs), training, and access to written records. Employers or producers of chemicals may withhold certain trade secret information, except when there is a medical emergency or when a health professional makes a written request explaining why information is needed. Requirements are detailed in the regulations. The manufacturers of the substances usually produce the MSDSs and provide them to purchasers.

Access to Medical and Exposure Records. OSHA regulations also provide for access to medical and exposure records (29 CFR 1910.20). Under these regulations, OSHA, workers, the workers' representatives, or health professionals may obtain medical records, records of exposure to workplace hazards, and records or any analysis of these records that are maintained by the workers' employers or former employers. Employers may withhold information they consider to be trade secrets.

Community Right to Know. Under the Superfund Amendments and Reauthorization Act (SARA, Title III), firms are required to provide the Environmental Protection Agency (EPA) with information about the release of toxic chemicals into the environment, inventories of toxic chemicals, and other information needed for emergency planning. This information is publicly available from a computerized storage and retrieval file (TRI) described below. Several states and cities also have enacted legislation granting communities and employees the right to acquire information about workplace and environmental hazards.

Information about Pesticides. Under FIFRA, pesticides—including insecticides, fungicides, herbicides, and rodenticides—must be labeled in accordance with regulations issued by the EPA.

Labor Unions' Right to Information. Under a precedent-setting case (*Oil Chemical and Atomic Workers v. National Labor Relations Board,* 711 F2d 348 [1983]), labor unions have an independent right to information about hazardous materials that is necessary for collective bargaining.

Reference Texts

Burgess WA. *Recognition of Health Hazards in Industry: A Review of Materials and Processes.* New York, NY: John Wiley & Sons; 1981.

Checkoway H, Pearce NE, Crawford-Brown DJ. *Research Methods in Occupational Epidemiology.* New York, NY: Oxford University Press; 1989.

Clayton GD, Clayton FE, eds. *Patty's Industrial Hygiene and Toxicology.* 3 vols. 3rd ed. New York, NY: John Wiley & Sons; 1978.

LaDou J. *Occupational Medicine.* Norwalk, Conn: Appleton & Lange; 1990.

Levy BS, Wegman DH, eds. *Occupational Health: Recognizing and Preventing Occupational Diseases.* 2nd ed. Boston, Mass: Little, Brown; 1988.

Monson RR. *Occupational Epidemiology.* 2nd ed. Boca Raton, Fla: CRC Press; 1990.

Parmeggiani L, ed. *Encyclopedia of Occupational Health and Safety.* 3rd ed., rev. Geneva, Switzerland: International Labour Office; 1983.

Plog BA, ed. *Fundamentals of Industrial Hygiene.* 3rd ed. Chicago, Ill: National Safety Council; 1988.

Proctor NH, Hughes JP, Fischman ML. *Chemical Hazards of the Workplace.* 2nd ed. Philadelphia, Pa: JB Lippincott; 1988.

Raffle PAB, ed. *Hunter's Diseases of Occupations.* London, England: Hodder & Stoughton; 1987.

Rom WN, ed. *Environmental and Occupational Medicine.* Boston, Mass: Little, Brown; 1983.

Rosenstock L, Cullen MR. *Clinical Occupational Medicine.* Philadelphia, Pa: WB Saunders; 1986. Blue Book Series.

Sax NI. *Dangerous Properties of Industrial Materials.* 6th ed. New York, NY: Van Nostrand Reinhold; 1984.

Waldron HA. *Occupational Health Practice.* 3rd ed. London, England: Butterworths; 1989.

Zenz C, ed. *Occupational Medicine: Principles and Practical Applications.* Chicago, Ill: Year Book Medical Publishers; 1988.

Key Scientific Journals

(In Occupational Medicine, Industrial Hygiene, and Occupational Injuries, with city of publication)

Accident Analysis and Prevention, Oxford, England

American Association of Occupational Health Nursing (AAOHN) Journal, Thorofare, NJ

American Industrial Hygiene Association Journal, Akron, OH

American Journal of Industrial Medicine, New York, NY

Annals of Occupational Hygiene, Oxford, England

Applied Industrial Hygiene, Akron, OH

Archives of Environmental Health, Washington, DC

Audiology, Basel, Switzerland

British Journal of Industrial Medicine, London

Contact Dermatitis, Copenhagen

Ear and Hearing, Baltimore, MD

Environmental Health Perspectives, Research Triangle Park, NC

Injury, Bristol, England

International Archives of Occupational and Environmental Health, Berlin, Germany

Journal of the Air Pollution Control Association, Pittsburgh, PA

Journal of Environmental Health, Denver, CO

Journal of Occupational Accidents, Oxford, England

Journal of Occupational Medicine, Baltimore, MD

Journal of the Society of Occupational Medicine, Edinburgh, Scotland

Journal of Trauma, Baltimore, MD

Occupational Health, London, England

Occupational Health Review, Ottawa, Ontario, Canada

Occupational Health and Safety, Waco, TX

Scandinavian Journal of Work, Environment, and Health, Helsinki, Finland

State of the Art Reviews: Occupational Medicine, Philadelphia, PA

The Bureau of National Affairs (BNA) in Washington, DC, publishes several serials relating to occupational and environmental regulations, court decisions, and policy matters. These publications include the *Occupational Safety and Health Reporter, Environmental Reporter, Chemical Regulation Reporter, Product Safety and Liability Reporter*, and *Toxic Law*

Reporter. Court decisions and full opinions on cases brought before the Occupational Safety and Health Review Commission, federal district courts, and the Supreme Court are routinely published.

Abstracting Services

Abstracting services review published literature and classify and publish abstracts.

Abstracts on Health Effects of Environmental Pollutants
Biosciences Information Services
2100 Arch St.
Philadelphia, PA 19103

Excerpta Medica
Occupational Health and Industrial Medicine
Environmental Health and Pollution Control

Journal Information Center
Elsevier Science Publishing Co., Inc.
52 Vanderbilt Avenue
New York, NY 10017
212-370-5520

Elsevier Science Publishers BV
P.O. Box 211
1000 AE Amsterdam
The Netherlands
020-5803911

Industrial Hygiene Digest
Industrial Health Foundation
5231 Centre Avenue
Pittsburgh, PA 15232

Computerized Data Bases

Computerized data sources provide interactive search and retrieval of bibliographic information, information about organizations, or information about chemicals to registered subscribers with a personal computer. The service is provided either over phone lines or by CD-ROM (compact disk-read only memory). In most instances, searches on specific topic areas can be arranged by contacting the data base manager or a commercial vendor. Services by the National Library of Medicine (NLM), NIOSH, and the International Labour Organization (ILO) are described below.

National Library of Medicine (U.S.). The NLM provides computerized search and retrieval of worldwide medical literature 24 hours a day, except for a brief daily maintenance period. Communication software (called Grateful Med, available for both DOS and Apple systems) and more information about any of the sources listed below, including training in their use, are available from

Medlars Management Section
National Library of Medicine
Bethesda, MD 20894
800-638-8480, or 301-496-6193 in Maryland.

The major files are MEDLINE and TOXLINE (see below). Files include the following, in alphabetical order:

CCRIS is the Chemical Carcinogenesis Research Information System developed and maintained by the National Cancer Institute (NCI). It contains about 1200 references to scientific journals, reports by the NCI, and a special core set of sources. Access and searching for references is managed through the TOXNET system (see below).

ChemID (Chemical Identification) is a chemical dictionary and thesaurus useful for identifying the names of chemicals. It contains about 200 000 records with information on the most commonly occurring substances of biomedical and regulatory interest. This data file can be searched by the Chemical Abstracts Services (CAS) registry number, other identifying numbers, molecular formulas, and names of substances. References can then be used to guide searches through other data files.

CHEMLINE (Chemical Dictionary Online) is a chemical dictionary file using data from a variety of services, including CAS. It enables search and retrieval of information for nearly 1 million chemical substances. The file contains CAS registry numbers, molecular formulas, chemical index nomenclature, generic and other names, classification codes, and a locator designation that leads to other files in the NLM system and to the inventory maintained by EPA.

CHEMLEARN is an interactive training service designed to teach users how to use CHEMLINE.

The Directory of Biotechnology Information Resources (DBIR) is a multicomponent data bank containing information related to biotechnology, including online data bases, publications, organizations, collections, and repositories of cells.

The Directory of Information Resources Online (DIRLINE) is an online data base concerned with information resource centers. Organizations include federal, state, and local government agencies; poison control centers; information and referral centers; professional organizations; self-help and support groups; voluntary associations; academic and research institutions; hospitals; libraries; and museums. It contains about 15 000 records.

ETIC (Environmental Teratology Information Center Database for current references) and ETICBACK (the backfile for ETIC, containing older references from 1950) are bibliographic data bases covering teratology and developmental toxicology. There are about 50 000 references included. Access is managed through the TOXNET system (see below) and can be searched in the same way as MEDLINE records, as well as by chemical names and CAS registry numbers.

HSDB (Hazardous Substances Data Bank) is a factual, nonbibliographic data base covering about 4200 chemical substance records. Information is peer reviewed and derived from a core set of standard texts, government documents, technical reports, and journal literature. HSDB includes emergency handling procedures, occurrence in the environment, extent of human exposure, detection methods, and regulatory requirements. All data are referenced. Access is managed through

TOXNET (see below). HSDB can be used to prepare MSDSs as required by OSHA and the Emergency Planning and Right-to-Know Act of 1986.

MEDLINE (MEDical information onLINE) is a bibliographic data base including about 6 million references in the scientific literature covering the fields of medicine, nursing, dentistry, veterinary medicine, and preclinical sciences. It contains all citations published in *Index Medicus, International Nursing Index,* and the *Index to Dental Literature.* It is updated weekly. Each entry includes author(s), title, and journal citation. Abstracts are included for almost all entries since 1986 and for many works published in prior years. It is possible to search by author, title, medical subject (employing MeSH [Medical Subject Headings], the controlled vocabulary used internationally), key words, journal, date of publication, and 21 other characteristics.

RTECS (Registry of Toxic Effects of Chemical Substances) is a file containing acute and chronic toxic effects of more than 100 000 chemicals. Regulatory requirements and exposure limits are included, and references are provided for all data. RTECS is developed and maintained by NIOSH, which has also published RTECS as books and microfiches since 1971. RTECS is part of TOXNET (see below). Data are arranged in four broad subject categories: identification of a substance, toxic effects, reviews of toxicology and carcinogenesis, and exposure standards and regulations. Administrative information, such as the last update and the record length, is also included.

TIP (Toxicology Information Program) is the name of the NLM program that develops and maintains CHEMLINE, DIRLINE, TOXLINE, TOXLIT, TOXNET, HSDB, RTECS, and CCRIS. TIP Files Demo Disk is an interactive demonstration of these services.

TOXNET (Toxicology Data Network) is a computerized system of files oriented to toxicology and related areas. It is an integrated system with sophisticated but user-friendly search and retrieval features. The component files are HSDB, RTECS, CCRIS, DBIR, ETIC, and EMICBACK. TOXNOTE is an on-line bulletin board that alerts users to current meetings, courses, and publications in toxicology.

TOXLINE, TOXLIT, and their respective backfiles, TOXLINE65 and TOXLIT65 (Toxicology Literature from Special Sources), are bibliographic data bases of nearly 3 million citations covering the pharmacological, biochemical, physiological, and toxicological effects of drugs and other chemicals. These files may be searched in the same way as MEDLINE files, as well as by chemical name, CAS registry number, and other features. Subfiles include ANEUPL (Aneuploidy File), ETIC, EPIDEM (Epidemiology Information System), HMTC (Hazardous Materials Technical Center), CIS (International Labour Organization), IPA (International Pharmaceutical Abstracts), NIOSHTIC, PESTAB (Pesticides Abstracts), PPIB (Poisonous Plants Bibliography), TOXBIB (Toxicology Bibliography), NTIS (Toxicology Document and Data Depository), BIOSIS (Toxicological Aspects of Environmental Health), CRISP (Toxicology Research Projects), and TSCATS (Toxic Substances Control Act Test Submissions).

TOXLEARN is an interactive training program designed to teach users how to use TOXLINE effectively.

TRI (Toxic Chemical Release Inventory) contains information on the annual estimated release of toxic chemicals into the environment based on data collected by the EPA. This file was mandated by SARA in 1986. Entries include the names of some 300 toxic chemicals; the names and addresses of facilities that manufacture, process, or use them; and the amounts of the chemicals released into the environment or transferred to waste sites.

National Institute for Occupational Safety and Health (NIOSH). NIOSHTIC is the computerized bibliographic data base developed and maintained by NIOSH. It contains over 150 000 citations to workplace safety and health literature and is updated quarterly with about 1500 new citations. It covers toxicology, epidemiology, occupational medicine, pathology, histology, physiology, metabolism, sampling and analytical methods, chemistry, industrial hygiene, health physics, control technology, engineering, behavioral sciences, ergonomics, safety, hazardous waste, occupational safety and health programs, and education and training. Entries can be searched by subject, chemical name, or CAS registry number, or by keywords; NIOSHTIC does not use a controlled vocabulary such as MeSH. It includes references not only to the scientific literature but also to NIOSH publications, including Criteria Documents, Current Intelligence Bulletins, Health Hazard Evaluation Reports, Technical Assistance Reports, conference proceedings, and selected references from CIS (see below). It is available on-line and in CD-ROM format. Entries since 1984 are available through the NLM, and the entire file is available through several vendors. Contact NIOSH for more information:

4676 Columbia Parkway
Cincinnati, OH
513-533-8302

NIOSHTIC is available in Canada, Latin America, and Europe as follows:

Canadian Centre for Occupational Health and Safety
250 Main Street East
Hamilton, Ontario L8N 1H6
800-263-8276

Pan American Health Organization
Biblioteca de ECP
Apartado Postal 37-473
06696 Mexico, DF

National Institute of Occupational Health
S-171 84 Solna, Sweden
46-8-7309000

International Labour Organization (ILO). CIS (International Occupational Safety and Health Information Center) maintains a computerized bibliographic data base covering the world literature on the recognition, evaluation, and control of occupational health and safety hazards, occupational medicine, toxicology, epidemiology, and industrial safety. It includes references to books, articles, conference proceedings, chemical information sheets, training materials, safety posters, films, laws, and regulations. Information is submitted by government agencies, trade unions, industries, and private publishers from over 50 countries in more than 30 languages. The CIS data base is available on-line or in CD-ROM format through commercial vendors. It is also available in printed versions and in microfiche. Information about subscribing can be obtained from

International Labour Organization
CH-1211 Geneva 2,
Switzerland
telephone: 41 22 799 67 40

CIS is available from the Canadian Centre for Occupational Health and Safety (see above) and from the Health and Safety Executive, UK.

Legislation and Case Law. CHEMLAW is a computerized data service pertaining to statutes and regulations concerned with chemical substances. Data include CAS registry numbers, chemical names, full texts, and summaries of regulations. A subfile of CIS, it can also be accessed through DIALOG, a commercial vendor.

LABORLAW is a computerized data service that includes legal decisions on occupational safety and health. Also included are decisions of the Occupational Safety and Health Review Commission, the Mine Safety and Health Review Commission, the National Labor Relations Board, federal and state courts, and arbitration awards. The vendor is DIALOG.

Worker Education and Training

Training is an essential element of any successful workplace hazard control program. Yet training alone is not an adequate prevention strategy.

Some believe that training alone can prevent injuries and illnesses, but this assumes that occupational disease and injuries are the result of worker ignorance. This idea was popular in the early part of this century. According to early theories, developing safety consciousness in workers was a primary solution to workplace injuries. This notion remains popular, with ongoing debates as to what percentage of accidents is due to unsafe acts as opposed to unsafe conditions.

Yet modern research points to a multiple causation theory of accidents. This theory suggests the need for multifaceted prevention strategies, in which training is but one component. Taken together, the four major elements include:

- management commitment,
- workplace analysis,
- hazard prevention and control, and
- safety and health training.

Training is most effective when placed in this context of a comprehensive prevention strategy. The Occupational Safety and Health Administration (OSHA) has proposed voluntary guidelines for the management of comprehensive safety and health programs in the workplace.

Objectives of Training

In its narrowest form, training can provide the skills and motivation for workers to use personal protective equipment properly and follow safety procedures. Safety and health training should be an integral part of job skills training. A competent worker is more likely to be a safe worker. But, in its broadest form, training can go beyond affecting individual compliance. It can also prepare workers to take an active role in recognizing and controlling hazards.

Workers have an invaluable role to play in preventing injury and illness in the workplace. Technical experts who do not know a job from daily experience cannot anticipate the full range of potential hazards or potential solutions. Good training can unleash the wealth of knowledge held by workers—the practical experts—and provide the opportunity for them to share their knowledge with the technical experts, ideally before problems occur.

Informed and active workers not only protect themselves but benefit management as well, both by reducing lost time and compensation costs and by increasing productivity. They also protect society as a whole. Costly government inspection and enforcement programs have a much smaller role to play in a society in which effective cooperation is taking place between informed labor and management. And, of course, workplace health and safety problems do not stop at the plant gate: Toxic substances that are not properly controlled in the workplace can become a community problem.

Unfortunately, much of the worker training offered in the workplace is not intended to empower workers and use their expertise. Some training even encourages passivity in workers by seeking compliance with work rules without allowing for worker input. Worker passivity, however, is dangerous because it tolerates hazards rather than controlling them.

Training to overcome passivity should include how to identify and control hazards as well as how to get problems corrected (using company procedures or legal enforcement). A comprehensive workplace prevention policy would be one that requires this type of action-oriented training and in turn uses workers as full participants in the process of making the workplace safe (through the health and safety committees, union safety representatives, protection for whistle-blowers, or other measures). Model worker participation programs such as this exist in several countries, including Sweden, Australia, and Canada.

A Quality Training Process

Despite the lack of any such policy in the United States, it is possible to design and provide high-quality training that empowers workers and encourages participation. The first step toward designing successful training is to understand that training is a continuing process, not an event. It is a process that requires careful and skillful planning through each major stage.

Step One: Assess Needs. A thorough needs assessment forms the foundation for the entire planning process. This includes both a hazards assessment—identifying high-priority problems to be addressed—and a profile of the target population. The target population profile attempts to answer a broad set of questions: Who can most benefit from training? What training has the target population already received? What knowledge and experience will the trainees bring to the process? What is the ethnic and gender makeup of the work force? What is the literacy level of the workers and what languages do they speak? Whom do they respect and whom do they mistrust? Needs assessment can be based on questionnaires, review of documents, observation in the workplace, and interviews with workers and their union representatives.

Step Two: Gain Support. Successful training also relies on identifying and involving key actors. The target population of training must be involved in the planning process; it is difficult to gain their trust without having sought their input. Who are the other key actors? Is it high-level management, who must make the necessary resources available? Is it the union, which must provide access to the work force or give the training credibility? Can a government or community organization provide the training, support, or follow-up activities? Whoever has a key role to play must be involved in the process through cosponsorship, participation on a planning committee, or other means.

Step Three: Establish Training Objectives. Using information from the needs assessment, the planning team can identify objectives. A common mistake is to assume that the objective of training is to present information. What is *presented* matters less than what the target population *receives*. In addition, one must consider not just what the learner should *know* (knowledge objectives) but also what the learner should *believe* (attitude objectives) and do (behavior objectives) as a result of training.

There is a hierarchy of these objectives. Knowledge objectives are the easiest to achieve (but by no means easy); attitude objectives are more challenging; and behavior objectives are the hardest. For example, it is possible to communicate the risks of asbestos to steamfitters. It is more difficult to change what steamfitters believe (eg, to convince them that they and their fellow workers are at risk and that something can be done about it). And the real challenge is to change their behavior—to get them to insist on using less hazardous substitute materials or to be sure that all necessary safety procedures are observed when they are working with asbestos.

HIERARCHY OF TRAINING OBJECTIVES

BEHAVIOR

ATTITUDES

KNOWLEDGE

To summarize before going on to the actual design of training sessions: (a) The goal of training must be to empower workers to act intelligently, not to control them. (b) Workers must be involved in planning their own training. (c) Training should be aimed at providing what workers most need to know, believe, or do. (d) For both ethical and practical reasons, it is essential that learners agree with the learning objectives.

Step Four: Select Training Methods. It is important to select the right methods for the chosen objectives. In general, the more ambitious the objectives, the more intensive the methods must be. Selection of appropriate methods should be based on basic principles of adult education, which include the following:

- People learn best by building on what they already know, by incorporating new ideas into their already vast reservoir of learning. Adults wish to be respected for their experience in life. Therefore, effective methods are those that draw on participants' own knowledge.

- People learn in different ways. Each person has a particular learning style. In a group, some will learn best by reading, some by listening, and some by practicing. It is important to offer training in more than one way. Variety not only ensures that each cognitive style is addressed but also provides repetition to reinforce learning and, of course, combats boredom.
- People learn better through active, participatory methods than through passive measures. Lectures and written materials have their place in a full repertoire of methods. But case studies, role plays, hands-on simulations, and other small-group activities that allow each individual to be involved are more likely to result in the retention and application of new learning. Participatory methods require more training time, smaller groups, and perhaps different skills than those that many trainers currently possess. But to increase the *impact* of training, participation is essential.

Whatever methods are selected, the profile of the work force must be considered. For example, if literacy is low, the trainer should use oral methods or highly graphic visuals. If the target population speaks a variety of languages, the trainer should use a multilingual approach.

Step Five: Implement Training. Actually conducting a well-planned training session becomes the easiest part of the process; the trainer simply needs to carry out the plan. The trainer is a facilitator who takes the learners through a series of activities designed to (a) explore new ideas or skills, (b) share their own thoughts and abilities, and (c) combine the two. If the session is planned well, including logistics (a comfortable facility, good outreach to get people there, etc), everything should run itself.

Step Six: Evaluate and Follow Up. Evaluation is often the forgotten step in the training process, but it serves several essential purposes. It allows the *learner* to judge his or her progress toward new knowledge, attitudes, or abilities; it allows the *trainer* to judge the effectiveness of the training to decide what has been accomplished; and it can document the success of training to justify future expenditures of resources. Follow-up is also critical. It involves planning how to reinforce new learning and how to support the application of new knowledge, inspiration, or skills resulting from training. For example, follow-up sessions to explore the application of learning can be scheduled; individual or small group consultations may be offered.

SUMMARY OF THE TRAINING PROCESS

- ASSESS NEEDS
- GAIN SUPPORT
- SET OBJECTIVES
- SELECT TRAINING METHODS
- IMPLEMENT TRAINING
- EVALUATE AND FOLLOW UP

Making a Commitment

A thorough, step-by-step training process requires a significant commitment of time and resources. All too often, training is conducted in the most expedient manner to meet minimal legal requirements (*see* box) rather than to educate the work force effectively. Some steps that could be taken to improve the quality of worker health and safety training include the following:

- education of the management community as to the need for worker training in its broadest form;

- enhanced government regulations, including a generic training standard that applies to all workers as well as a right for workers to participate fully in workplace health and safety programs;
- greater inclusion of educators in health and safety programs; and
- innovative programs to reach workers—especially nonunionized workers, women, minorities, and young workers—who are not getting the training they need in the workplace (eg, young workers could be reached through high school curricula addressing workplace health and safety).

Further Reading

American Conference of Governmental Industrial Hygienists. *Threshold Limit Values and Biological Exposure Indices for 1990-1991.* Cincinnati, Ohio: American Conference of Governmental Industrial Hygienists; 1990.

Boden LI. Government regulation of occupational safety: underground coal mine accidents, 1973-75. *Am J Public Health.* 1985;75(5):497-501.

Boden LI, Hall JA, Levenstein C, Punnett L. The impact of health and safety committees: a study based on survey, interview, and Occupational Safety and Health Administration data. *J Occup Med.* 1984;26(11):829-834.

Freeman RB, Medoff JL. *What Do Unions Do?* New York, NY: Basic Books; 1984.

Markowitz G, Rosner D. More than economism: the politics of workers' safety and health, 1932-1947. *Milbank Q.* 1986;64(3):331-354.

Mendeloff J. The role of OSHA violations in serious workplace accidents. *J Occup Med.* 1984;26(5):353-360.

OSHA Training Requirements

OSHA has no generic training requirements; rather, it has "Voluntary Training Guidelines." However, more than 100 of OSHA's current standards require training. Some of the standards with significant worker training requirements are listed below.

1. Hazard Communication (referring broadly to toxic substances) (29 CFR Part 1910, Subpart 2)
2. Specific Toxic and Hazardous Substances (including asbestos, lead, vinyl chloride, inorganic arsenic, cotton dust, coke oven emissions, and ethylene oxides; and other substances) (29 CFR, Part 1910 Subpart 2)
3. Hazardous Waste Operations and Emergency Response (29 CFR, Part 1910 Subpart 14)
4. Special Industries (including pulp, paper, and paperboard mills; laundry machines and operations; sawmills; and telecommunications) (29 CFR, Part 1910 Subpart R)
5. Machinery and Machine Guarding (29 CFR, Part 1910 Subpart O)
6. Welding, Cutting, and Brazing (29 CFR, Part 1910 Subpart Q)
7. Farm Equipment (29 CFR, Part 1928 Subparts C & D)
8. Construction Safety Orders (including an accident prevention program and a general safety and training requirement) (29 CFR, Part 1926)
9. Machine Training Requirements (29 CFR, Parts 1915, 1917, 1918)
10. Federal Employee Programs (including general training requirements for top management; supervision, safety, and health specialists and inspectors; health and safety committee members; employees; and employee representatives) (29 CFR, Part 1960 Subpart 14)

A complete guide to "Training Requirements in OSHA Standards and Training Guidelines" is available from the OSHA Publications Office, U.S. Department of Labor, 200 Constitution Ave., NW, Room N3101, Washington, DC 20210. This publication also includes OSHA's "Voluntary Training Guidelines for Employers."

Robertson LS, Keeve JP. Workers' injuries: the effects of workers' compensation and OSHA inspections. *J Health Polit Policy Law*. 1983;8(3):581-597.

Ruttenberg R. *The Role of Labor Management Committees in Safeguarding Worker Safety and Health*. Washington, DC: U.S. Department of Labor, Bureau of Labor-Management Relations and Cooperative Programs; 1988:i. BLMR 121.

Stone KVW. The post-war paradigm in American labor law. *Yale Law Journal*. 1981;90(7):1509-1580.

Weeks JL, Fox MB. Fatality rates and regulatory policies in bituminous coal mining, United States, 1959-1981. *Am J Public Health*. 1983;73(11):1278-1280.

Weil D. *Government and Labor at the Workplace: The Role of Labor Unions in the Implementation of Health and Safety Policies*. Cambridge, Mass: Harvard University; 1987. Dissertation.

Part 2
Occupational Diseases and Injuries

Acne and Other Hair Follicle Abnormalities

ICD-9 706.0, 706.1, 704

Identification

Acne vulgaris is characterized by open and closed comedones (blackheads and whiteheads), erythematous papules and nodules, pustules, and large cysts. Acne vulgaris may be aggravated by occupational factors such as exposure to cutting oils and machine grease. Other pilosebaceous follicle abnormalities appear as (1) follicular papules (round elevated lesions < 1 cm in diameter around hair follicles), pustules, and nodules (oil folliculitis); or (2) comedones (follicular impactions of keratin and lipids), hyperpigmentation, and oil cysts (chloracne).

Occurrence

The epidemiology of pilosebaceous follicle abnormalities depends on the particular agent responsible and the resultant disorder. (*See* Contact Dermatitis, Irritant, for data on incidence rates for occupational skin diseases.) Machine-tool operators who are exposed to heavy oils, such as insoluble cutting oils and greases, may be at risk for idiopathic acne or for developing oil folliculitis. Workers at risk for chloracne include those exposed to chlorinated hydrocarbons, including herbicide manufacturers.

Causes

Both acne and folliculitis can be caused or aggravated by oil blockage of follicular orifices.

Pathophysiology

The blockage of follicular orifices may be due to the physical presence of the oil or grease within the follicular orifice or to changes in the epidermal cells lining the follicular orifices. Follicular blockage may cause bacterial overgrowth, accumulation of sebum within the follicle, follicular rupture, and inflammation.

Prevention

Exposure to the responsible agents may be minimized or avoided by using gloves, protective clothing, adequate ventilation, and closed systems.

Other Issues

Although chloracne is not itself a disabling illness, it is important evidence of percutaneous absorption of chloracnogens, which have been associated with hepatic damage and malignancies in animals.

MB, KA

Further Reading

Adams RM. *Occupational Skin Disease.* Philadelphia, Pa: WB Saunders; 1990.

Arndt KA, Bigby M. Skin disorders. In: Levy BS, Wegman DH, eds. *Occupational Health: Recognizing and Preventing Work-Related Disease.* 2nd ed. Boston, Mass: Little, Brown; 1988:371-385.

Suskind RR. Chloracne, the hallmark of dioxin intoxication. *Scand J Work Environ Health.* 1985;11:165-171.

Acquired Immunodeficiency Syndrome (AIDS)—*See* Human Immunodeficiency Virus (HIV) Infection/Autoimmune Deficiency Syndrome (AIDS)

Angiosarcoma, Hepatic—*See* Hepatic Angiosarcoma

Aplastic Anemia

ICD-9 284.8
SHE/O

Identification

Aplastic anemia is a bone marrow stem cell disorder characterized by reduced hematopoietic tissue and consequent deficiencies in circulating white cells, red cells, and platelets. Deficiency may involve only one or two of the blood elements, and varying degrees of pancytopenia may precede frank bone marrow failure. Some cases evolve into leukemia. The case-fatality rate is approximately 50% within 1 year of onset, owing primarily to infection and hemorrhage.

Occurrence

Annual incidence is 5 to 10 per million persons in Western nations, with a rising age-specific incidence from 4 per million in children to 60 per million in people over age 65. This low rate of occurrence makes it unusually difficult to conduct epidemiological studies with adequate power to detect increased risks for specific exposures, and thus much etiologic knowledge is based on case reports. In the U.S., approximately 50% of cases are attributed to a defined agent, most often a therapeutic or infectious agent. Rates of occurrence have not been calculated for occupational groups known to be at risk. Many workers in the petrochemical and pharmaceutical industries have potential exposure to benzene, although compliance with the OSHA standard of 1 ppm TWA exposure should usually keep exposures below the range in which cases have been reported. Other groups exposed to benzene include garage, gas station, rubber, and electronics workers. Various glycol ethers have been associated with aplasia in animals and humans, and although they are not as firmly established as causes of aplastic anemia as benzene is, their widespread use as solvents in paints, coatings, and other products potentially puts many people at risk.

Causes

Occupational

Benzene. This is the classic occupational cause of aplastic anemia. There is an apparent relationship between exposure concentration of benzene and development of aplastic anemia. Prior to development of the disorder, lesser degrees of reversible depressions in one or more cell lines have sometimes been observed, such that repeated inhalations of benzene in excess of 100 ppm may be expected to cause peripheral cytopenia of at least one blood element in most individuals. In highly exposed groups, the incidence of aplastic anemia has been estimated at about 3% to 4%. There is probably a threshold below which aplastic anemia does not occur (<10 ppm TWA), although subclinical marrow toxicity (based on animal studies) and carcinogenic effects do not show such a threshold.

Ionizing Radiation. Aplastic anemia from ionizing radiation is also characterized by a dose-response relationship. There is a threshold for marrow toxicity of approximately 1.25 grays (1 Gy = 100 rads of energy absorbed), with a dose of 5.0 Gy causing 50% mortality from aplasia. More intense exposure also increases toxicity.

Pesticides. There are multiple case reports of aplastic anemia following relatively high exposures to Lindane (hexachlorobenzene). Animal data have not been confirmatory. Chlordane/heptachlor termiticides also have been implicated in case reports.

Ethylene glycol ethers. Ethylene glycol monomethyl ether and other glycol ethers have been reported to cause cytopenia and hypoplastic bone marrow in groups of workers, sometimes with relatively low exposure.

Arsenic. Therapeutically used organic arsenicals are documented causes of bone marrow depression. Inorganic arsenic exposure may also cause pancytopenia, with recovery expected on withdrawal from exposure. Pancytopenia is unusual without other signs of significant inorganic arsenic poisoning.

Trinitrotoluene. Large outbreaks of aplastic anemia have been reported in the munitions industry, often accompanied by hemolytic anemia, methemoglobinemia, dermatitis, and hepatitis.

Nonoccupational

Chloramphenicol. For this classic example and most other pharmaceutical agents, the mechanism of disease is idiosyncratic rather than dose dependent, occurring about once in 50 000 patients. This idiosyncratic reaction should not be confused with the predictable and mild suppression of erythropoiesis (production of red blood cells), which commonly accompanies treatment with this agent. There is no threshold dose for occurrence of aplastic anemia. Approximately 100 other drugs also have been reported to cause aplastic anemia, including phenylbutazone, sulfa drugs, and gold compounds.

Systemic Illness. Viral hepatitis, other viral infections, miliary tuberculosis, and systemic lupus erythematosus have all been demonstrated to result in aplastic anemia.

Chemotherapeutic Agents. Alkylating agents, antimetabolites, mitotic inhibitors, and anthracyclines (alone or with ionizing radiation) predictably cause marrow aplasia in a dose-dependent fashion. Risk is potentially present for pharmaceutical workers, although the apparent dose dependence of this toxicity probably offers protection.

Pathophysiology

Pathogenesis of aplastic anemia is complex and not completely understood. The disorder is most commonly thought to be caused by injury or destruction of a common pluripotential stem cell, which thus affects all three cell populations (red cells, white cells, and platelets). Immune-mediated marrow suppression has also been identified as a component, although not so much in occupational cases. In recognized occupational cases, aplastic anemia has been diagnosed soon after the responsible exposure. In contrast, in recognized cases of occupational leukemia that are associated with some of the same exposures, there is a longer delay between exposure and diagnosis of disease. Only for benzene, ionizing radiation, and cytostatic drugs is the pancytopenia established to be dose dependent, although even for benzene exposure, only a fraction of those people who have been highly exposed progress to frank aplastic anemia. Intermittent high exposures on a background of lower exposures have been implicated in animals and humans as potentially having etiologic importance. Recovery from aplastic anemia associated with benzene and other chemicals is substantially more frequent than when the disorder is due to idiopathic and idiosyncratic causes.

Prevention

For the idiosyncratic causes of aplastic anemia, prevention is problematic. The two major occupational causes, benzene and ionizing radiation, have well-developed primary and secondary preventive approaches. Because both agents are also carcinogenic, exposure is regulated more strictly than would be required for prevention of aplasia alone.

For ionizing radiation, strict monitoring of dose with environmental and personal sampling is indicated. The same is true for benzene. Biological monitoring of blood for benzene, or of urine for the benzene metabolite phenol, are not practical or effective, especially at the current, relatively low day-to-day exposures in the U.S. However, measurement of urinary phenol may allow the retrospective verification of acute benzene overexposures. Breath sampling for exhaled benzene and other volatile organic compounds is being developed as a research tool.

Serial monitoring with complete blood counts is of limited sensitivity at current levels of exposure. If aggregate blood counts in a population or individual were depressed, this would certainly indicate a myelotoxic or potentially myelotoxic problem. However, because of the apparent threshold discussed above, this effect is not likely to occur until exposures reach levels between one and two orders of magnitude above current exposure levels (PEL 1 ppm TWA). Thus, normal counts are reassuring in terms of aplastic anemia, but they are not adequate to ensure compliance with the standard and to reduce carcinogenic risk to a minimum.

Other Issues

Prognosis is poor, with a median survival of 3 months and a 50% case-fatality rate at 1 year. Quick removal from exposure of a patient with developing aplasia may enhance the chances of recovery. Progress is being made with immunologic therapies and bone marrow transplantation. Both during recovery and later, there is a substantially increased risk for development of acute leukemia (1% to 5%).

HK, BG

Further Reading

Cullen MR, Rado T, Waldron JA, Sparer J, Welch LS. Bone marrow injury in lithographers exposed to glycol ethers and organic solvents used in multicolor offset and ultraviolet curing printing processes. *Arch Environ Health.* 1983;38:347-354.

Epstein S, Ozonoff D. Leukemias and blood dyscrasias following exposure to chlordane and heptachlor. *Teratogenesis, Carcinog Mutagen.* 1987;7:527-540.

Jandl JH. Aplastic anemias. In: *Blood: Textbook of Hematology.* Boston, Mass: Little, Brown; 1987:115-152.

Laskin S, Goldstein BD, eds. Benzene toxicity: a critical evaluation. *J Toxicol Environ Health.* 1977;(suppl 2).

Arrhythmias/Sudden Death

ICD-9 427, 798.1

Identification

Arrhythmias are irregular heart rhythms that may be identified on a routine physical examination or electrocardiogram (ECG). Symptomatic individuals may complain of thumping in the chest, irregular heartbeat, or pause, which may produce anxiety and fatigue. If cardiac function is sufficiently compromised, individuals may present with syncope (fainting), dyspnea (shortness of breath), or angina (chest pain). Symptoms of occlusion of blood vessels, such as occur with a stroke, may result from blood clots (formed because of arrhythmias) that travel from the heart.

Physical examination, chest x-ray, and echocardiogram may indicate the underlying disease. The diagnosis may be made with a resting ECG or with further testing, such as stress testing or long-term ECG (Holter) monitoring. Invasive electrophysiological studies, with actual stimulation of the arrhythmia, may be required.

There are many different types of arrhythmias, with characteristic presentations and different prognoses. Major categories include deviations of sinus rhythms, supraventricular arrhythmias, sick sinus syndrome, ventricular arrhythmias, heart block, atrioventricular junctional variants, and escape rhythms.

"Sudden death" may be defined as a rapid, unexpected death occurring in an apparently healthy individual without a history of pertinent disease or in an individual with a stable, underlying, chronic disease. There is no one definition on which most investigators agree. Variations in definition, such as time interval from onset to death and inclusion of unwitnessed deaths, as well as the detail with which autopsy is performed, preclude a consensus on pathological findings.

Occurrence

There are an estimated 2 million hospital discharges annually in the U.S. with arrhythmias on the discharge summary. For 25% of these discharges, an arrhythmia is listed as the primary cause of admission. There are 2 million physician office visits in the U.S. per year for these conditions. The prevalence of arrhythmias in the general population is high. If all arrhythmias are included, abnormalities have ranged as high as 28% in normal children, 50% to 60% in college and graduate students, and approximately 90% in individuals over the age of 60. Significantly lower percentages are reported for arrhythmias with poor prognoses.

"Sudden death" is the single largest cause of death among middle-aged men in the U.S. It is responsible for approximately 50% of all deaths due to ischemic heart disease, and its incidence is similar to that of ischemic heart disease. Of all deaths that occur at work, "sudden death" is the leading cause, outnumbering traumatic fatalities by two to one in one descriptive epidemiological study.

Causes

The most common underlying cause of arrhythmias and sudden death is atherosclerosis. Other, less common causes include congenital malformations, connective tissue diseases with arteritis, infectious diseases, heart tumors, alcoholism, iatrogenic disorders, and electrolyte disorders.

Exposure to carbon monoxide (CO) and to many different types of solvents has been associated with both arrhythmias and sudden death. No particular type of arrhythmia is specific to chemical exposures. Solvents that have been reported in case reports to cause both arrhythmias and sudden death include fluorocarbons, benzene, chloroform, trichloroethylene, 1,1,1-trichloroethane, perchloroethylene, toluene, phenol, and gasoline. Methylene chloride, a weak sensitizing agent, is converted into CO in the body. There are no data on the incidence of arrhythmogenic effects from these widely used substances. With regard to sudden death, atherosclerotic change contributes significantly to the cause of the final arrhythmia. Most autopsies of individuals after sudden death do not reveal myocardial infarction.

Pathophysiology

Arrhythmias involve a disturbance in the normal sequence of cardiac activation and/or a change in rate or regularity beyond conventionally defined limits of normal.

Abnormalities in impulse formation or conduction underlie the different types of arrhythmias. The sinus node, from which impulses originate in the normal heart, is under the influence of the nervous system. Special ionic mechanisms and differentials are present in cells that initiate and conduct electrical impulses. Damage occurs to these cells from ischemic changes and, less frequently, from congenital, infectious, or immune abnormalities. The type of arrhythmia depends on the particular nodal or conduction cells that are affected.

Solvents have been shown to sensitize the heart to epinephrine, which lowers the initiation threshold, and to have a direct negative inotropic effect. Carbon monoxide has both a direct and indirect ischemic effect and a direct effect on mitochondrial enzymes.

Prevention

Primary prevention measures include reducing exposures to substances that cause ischemic heart disease, hypertension, or direct arrhythmogenic effects. The best-accepted approach would probably be reducing workplace exposures, combined with a program to reduce personal life-style habits that are risk factors for ischemic heart disease.

Whether individuals with underlying heart disease should be allowed to work in jobs where significant exposure to solvents or CO occurs, even within legal limits, needs to be assessed on an individual basis. The adequacy of current standards to protect against arrhythmias needs further investigation. Because only about half of all cardiac deaths have a clinical history of heart disease, medical screening may be only partially effective for identifying potential victims.

Other Issues

Individuals with underlying ischemic heart disease are theoretically more at risk than others to manifest arrhythmias or die suddenly with exposure to CO and/or solvents. Carbon monoxide levels from all sources, including cigarette smoking, are additive to CO inhaled at the workplace.

Exposure to levels of CO and/or solvents that are high enough to cause sudden death is most likely to occur in enclosed spaces, such as in a reactor vessel.

The following principles from animal studies need to be considered: (a) The threshold for initiation is independent of duration but dependent on dose. (b) The heart remains sensitized until the solvent level in the blood is cleared below the threshold of initiation, not when exposure in the air ceases. (Therefore, arrhythmias after work may be secondary to elevated levels of chemicals in the blood from work exposures.) (c) Halogenated solvents are more active than aliphatic solvents. (d) Other stresses, such as noise, lower the initiation threshold.

KR

Further Reading

Abedin Z, Cook RC, Milberg RM. Cardiac toxicity of perchloroethylene (a dry cleaning agent). *South Med J.* 1980;73:1081-1083.

Bass M. Sudden sniffing death. *JAMA*. 1970;212:2075-2079.

Magos L. The effects of industrial chemicals on the heart. In: Balazs T, ed. *Cardiac Toxicology.* Boca Raton, Fla: CRC Press; 1981;2:203-275.

Robinson CC, Kuller LH, Perper J. An epidemiologic study of sudden death at work in an industrial county, 1979-1982. *Am J Epidemiol.* 1988;128:806-820.

Sheps DS, Herbst MC, Hinderliter AL, Adams KF, et al. Production of arrhythmias by elevated carboxyhemoglobin in patients with coronary artery disease. *Ann Intern Med.* 1990;113:343-351.

Speizer FE, Wegman DH, Raminey A. Palpitation rates associated with fluorocarbon exposure in a hospital setting. *N Engl J Med.* 1975;292:624-626.

Zakhari S, Aviado DM. Cardiovascular toxicology of aerosol propellents, refrigerants and related solvents. In: Van Stee EW, ed. *Cardiovascular Toxicology.* New York, NY: Raven Press; 1982:281-314.

Asbestos-Related Diseases

Five different diseases are caused by exposure to asbestos: asbestosis, nonmalignant pleural disease, respiratory tract cancer (cancer of the lung, larynx, and pharynx), mesothelioma, and certain cancers of the gastrointestinal tract. Some studies have also associated cancer of the kidney with asbestos exposure, but other studies have not. Adverse effects of asbestos have been known since about the late 19th century. The purpose of this introduction is to describe disease control measures common to all asbestos-related diseases. (*See* Asbestosis, Laryngeal Cancer, Lung Cancer, Mesothelioma, and Pleural Diseases, Asbestos-Related.)

Asbestos is a naturally occurring class of silicate fibers mined primarily by open-cast mining. More than half of the world's production comes from mines in the Soviet Union and Canada. Other producing countries include South Africa, Zimbabwe, the U.S., Italy, China, and Australia. Worldwide production peaked at about 6 million tons in 1973.

Properties and Uses

Asbestos has very high tensile strength and durability. It resists both physical and chemical corrosion. It has been sprayed, woven, and molded. Because of its properties, it has been and is being used in many ways: for fire and electrical insulation, as filler in cement and other products (eg, pipes, floor tiles, wall coverings, roofing) to add durability, and in friction products such as brake shoes and clutch plates. It also has been used in filters.

Exposure

The route of exposure is primarily by inhalation of airborne fibers and secondarily by ingestion. Following inhalation, fibers are transported by the lung's clearance mechanisms. They may then enter the gastrointestinal tract by being coughed up and swallowed. They may also be ingested when contaminated items are consumed. Consequently, any process or activity that generates airborne asbestos dust—spraying, grinding, demolition, milling, drilling, sawing, packaging, etc.—creates a risk of exposure.

Occupational exposure has occurred among miners and millers of asbestos, makers of asbestos products, construction workers (eg, insulation workers, plumbers, and pipe fitters), electricians, sheet metal workers, and shipbuilders. In the past, major clusters of exposure and of asbestos-related disease have occurred in building construction workers, makers of asbestos products, and shipbuilders. There are lower rates of asbestos-related disease in miners than there are in people who work later in the cycle of mining, milling, manufacture, use, renovation, and disposal. Asbestos insulation was sprayed onto buildings during construction for fire protection until the EPA banned spray-on asbestos insulation in 1973. Asbestos has been commonly used as insulation on steam pipes and boilers. In shipbuilding, asbestos was used as thermal and electrical insulation on pipes and conduits and has been sprayed onto bulkheads. Current exposure can occur during building demolition and renovation, asbestos removal, ship breaking and repair, and motor vehicle maintenance (eg, brake shoes and clutch plate repair).

Use of asbestos in the U.S. was very common as a result of shipbuilding during World War II; however, it reached a peak in 1973 and declined steadily thereafter. Thus, although occupational exposure to asbestos in the U.S. was widespread, it is now decreasing. As of 1990, approximately 5 million workers were exposed.

Exposure also has occurred secondary to occupational exposure when asbestos workers have come home contaminated with fibers in their clothing or hair. Children, spouses, and others have been exposed in this way, as well as by living near establishments that use asbestos. Exposure has been reported near asbestos mines in Canada and California and near surface deposits in several locations, including Turkey and Corsica.

Because of their long latency, asbestos-related cases of mesothelioma and lung cancer are expected to persist into the next century. Past exposure may account for 2000 mesothelioma deaths and 4000 to 6000 lung cancer deaths per year in the U.S.

Fiber Types

There are two main types of asbestos: serpentine and amphibole. These differ primarily in their shape and molecular structure. Serpentine, named because of the curled, snakelike shape of the fibers, includes the subtype chrysotile, also called white asbestos. It typically occurs in bundles and fragments about 10 micrometers or more in length. Chrysotile asbestos accounts for more than 90% of consumption of commercial asbestos, so exposure to chrysotile is widespread. Amphiboles include the subtypes of amosite (brown asbestos), crocidolite (blue asbestos), actinolite, tremolite, and anthophyllite (the latter three of which occur in fibrous and other forms as a result of different patterns of crystalline growth). Amphiboles, which are straight and needle-like, are found in some deposits of chrysotile, talc, vermiculite, and other minerals. Amosite and actinolite are more common in North America, and crocidolite, anthophyllite, and tremolite are more common in Britain. Amosite is also found in soil in Turkey. Tremolite and anthophyllite fibers are not used commercially but frequently contaminate other types.

Although both serpentine and amphiboles can apparently cause cancer, there is some evidence that amphiboles are more likely to produce mesothelioma and lung cancer. The evidence for this hypothesis is indirect. In some epidemiological studies, the occurrence of mesothelioma is more frequent among workers exposed primarily to amphibole fibers than to chrysotile. However, in other investigations, mesothelioma also occurs among workers exposed only to chrysotile. Both fiber types produce mesothelioma in laboratory animals exposed by inhalation. Some laboratory studies show that chrysotile fibers may be more toxic than amphiboles.

Other evidence for this hypothesis comes from the distribution of fiber types in lung tissue. Amphiboles are more common in lung tissue of people autopsied for mesothelioma. Asbestos bodies (ie, iron-rich deposits around an asbestos core, also called ferruginous or coated bodies) are also found in lung tissue and are more likely to contain a core of amphiboles than of chrysotile. Uncoated fibers are more numerous than asbestos bodies in lung tissue. The reason for these differences in the occurrence of coated and uncoated fibers may be that chrysotile fibers are physically cleared more efficiently or are dissolved in lung tissue.

Either serpentine or amphibole asbestos can also produce asbestosis and nonmalignant pleural disease. (There is little information about differential associations with fiber type and risk of gastrointestinal cancers.) The mechanisms whereby these fibers produce fibrosis, nonmalignant pleural disease, or malignancy is unknown. Either shape or surface characteristics of the fibers may play an important role. Based on investigations in animals, which involved both inhalation and intrapleural implantation, carcinogenesis is influenced by fiber length and width. Thus, long fibers (8 um or longer) with a high aspect ratio (length to width) tend to be more carcinogenic than shorter (5 um or less) and thicker fibers. An alternative mechanism attributes carcinogenicity to fiber surface characteristics and immunologic mechanisms. If shape is the most important factor for producing disease, then substitute products such as fibrous glass or mineral wool, which have a similar shape, could pose a similar risk.

Prevention

Primary prevention of asbestos-related disease can be achieved by eliminating the use of this substance. Substitutes exist for most applications. Given the risk of disease and death even with apparently trivial exposures, reliance on even the most fastidious industrial hygiene techniques does not entirely eliminate risk. Additional use of asbestos has been banned or significantly restricted in Sweden, the U.S. (by the Environmental Protection Agency [EPA]), and other countries.

When elimination is not possible—as, for example, during asbestos removal from buildings, demolition of buildings containing asbestos, or ship breaking or repair—careful attention must be paid to hygienic practices. These should include total enclosure of an area under negative pressure, use of specially designed disposable clothing, wetting and other dust control techniques, and vacuum removal. Asbestos-removal protocols have been developed by the EPA and the National Institute of Building Sciences.

An important part of the successful control of exposure in the U.S. has been achieved by litigation. Many former asbestos workers with asbestos-related disease have sued asbestos manufacturers as third parties (*see* Government Regulation in chapter 2). Manufacturers have been found negligent for not warning asbestos workers of health risks about which the manufacturers knew or should have known. When these suits have been successful, awards have been significantly larger than any amounts that would have been paid under workers' compensation claims. Adverse publicity, the magnitude of awards, and projected future liabilities have prompted some asbestos manufacturers to claim bankruptcy because projected liabilities exceeded their net worth. Although litigation has done nothing to prevent disease for those people exposed in the past, it has served as an important object lesson and incentive for makers and users of asbestos both to inform and educate workers and to control exposure now in order to avoid similar liabilities in the future.

Opportunities for secondary prevention based on medical screening for early signs of asbestos-related disease are limited because treatment is often ineffective. Medical screening of workers known to have been or suspected of having been exposed to asbestos should focus on early signs of nonmalignant disease (pleural thickening, pleural plaques, and fibrosis). Workers with pleural thickening, pleural plaques, pleural calcification, or early signs of parenchymal fibrosis are at increased risk of mesothelioma, lung cancer, and progressive fibrosis. Bilateral pleural thickening on chest x-ray is a reasonably sensitive and specific indicator of asbestos exposure. In one investigation, the predictive value of bilateral pleural thickening for past exposure to asbestos was 80% if unilateral thickening and nonasbestos causes of pleural thickening (chronic renal failure, chest surgery or trauma, severe rheumatoid arthritis, bilateral empyema) were eliminated. Recovery of asbestos fibers or asbestos bodies in bronchoalveolar lavage fluids, though not a screening test, is also an indicator of past exposure. Successful prevention for workers with positive evidence of exposure is limited to restricting additional exposure to asbestos and other pulmonary hazards and advising workers of the significantly added risk of lung cancer if they smoke. Only smoking cessation has been shown to be helpful.

Because knowledge about how health effects differ with fiber type is incomplete, prevention strategies should be uniform for exposure to all types of asbestos. Exposure to a homogeneous type of fiber is rare in any case; most often, one encounters a mixture. Both OSHA and NIOSH recommend an exposure limit of 0.1 fibers per cubic centimeter (cc) (100 000 fibers per m^3), regardless of fiber type. The ACGIH, however, recommends a TLV of 0.5 fibers per cc for amosite, 0.2 for crocidolite, and 2.0 for all other forms. The OSHA standard (29 CFR 1910.1001), in addition to setting an exposure limit, also requires exposure monitoring, medical surveillance, worker training and education, shower and change facilities, and use of personal protective equipment.

In the past, limits on exposure were much more lenient. Prior to 1972, occupational exposure to 12 fibers per cc of air was permitted. OSHA reduced the limit to 2 fibers per cc in 1972 to prevent asbestosis. Then in 1985, the limit was reduced to 0.1 fibers per cc. Some malignancies will occur even at this exposure level.

Industrial Hygiene Measurement and Identification of Exposure

Airborne dust concentrations are measured not gravimetrically, as are most other dusts, but by counting the number of fibers for a given volume of air. Specific mineralogical characterization of asbestos fibers in samples usually requires both transmission or scanning electron microscopy to identify the shape, and x-ray dispersive analysis to identify the chemical composition. Fibers are difficult to identify positively with the ordinary light microscope, but phase-contrast light microscopes are often used for screening purposes.

JLW

Further Reading

Albelda SM, Epstein DM, Gefter WB, Miller WT. Pleural thickening: its significance and relationship to asbestos dust exposure. *Am Rev Respir Dis.* 1982;126:621-624.

American Thoracic Society. The diagnosis of nonmalignant diseases related to asbestos. *Am Rev Respir Dis.* 1986;134:363-368.

Becklake MR. Asbestos-related diseases of the lung and pleura: current clinical issues: review. *Am Rev Respir Dis.* 1982;126:187-194.

Castleman BI. *Asbestos: Medical and Legal Aspects.* New York, NY: Prentice-Hall; 1987.

Cullen MR. Controversies in asbestos-related lung cancer. In: Rosenstock L, ed. Occupational lung disease. *State of the Art Reviews: Occupational Medicine.* 1987;2:259-272.

Merchant JA, ed. Occupational Respiratory Diseases. Washington, DC: U.S. Government Printing Office; 1986. DHHS (NIOSH) publication 86-102.

National Institute of Building Sciences. *Model Guide Specifications: Asbestos Abatement in Buildings.* Washington, DC: National Institute of Building Sciences; 1986.

Rosenstock L, Hudson LD. The pleural manifestations of asbestos exposure. In: Rosenstock L, ed. Occupational lung disease. *State of the Art Reviews: Occupational Medicine.* 1987;2:383-407.

U.S. Environmental Protection Agency. Guidance for Controlling Asbestos-Containing Materials in Buildings. Washington, DC: U.S. Government Printing Office; 1985. EPA 560/5-85-024.

Asbestosis

ICD-9 501
SHE/O

Identification

Asbestosis is a pneumoconiosis produced by inhalation of asbestos fibers. It is a chronic disease with slow onset that usually requires several years of exposure, depending on the intensity of exposure. Clinically, it is characterized by diffuse interstitial pulmonary fibrosis, often accompanied by thickening and sometimes calcification of the pleura. Shortness of breath on exertion is the most common presenting symptom. A chronic dry cough is common, but the cough may be productive, especially among smokers. Finger clubbing may appear in advanced cases.

In most cases, the first and often the only physical sign is crackles, also known as rales, usually detected near the end of a full inspiration. Chest x-rays reveal small, irregular opacities commonly distributed in the middle and lower lung fields. With disease progression, all lung zones may be affected, and honeycombing, especially in the lower zones, is not unusual. Late manifestations can include an irregular diaphragm and cardiac border, which is associated with pleural plaques and diffuse pleural thickening.

Chest radiographs are often classified according to a standardized method developed by the International Labour Organization. Chest x-ray findings may be interpreted as normal in close to 20% of asbestos workers with fibrosis on microscopic examination of lung tissue.[1]

In diffuse interstitial fibrosis, particularly cases of low levels of profusion, other etiologies should be considered. These include desquamative interstitial pneumonitis, sarcoidosis, scleroderma, and lipoid pneumonia, all of which can present with irregular opacities in the lower lung fields. If nodular opacities are prevalent, the possibility of exposure to silica or coal dust should be considered.

Lung function tests can help to quantify the level of pulmonary dysfunction. In advanced asbestosis, vital capacity, functional residual capacity, and, hence, total lung capacity are reduced. Diffusion capacity is also decreased, and, in more advanced cases, resting arterial oxygen concentration is reduced. Earlier in the course of disease, arterial oxygen concentration may be reduced only with exercise. Carbon dioxide exchange is usually not affected. There may be a mixed restrictive and obstructive pattern in lung function, even in the absence of smoking. This is consistent with pathological observations that show peribronchiolar fibrosis early in the course of asbestosis, with distortion of the airways in advanced cases.

A positive identification usually requires a history of exposure to asbestos established by work history, by the presence of pleural thickening or plaques, or by asbestos bodies in sputum, and by sufficient latency (usually 15 or more years), combined with some or all of the following: (a) rales or crackles, (b) positive chest x-ray findings for fibrosis (which may only appear later in the course of disease), (c) reduced lung function (vital capacity, total lung capacity, or gas transfer), and (d) shortness of breath on effort.

Occurrence

The incidence of asbestosis depends on both duration and intensity of exposure. Prevalence increases among stable groups of asbestos workers with increasing length of exposure. Higher exposure produces asbestosis earlier. Estimates of prevalence have been reported as 2.5% to 4% with exposure between 50 and 99 fiber-years/cc (fibers/cc x years exposed) and as 6% to 8.5% between 100 and 149 fiber-years/cc. Even brief exposure can result in disease years later.

Prevalence studies have reported variable rates of pulmonary parenchymal fibrosis among asbestos-exposed workers. A 1965 cross-sectional survey of asbestos insulation workers found a 50% prevalence of radiographically determinable pulmonary fibrosis overall, with increasing rates of fibrosis associated with longer employment. Surveys of asbestos textile workers have found that between 6% and 40% of workers reveal fibrosis on x-ray, with higher rates associated with longer periods between first exposure and examination. Surveys of asbestos miners, millers, cement workers, railroad repair workers, plumbers, pipe fitters, and maintenance workers have also demonstrated significant levels of asbestosis.

In 1986, there were 180 deaths in the U.S. with asbestosis listed as the primary cause on the death certificate and 382 deaths with asbestosis listed as a contributing cause. These are most likely underestimates. Among insulation workers in New York and New Jersey, 7.7% of all deaths were due to asbestosis, with an average latency of 45 years. From 1965 to 1974, there were 877 compensation awards and 39 deaths due to asbestosis in Canada.

Causes

Asbestosis is caused by exposure to airborne asbestos dust, including both serpentine (ie, chrysotile) and amphibole fibers. There are no other causes, although the diagnosis can be mistaken, as indicated above.

Pathophysiology

Although the pathogenesis of asbestosis is unknown, a variety of pathological and physiological changes have been identified as a result of asbestos fiber inhalation in laboratory animals and in humans. Asbestos fibers deposited in the air exchange units are ingested by macrophages, resulting in release of inflammatory mediators, inflammation, and, eventually, scar formation. The pattern of scar formation is indistinguishable from that of other fibrotic lung diseases except for the presence of asbestos bodies, which are seen as markers of exposure. With progressive fibrosis, there is destruction of the normal lung architecture. Fibrosis is generally concentrated at the lung bases but may spread to include the entire lung.

Prevention

See Asbestos-Related Diseases.

Other Issues

Asbestos exposure has also been associated with respiratory tract cancer, mesothelioma, cancer of the gastrointestinal tract, and, in some reports, kidney cancer. These conditions are discussed elsewhere. There is no apparent interaction between asbestosis and tuberculosis.

There has been considerable litigation and contention concerning all aspects of asbestos-related disease. Controversy has spilled over into the scientific literature.[2-5] The ambiguity that results from disagreement among qualified experts should not be used to discourage prudent public health actions in favor of health protection.

DCC

References

1. Kipen HM, et al. Pulmonary fibrosis in asbestos insulation workers with lung cancer: a radiological and histopathological evaluation. *Br J Indus Med.* 1987;44:96-100.
2. Mossman BT, Gee JBL. Asbestos-related diseases. *N Engl J Med.* 1989;320:1721-1730.
3. Kern DG, Gee JBL, Corn M. Asbestos-related diseases. *N Engl J Med.* 1991;324:195-197. Letter.
4. Mossman BT, et al. Asbestos: scientific developments and implications for public policy. *Science.* 1990;247:294-301.
5. Nicholson WJ, Johnson EM, Melius JM, Landrigan PJ. The carcinogenicity of chrysotile asbestos. *Science.* 1990;248:795-799. Letter.

Further Reading

American Thoracic Society. The diagnosis of nonmalignant diseases related to asbestos. *Am Rev Respir Dis.* 1986;134:363-368.

Asbestos Working Group. *Asbestos-Related Diseases: Clinical, Epidemiologic, Pathologic, and Radiologic Characteristics and Manifestations.* Richmond, Va: American College of Radiology; 1982.

Dement JM, Merchant JA, Green FHY. Asbestosis. In: Merchant JA, ed. *Occupational Respiratory Diseases.* Washington, DC: U.S. Department of Health & Human Services; 1983:287-327.

Rom WN. Asbestos and related fibers. In: Rom WN, ed. *Environmental and Occupational Medicine.* Boston, Mass: Little, Brown; 1983:157-183.

Rom WN, Travis WD, Brody AR. Cellular and molecular basis of the asbestos-related diseases. *Am Rev Respir Dis.* 1991;143:408-422.

Selikoff IJ. Asbestos-associated diseases. In: Last JM, ed. *Maxcy-Rosenau: Public Health and Preventive Medicine.* 12th ed. Norwalk, Conn: Appleton Century Crofts; 1986;523-525.

Selikoff IJ, Lee DKH. *Asbestos and Disease.* New York, NY: Academic Press; 1978.

Asphyxiation

ICD-9 799.0, 986

Identification

Asphyxiation is any event that results in insufficient tissue oxygenation. Tissues can become oxygen starved if not enough oxygen is inspired or if cells are inhibited from using available oxygen. Thus, there are two broad categories of asphyxiants: simple and chemical.

Simple asphyxiants cause inadequate tissue oxygenation primarily by displacing oxygen within an environment—that is, the higher the concentration of a simple asphyxiant, the lower the concentration of oxygen. At some level, this begins to have adverse physical consequences (*see* Table 1). Therefore, it is the concentration of oxygen in air that is ultimately important. Simple asphyxiants, such as carbon dioxide, methane, ethane, argon, and nitrogen, are usually physiologically inert.

A chemical or toxic asphyxiant (carbon monoxide, cyanide, acrylonitrile, or hydrogen sulfide) exerts its effects by interfering with cellular metabolism, thereby causing cells to become starved for oxygen. Toxic asphyxiants, such as hydrogen cyanide, carbon monoxide, and hydrogen sulfide, work through a variety of mechanisms, discussed separately below.

Occurrence

Occupational deaths from asphyxiation are not common, although morbidity and mortality associated with exposure to confined spaces, oxygen-deficient environments, and toxic asphyxiants are probably substantially higher than are currently reported. Deaths due to asphyxiation usually occur rapidly as a result of unanticipated exposure. Most victims are men, and younger men may be at greater risk, possibly because they are given more hazardous tasks and/or because they lack experience.

Accurate incidence data for morbidity and mortality from asphyxiation are lacking. One group of investigators, using OSHA data, documented 423 occupational deaths from asphyxiation and poisoning during a 3-year period.[1] Given well-documented problems with current occupational data bases, this figure is probably a significant underestimation of the true magnitude of the problem. The most common cause of asphyxiation is mechanical, such as in a trench cave-in or immersion in a suffocating material such as grain. In the study cited above, 53% of all occupational deaths (223 of 423) fell into this category. Simple asphyxiants and oxygen-deficient environments accounted for 15% of the 423 deaths reported to OSHA between 1984 and 1986.

Manufacturing, oil and gas industries, construction, and utilities appear to account for most fatalities due to asphyxiation. Asphyxiation also has occurred in a wide range of occupations, including agricultural workers and emergency response workers.

Death due to simple asphyxiants most commonly occurs in relation to entry into a confined space. NIOSH defines a confined space as one that "by design has limited openings for entry and exit; unfavorable natural ventilation which could contain or produce dangerous air contaminants; and which is not intended for continuous worker occupancy."[2] In confined spaces, oxygen can be replaced by other gases, leading to an oxygen-deficient atmosphere or to an atmosphere with high concentrations of toxic gases. Workers can be overcome, and death can occur upon entry unless adequate fresh air or a self-contained breathing apparatus is available. Would-be rescuers are often overcome as well, leading to multiple fatalities.[3]

Causes and Pathophysiology

Simple Asphyxiants. At room temperature and sea level, normal inspired air consists of 21% oxygen by volume. As a class, simple asphyxiants—which include argon, nitrogen, hydrogen, helium, methane, ethane, and carbon dioxide—are biologically inert; they affect health by decreasing the percentage of oxygen in inspired air. This displacement causes the partial pressure of oxygen in the alveoli to decrease, and, ultimately, less oxygen is delivered to tissues. Many of the clinical effects of an oxygen-poor atmosphere cluster around the central nervous system (*see* Table 1).

The adverse health effects of simple asphyxiants also depend on the presence or absence of other variables. For example, the effects of asphyxiants can be accelerated or exacerbated by increased work pace, increased duration of exposure, underlying medical conditions of the exposed worker (eg, chronic lung disease), absence or improper use of personal protective equipment, inadequate ventilation, and other environmental conditions of the workplace (eg, high altitude, high temperature). Ultimately, all these factors operate either to decrease the amount of oxygen available to tissues or to augment the tissues' oxygen requirements, which cannot be matched by the available oxygen in the inspired atmosphere.

Table 1.—Physiological Effects of Oxygen Deficiency

Oxygen Concentration (% inspired)	Clinical Effects
16 – 21	Asymptomatic
12 – 16	Tachypnea, tachycardia, motor incoordination
10 – 14	Emotional lability, fatigue, exertional dyspnea
6 – 10	Nausea/vomiting, lethargy, possible unconsciousness
<6	Seizure, apnea, asystole

Table 1 implies that levels of inspired oxygen of 16% to 21% can be tolerated without adverse effects. However, this may not hold true for situations in which increased tissue oxygen requirements exist (eg, a heavy workload). Consequently, recognizing that oxygen requirements increase with exertion and are influenced by environmental factors such as altitude and heat, OSHA has set a standard that requires 19% oxygen as a minimum.

Chemical Asphyxiants. Chemical asphyxiants produce their symptoms by interfering with cellular oxygen utilization. The mechanisms by which each toxic asphyxiant interrupts cellular metabolism vary and will be addressed individually. However, as with simple asphyxiants, other variables influence the severity of clinical effects of any given toxic asphyxiant. Concentration of the gas, length of exposure, and adequacy of ventilation are important environmental conditions that influence toxicity. Individual factors, such as baseline health status (eg, cardiac or pulmonary disease) and improper use or nonuse of personal protective equipment (*see* Prevention below), can modify toxicity as well. Finally, concurrent exposure to other asphyxiants or gases may augment the potential toxicity of any given asphyxiant. Such is the case for fire fighters, who are exposed to high levels of carbon monoxide and hydrogen cyanide in fires in which pyrolysis of plastics occurs.

Carbon monoxide (CO): OSHA PEL = 50 ppm (8-hour TWA)
NIOSH REL = 35 ppm (10-hour TWA)

Carbon monoxide is a colorless, odorless, and tasteless gas produced as a by-product of incomplete combustion, such as in car exhaust. It is ubiquitous in our environment. Although exposure to CO is likely to be found in any industrial setting, it can also be present in agricultural and commercial worksites. Some occupations with well-recognized CO exposure are fire fighters, tollbooth operators, traffic police, coal miners, coke oven workers, and smelter workers.

It is estimated that more than 3500 deaths due to CO poisoning occur annually, including occupational and nonoccupational deaths.[4] Yet estimates of annual mortality and morbidity due to CO exposure, along with those due to other asphyxiants, are notoriously inadequate and probably grossly underreported. The majority of deaths attributed to smoke inhalation are probably caused by CO exposure.

Carbon monoxide has 250 times the affinity for hemoglobin that oxygen has, forming a compound referred to as carboxyhemoglobin (Hb_{CO}). A consequence of this avidity is that very small concentrations of CO in inspired air can have significant clinical effects (*see* Table 2). Carbon monoxide causes a displacement of oxygen from hemoglobin and subsequently inhibits further binding of oxygen to the hemoglobin molecule. In addition, as the blood concentration of carboxyhemoglobin increases, the oxygen dissociation curve shifts, making it progressively more difficult for oxygen to be released from hemoglobin for tissue uptake. Carbon monoxide also interferes with the cytochrome oxidase system, the core enzyme system for cellular oxidative metabolism, although the clinical significance of this has not yet been defined.[4]

Table 2.—Relationship of Carbon Monoxide Levels and Symptoms*

Intensity	Airborne CO Concentration	$\%Hb_{CO}$	Symptoms
Mild	.002 – .007	< 20	Nausea, tinnitus, dyspnea on exertion, headache
Moderate	.011 – .035	20 – 40	Fatigue, poor judgment, mental obtundation
Severe	> .035	> 40	Arrhythmias, death

*Adapted from Rosenstock, 1987.

The clinical effects of CO exposure depend on the degree of carboxyhemoglobinemia, which itself depends on the concentration of inspired CO, the duration of exposure, and the activity level of the exposed individual. At 200 ppm (or 0.02% inspired CO by volume), headache, tinnitus, and a sense of discomfort are frequent after several hours of exposure. At 800 ppm (0.08% inspired CO), frontal headache, nausea, and dizziness occur within an hour. Concentrations above 1600 ppm (0.16% inspired CO) can lead to narcosis, coma, and, eventually, death. Table 2 indicates at what Hb_{CO} level various clinical effects are expected. It is important to recognize that carbon monoxide's affinity for hemoglobin means that, even at low concentrations in inspired air, CO can have drastic clinical effects. The correlation between blood Hb_{CO} level and symptoms is not consistent. Severity of poisoning is judged clinically rather than through the determination of Hb_{CO} level.

Cigarette smokers often have high baseline levels of Hb_{CO} (up to 8%) and may be more susceptible than nonsmokers to adverse health effects of ambient CO exposure at lower concentrations. Exposure to CO at levels of activity that lead to an increase in minute ventilation, as occurs with heavy labor or high-altitude work, also increases Hb_{CO} levels and can produce symptoms and signs more rapidly. Despite some debate, experimental animal data support the position that coronary artery disease can be accelerated by CO exposure. Many case reports and some human and animal experimental data have shown that individuals with known coronary disease may have symptoms provoked by low levels of inspired CO.[5]

Hydrogen cyanide (HCN): OSHA PEL = 10 ppm (8-hour TWA)
NIOSH REL = 4.7 ppm
(10-minute Ceiling Limit)

Cyanide is a systemic poison with a long and notorious history. By binding with mitochondrial cytochrome oxidase, cyanide interrupts aerobic metabolism and forces cells to produce adenosine triphosphate through anaerobic (ie, lactate) metabolism. The clinical consequence of cyanide exposure is rapid cessation of cellular respiration, profound metabolic acidosis, and death. Cyanide can be inhaled as a gas (eg, HCN, acrylonitrile), ingested (eg, Laetrile), or absorbed cutaneously. There are a variety of compounds with CN^- as the leaving group.

Industrial uses for cyanide are found in electroplating, metal extraction (eg, gold from ore), fumigants, metal (steel) hardening, photography, blast furnaces, nitrile preparation, chemical manufacture, and pharmaceuticals.

Thermal degradation of polyurethanes and plastics can release HCN, representing a risk to fire fighters as well as to aircraft crew and passengers during on-board fires.[4]

Hydrogen sulfide (H_2S): OSHA PEL = 20 ppm (Ceiling Limit)
OSHA PEL = 50 ppm (peak <10 minutes)
NIOSH REL = 10 ppm (<10 minutes)

Descriptions of diseases afflicting miners and sewer workers suggest that H_2S is an industrial hazard with a long history. However, only with the advent of oil drilling and refining in the 20th century has widespread industrial exposure to H_2S occurred. Hydrogen sulfide is a colorless, dense, malodorous gas produced through the anaerobic decomposition of organic matter, such as might naturally be found in oil drilling, mining (where it is known as "stink damp"), and wastewater treatment facilities. It is also used in or given off as a by-product of a variety of industrial processes. Industries in which H_2S is commonly found include the oil and gas industry, heavy water production, sewage treatment, rayon manufacturing, leather tanning, mining (coal and metal), and rubber vulcanization.

Accurate current figures on the annual incidence of H_2S intoxication, asphyxiation, and fatalities are not available. However, there are a number of industrial case series documenting H_2S intoxication and fatalities and several well-known accidental community overexposures to H_2S. For example, in 1950, 320 people were hospitalized and 22 died in Poza Rica, Mexico, when a local oil refinery released a large cloud of H_2S.[6] Yet one group of investigators found only 14 documented case reports in the English literature between 1960 and 1974, which suggests that the annual incidence of fatal asphyxiations is probably small.[7]

At minute (<0.025 ppm) concentrations, H_2S is a potent respiratory and mucous membrane irritant, with an odor invariably characterized as that of rotten eggs; at higher concentrations, H_2S can be a rapid and deadly asphyxiant (*see* Table 3).

But despite its potent odor, smell is an unreliable means of warning workers in danger from H_2S. The brain's olfactory center becomes desensitized to the characteristic odor as the concentration of H_2S increases; therefore, it is not uncommon for workers who survive near-fatal asphyxiation to be unable to recall any "warning" odor. Subacute intoxication due to acute low-level (50 to 100 ppm) or chronic low-level exposure to H_2S has resulted in delayed pulmonary edema and death.

Table 3.—Relationship of Hydrogen Sulfide Levels and Clinical Effects*

Hydrogen Sulfide Concentration (in ppm)	Clinical Effect
0.1–0.2	Odor threshold
10–100	Eye and upper respiratory tract irritant
>200	Anosmia (delayed), pulmonary edema
>500	Hyperpnea, apnea
>1000	Respiratory paralysis, death

*Adapted from National Research Council, 1979; Deng, 1987.

Hydrogen sulfide exhibits its asphyxiant properties by binding to cytochrome oxidase and uncoupling oxidative phosphorylation. This results in an interruption of aerobic metabolism and in profound metabolic acidosis. Hydrogen sulfide also has a direct effect on the brain's respiratory center and the carotid body, initially causing respiratory stimulation, followed by depression, apnea, and death.[6,7] Hydrogen sulfide is denser than air and so has a tendency to accumulate at the bottom of pits, tanks, or other enclosed spaces. This fact has obvious ramifications for the occurrence and prevention of H_2S asphyxiation.

Prevention

Ideally, exposure to toxic asphyxiants could be controlled by appropriate environmental controls. Dilution ventilation and effective maintenance of machinery, pipes, ventilation, and exhaust systems could keep ambient levels of asphyxiants below levels at which clinical effects occur or could prevent asphyxiants from accumulating altogether. Because most asphyxiant deaths occur without warning, engineering and safety programs must anticipate potential asphyxiant hazards if they are to be effective in preventing accidents. For example, virtually every suffocation due to trench cave-ins could be prevented if correct trenching and sloping techniques were used in the first place. For another example, all potentially confined, poorly ventilated spaces should have available means of egress in the event of unanticipated asphyxiant exposures.

Confined, nonventilated spaces should never be entered without prior air monitoring using a number of readily available and accurate field devices to determine oxygen content and content of toxic or explosive gases in the air. If the levels of oxygen or toxic gases are dangerous, trained workers with air-supplied (not cartridge) respirators should be assigned. Finally, the "buddy system" with trained workers should always be in effect when confined spaces are entered or rescue efforts are attempted. However, fatalities occur even in companies that pay meticulous attention to worker education, hazard evaluation, and safety precautions.[6] If asphyxiants have been generated within a workplace, nonoccupational environment, or other enclosed space, adequate preparation for entry into these zones must include the above precautions. A proposed OSHA standard would require minimum standards for confined spaces and establish a system of permits to limit access to such spaces. The proposed standard addresses flammable, oxygen-deficient confined spaces and provides mechanisms to evaluate the potential presence of these spaces before entry.

For the most part, biological surveillance for asphyxiants is neither technically possible nor reasonable, because most deaths and overexposures occur through inadvertent exposure. However, for workers with known potential CO exposure, surveillance mechanisms can be instituted to monitor for CO intoxication. For example, postshift Hb_{CO} levels or expired alveolar air for CO are both reliable means of surveillance. If surveillance is instituted, adjustment for the potential confounding by tobacco smoke is necessary.

Once an individual is overcome by an asphyxiant, individualized treatment of the condition depends on the suspected exposure. In all instances, the individual should be removed from the asphyxiant atmosphere before resuscitation is attempted. Would-be rescuers must use proper protective equipment, such as air-supplied respirators. Oxygen (100%) should always be administered to victims in cases of suspected asphyxiation.

High concentrations of delivered oxygen, including hyperbaric oxygen, have been shown to decrease the half-life of CO elimination.[8] Some authors have questioned the value of oxygen therapy following asphyxiation with HCN or H_2S; they argue that H_2S and HCN prevent cellular oxygen use, so unless this action is counteracted, oxygen therapy is not useful in acute asphyxiation.[7,9] Nevertheless, laboratory animal data have shown an independent protective effect of oxygen when it is given as part of an overall treatment protocol for toxic asphyxiation.[9] Medical consensus is that oxygen should be provided in all instances.

Induction of methemoglobinemia through the administration of nitrites (4-DMAP, or sodium nitrite) has been shown to be effective in reversing the binding of cyanide and H_2S anions to cytochrome oxidase. Intravenous administration of sodium nitrite or inhalation of amyl nitrate are standard treatments of intoxication with H_2S or HCN. Some authors have cautioned that nitrites and DMAP can cause unpredictable levels of methemoglobinemia, which in itself can complicate resuscitation.[10] However, intravenous hydroxocobalamin has been shown to decrease the cyanide toxicity associated with nitroprusside administration by forming less toxic cyanocobalamin.[11] It has recently found a use in the treatment of cyanide intoxication, but it remains available in the U.S. on an investigational basis only. There is a field kit for the treatment of cyanide poisoning (Lilly, Cyanide Antidote Kit). Finally, industries in which HCN is a known hazard should provide appropriate first aid training, including the administration of amyl nitrate, to employees potentially exposed to HCN.

MG, ES

References

1. Suruda A, Agnew J. Deaths from asphyxiation at work in the United States 1984-86. *Br J Ind Med.* 1989;46:541-546.

2. National Institute for Occupational Safety and Health. Criteria for a recommended standard: working in confined spaces. Washington, DC: Department of Health, Education, & Welfare; 1979. Publication No. 80-106.

3. Fatalities attributed to methane asphyxia in manure waste pits: Ohio, Michigan, 1989. *MMWR.* 1989;38(33):583-586.

4. Rosenstock L, ed. Occupational pulmonary disease. *State of the Art Reviews in Occupational Medicine.* 1987;2(2):297-318.

5. Atkins EH, Baker EL. Exacerbation of coronary artery disease by occupational carbon monoxide exposure: a report of two fatalities and a review of the literature. *Am J Ind Med.* 1985;7(1):73-79.

6. National Research Council, Subcommittee on Hydrogen Sulfide. *Hydrogen Sulfide.* Baltimore, Md: University Park Press; 1979:48-65.

7. Stine RJ, Slosberg B, Beacham BE. Hydrogen sulfide intoxication. A case report and discussion of treatment. *Ann Intern Med.* 1976;85(6):756-758.

8. Myers RA, Snyder SK, Linberg S, Cowley RA. Value of hyperbaric oxygen in suspected carbon monoxide poisoning. *JAMA.* 1981;246(21):2478-2480.

9. Smith RP, Kruszyna R, Kruszyna H. Management of acute sulfide poisoning: effects of oxygen, thiosulfate, and nitrite. *Arch Environ Health.* 1976;31(3):166-169.

10. Heijst ANP, et al. Therapeutic problems in cyanide poisoning. *Clin Toxicol.* 1987;25(5):383-398.

11. Cottrell JE, et al. Prevention of nitroprusside-induced cyanide toxicity with hydroxocobalamin. *N Engl J Med.* 1978;298(15):809-811.

Further Reading

Arnold IM, Dufresne RM, Alleyne BC, Stuart PJ. Health implications of occupational exposures to hydrogen sulfide. *J Occup Med.* 1985;27(5):373-376.

Carbon monoxide intoxication: a preventable environmental health hazard. *MMWR.* 1982;31(39):529-530.

de Kort WL, Sangster B. Acute intoxications during work. *Vet Hum Toxicol.* 1988;30(1):9-11.

Deng JF, Chang SC. Hydrogen sulfide poisonings in hot-spring reservoir cleaning: two case reports. *Am J Ind Med.* 1987;11(4):447-451.

Olishifsku JB, ed. *Fundamentals of Industrial Hygiene.* 2nd ed. Washington, DC: National Safety Council; 1979:479-481.

Parmeggiani L, ed. *Encyclopedia of Occupational Health and Safety.* 3rd ed., rev. Geneva, Switzerland: International Labour Office; 1983:396-399.

Radford EP. Proceedings: carbon monoxide and human health. *J Occup Med.* 1976;18(5):310-315.

Rom WN, ed. *Environmental and Occupational Medicine.* Boston, Mass: Little, Brown; 1983.

Rosenstock L, Cullen MR, eds. *Clinical Occupational Medicine.* Philadelphia, Pa: WB Saunders; 1986.

Vogel SN, Sultan TR, Ten Eyck RP. Cyanide poisoning. *Clin Toxicol.* 1981;18(3):367-383.

Asthma

ICD-9 493, 506, 507.8
SHE/O

Identification

Occupational asthma is defined as reversible, generalized airway narrowing as a result of exposure to airborne dust, gases, vapors, or fumes in the work environment. No one definition of occupational asthma is uniformly accepted. For example, some definitions require a period of sensitization; others include irritant-induced bronchoconstriction. Diversity of opinion reflects difficulty in defining asthma in general (there is still no single epidemiological definition) and different pathogenic mechanisms.

Diagnosis is made by confirming the presence of asthma and establishing a relationship with work exposures. Patients may present with typical symptoms of episodic dyspnea, chest tightness, and wheezing associated with airflow limitation, which are reversible after administration of a bronchodilator. Patients may also present with recurrent attacks of bronchitis. A detailed work history is important, with particular attention to the periodicity of symptoms. Suggestive patterns are symptoms that occur only at work or in the evening of workdays, worsen at the start or in the course of the workweek, improve on weekends or vacations, or resolve with changes in the working environment.

Lung function tests may be normal or may show obstruction with improvement after inhaled bronchodilator is administered. In patients with normal pulmonary function, tests of airway reactivity can be done using methacholine, a very sensitive indicator of airway hyperresponsiveness. Spirometry can also be done before and after a workshift to determine the relationship of airflow obstruction to work; a 10% or greater decrease in forced expiratory volume (FEV) is considered significant. However, a smaller decrease, or no decrease at all, does not exclude work-related asthma. Although skin testing with common inhalants and food allergens can be used to define the atopic status, this has no relevance to most cases of occupational asthma.

With some exposures (eg, high molecular-weight compounds), skin tests may help to identify the responsible agent. They have been used in such exposures as flour, dust, coffee, and some animal products. Serological tests (eg, IgE antibodies, radioallergosorbent test [RAST], or enzyme-linked immunosorbent assay [ELISA]) to various occupational allergens can be done, but these can yield positive results after exposure alone, with no evidence of disease.

It is generally unnecessary to resort to specific inhalation challenge testing of workers using a suspected cause in order to diagnose occupational asthma. Challenge testing can be a valuable research tool that should only be performed in specialized centers with personnel experienced in recognizing and treating complications that can arise from such procedures, such as severe or delayed reactions. As with all testing, an informed decision must be made that the information that can be derived from the test is worth the risk to the test subject.

A case definition for surveillance purposes has been published by the CDC. It consists of "a physician diagnosis of asthma [as above], an association (patterns of which may vary) between symptoms of asthma and work and either (1) workplace exposure to an agent or process previously associated with occupational asthma or (2) significant work related changes in forced expiratory volume in 1 second [FEV_1] or peak expiratory flow-rate or (3) significant work-related changes in airways responsiveness as measured by nonspecific inhalation challenge or (4) positive response to inhalation provocation testing with an agent to which the patient is exposed at work." In addition, "the number of co-workers with exposures similar to those of the reported case-patient, and the number of co-workers with respiratory symptoms" should also be ascertained.[1]

Occurrence

About 5 million American adults are asthmatic. The overall prevalence of occupational asthma is unknown, but an estimated 2% to 15% of adult asthma cases in the U.S. are attributed to occupational exposure. There are numerous case reports in the medical literature. In Japan, 15% of all adult males with asthma have occupationally related cases. Prevalence of the disease depends on the nature of the agent—the prevalence of asthma among workers exposed to proteolytic enzymes is 50% to 66% compared with 5% for workers exposed to volatile isocyanates—and the level of exposure.

Workers exposed to agents with known or suspected allergic properties are at increased risk for developing asthma. Such occupations include animal handlers (prevalence rate of 3% to 30% for laboratory animal handlers); workers in snow crab and egg processing; grain handlers (4% to 11% of grain workers showed a postshift fall in FEV_1 of greater than 10%); bakers (20% of established bakers in West Germany show allergic symptoms); workers in factories manufacturing detergents containing biologic enzymes; workers exposed to isocyanates used in industry (5% to 10% of workers exposed to toluene diisocyanate develop asthma); workers involved in the manufacture of paints, plastics, and adhesives who are exposed to anhydrides in epoxy resins; red cedar workers (4% to 15% prevalence); workers exposed to metal salts or working in platinum refineries, jewelry making, nickel plating, manufacture of pigments, and the tanning industry (with chromium exposure); and workers exposed to soldering emissions (with colophony resin exposure). In Japan, cases of hard metal-induced asthma developed from cobalt sensitivity.

Causes

To date, more than 200 agents are reported to cause occupational asthma (*see* Table 1 at end of entry).

Pathophysiology

Pathological appearance of occupational asthma is not different from that of asthma of any cause. The bronchial lumen is narrowed and filled with mucus ("mucus plugging"). Both the mucus and the bronchial wall are infiltrated with inflammatory cells. The basement membrane is thickened and bronchial muscle hypertrophied. The mechanisms of occupational asthma include the following:

- Reflex bronchoconstriction—eg, by cold air, noxious gases, and irritants.
- Acute inflammation—eg, by irritant gases and vapors, like acid mists, or by pyrolysis products of polyvinyl chloride.
- Pharmacological reaction—eg, by isocyanates (partially explains mechanism).
- Allergic reaction (accounting for the greatest number)—by grain, baker's flour, detergent enzymes, isocyanates, wood, and metals, such as nickel, platinum, and chromium.

Asthmatic reactions can be immediate, late (delayed), or both (dual). The immediate reaction can be produced by nonallergic or allergic mechanisms. Nonallergic reactions, via reflex bronchial constriction, can occur in persons with preexisting bronchial hyperreactivity. Allergic asthma is mediated via IgE reaginic antibodies, which have a high affinity for membrane receptors of circulating basophils and tissue mast cells. These antibodies are the source of potent chemical mediators such as histamine, eosinophilic chemotactic factor of anaphylaxis, neutrophilic chemotactic factor of anaphylaxis, platelet-activating factor, and a number of arachidonic acid metabolites such as prostaglandins.

The late type of reaction can occur alone or as a sequel to the immediate reaction. When occurring as a sequel, it is part of a dual reaction. The late or dual reaction may be induced by allergens or by a number of low molecular-weight compounds.

The late reaction is also associated with an ongoing inflammatory process that probably causes the nonspecific bronchial reactivity that is seen in asthma induced by any of the above mechanisms.

Prevention

Aggressive measures need to be taken to prevent the development of occupational asthma in both atopic and nonatopic individuals. Preventive strategy should include (a) worker education; (b) product substitution, where possible, using less sensitizing materials; (c) efficient environmental control of exposures; (d) safer work practices, such as safe handling procedures, avoidance of spills, and good housekeeping; and (e) medical surveillance, including periodic questionnaires, chest examinations, and spirometry. Currently, there are no preemployment screening criteria that have proved to be very useful in predicting occupational asthma. Prophylactic desensitization has not been shown to be effective.

Secondary prevention requires early identification of workers who develop asthma and absolute control of future exposures. Failure to control exposure may result in the development of irreversible airways obstruction. Once sensitized to an agent, a worker may respond to minute quantities of it. For these workers, job change with rate retention may be the only reasonable preventive strategy.

Other Issues

Predisposing host factors may be important in the development of occupational asthma. Atopy is the capacity to develop immediate sensitivity after exposure to common environmental allergens. Atopic workers become sensitized more readily than nonatopic workers when high-molecular-weight compounds are used, such as in the enzyme detergent industry, in bakeries, and in industries in which animals are handled. However, because nonatopic workers also become sensitized at significant rates, it is inappropriate to exclude atopic individuals from employment in these industries.

The role of cigarette smoking in the development of occupational sensitization and asthma is unknown. Research findings have been contradictory, and there is little evidence to suggest that smokers are more predisposed to asthma. Most workers with symptomatic occupational asthma have nonspecific bronchial hyperreactivity, which is most likely a consequence of workplace exposures rather than a predisposing factor.

Work-relatedness of asthma may be obscured if a worker becomes sensitized and develops symptoms after a long asymptomatic period in the same workplace, or if a worker chronically exposed to a causative agent develops relatively fixed airways obstruction that does not lead to recognition of an association with work.

Airway obstruction may resolve over months to years following cessation of exposure, or it may remain fixed.

DCC

Reference

1. Occupational disease surveillance: occupational asthma. *MMWR*. 1990;39:119-123.

Further Reading

Chan-Yeung M, Lam S. Occupational asthma. *Am Rev Respir Dis.* 1986;133:686-703.

Grammer LC, Patterson R. Occupational immunologic lung disease. *Ann Allergy.* 1987;58:151-159.

Lam S, Chan-Yeung M. Occupational asthma: natural history, evaluation and management. In: Rosenstock L, ed. Occupational Lung Disease. *State of the Art Reviews: Occupational Medicine.* 1987;2:373-381.

Merchant JA, ed. Environmental and occupational asthma. *Chest.* 1990;98(5)(suppl):145S-252S.

Parkes WR. *Occupational Lung Disorders.* 2nd ed. London, England: Butterworths; 1982.

Salvaggio JE, Taylor G, Weill H. Occupational asthma and rhinitis. In: Merchant JA, ed. *Occupational Respiratory Diseases.* Washington, DC: U.S. Government Printing Office; 1986. DHHS (NIOSH) publication 86-102.

Table 1.—Hazards and Occupations Associated with Asthma

Hazard	Occupation	Prevalence
Substances of Animal Origin		
Hair, epidermal squamae, animal products, urine protein, insects, birds, laboratory animals, danders	Hairdressers, research lab workers, animal handlers, grain and poultry workers, veterinarians, laboratory workers	3% to 30%
Bee toxin and squamae, hairs, chitin	Entomologists, fish bait breeders, apiarists	?
Marine organisms	Oyster farmers, crab and prawn processors	16% to 35%

Hazard	Occupation	Prevalence
Silk hair and larva, butterfly squamae	Sericulture workers, silkwork cutters	?
Locusts, river flies, screwworm flies, sewage flies	Research lab workers, flight crews, outdoor workers	3% to 70%
Substances of Plant Origin		
Grain dust	Grain handlers, millers	?
Wheat/rye flour, buckwheat	Bakers, millers	20%
Coffee bean, caster bean	Coffee workers, oil extractors	?
Tea	Tea workers	?
Tobacco leaf	Tobacco workers	?
Hops *(Humulus lupulus)*	Brewery chemists, farmers	?
Wood dust	Carpenters, construction workers, sawmill workers, cabinetmakers	?
Cotton, flax, and hemp dust	Textile workers, weavers	?
Gum acacia	Printers	19%
Gum tragacanth	Gum manufacturers	?
Substances of Chemical Origin		
Toluene diisocyanate	Polyurethane industry, plastics, varnish workers	5% to 38%
Diphenylmethane diisocyanate, furfuryl alcohol	Foundry workers	5% to 10%
Hexamethylene diisocyanate, dimethyl ethanolamine	Spray painters	?

Hazard	Occupation	Prevalence
Phthalic trimetalic and tetra-chlorophthalic anhydrides	Epoxy resin, plastic, paint workers	29% to 36%
Hexachlorophene, formalin	Hospital workers	29%
Paraphenylene diamine, dioazonium salt	Dye workers	38%
Pyrethrins, organophosphates	Insecticide manufacturers, farmers	?
	Metals	
Platinum	Platinum refinery workers, jewelers, electroplating workers	57%
Nickel	Metal-plating workers, stainless steel workers	?
Chromium	Manufacturers of pigments, tannery workers, precision casters, stainless steel welders	?
Cobalt	Metalworkers, cobalt refinery or alloy workers	?
Vanadium	Metalworkers, mineral ore processors	33%
Tungsten carbide	Metalworkers	?
Nickel sulfate	Nickel workers, metal-plating workers	?
Aminoethyl ethanolamine and colophony soldering flux	Electricians and electronics manufacturing workers	?
	Biologic Enzymes	
B. subtilis	Detergent industry or laundry workers	?

Hazard	Occupation	Prevalence
Trypsin	Plastic and pharmaceutical workers	?
Pancreatin, pepsin, bromelin, flaviastase	Pharmaceutical workers	?
Fungal amylase	Manufacturing workers, bakers	?
Papain	Laboratory and food technicians, packagers	?
Penicillium casei	Cheese production workers	?
	Drugs	
Pencillins, Methyldopa, Cephalosporins, Spiramycin, Sabutamol intermediate, Phenylglycine acide chloride, Tetracycline	Pharmicists, nurses, physicians, factory workers	?
Piperazine hydrochloride	Chemists	?
Psyllium	Laxative manufacturers	?
Amprolium HCl	Poultry feed mixers	?
Sulphone choramides	Brewery workers	?

Ataxia

ICD-9 334.3, 781.3
SHE/O

Identification

Ataxia is characterized by an unsteady, wide-based gait and difficulty maintaining standing posture. In severe cases, falling occurs. Ataxia can result from weak legs and trunk muscles, loss of sensation in legs and feet, or poor coordination of movements.

Exposure to neurotoxins in the workplace can lead to ataxia, either transiently or persistently, depending on the chemical's neuropathological mechanism of action and the extent of the exposure. Transient ataxia may be due to intoxication by various central nervous system depressants, which may cause toxic encephalopathy after more prolonged exposure. In cases of myopathy, weak proximal muscles cause inability to fixate the hips and knees, resulting in unsteady walking. In peripheral neuropathy affecting the legs, lack of input of sensory information regarding limb position results in an uncertain gait. Ataxia due to sensory loss is worsened by poor lighting; it is helped by touching walls or furniture to gain the auxiliary sensory cues needed for stability of gait and posture. Ataxia due to poor coordination of movements may result from either cerebellar or vestibular dysfunction; cerebellar ataxia is due to impaired coordination functions of the cerebellum and its connecting systems. A sway of posture with eyes closed suggests that the posterior spinal columns are the anatomical site of dysfunction. Symptoms of vestibular dysfunction include vertigo, dizziness, and ataxia. These symptoms may progress from an unsureness of step, a tendency to drift to one side, or a slight unsteadiness that worsens in darkness to a staggering gait or an inability to stand.

Abnormalities on the physical exam can assist in determining the underlying disease process. Tests of proximal muscle weakness include evaluation of muscle flexion and extension through the major joints, with and without resistance, to measure muscle strength. Careful examination of sensation (ie, pinprick testing, vibratory stimulation, position sense testing, and tendon reflex elicitation) that shows diminished perception indicates that peripheral neuropathy is the probable cause of the sensory ataxia. A Romberg test that yields positive results when eyes are closed is also an indicator. Nerve conduction studies can be used to confirm the diagnosis of neuropathy.

Tests of cerebellar dysfunction include varying ability in performing standard clinical neurological tests, such as finger-to-nose test with eyes open and closed and with either hand; rapid alternating movements; and/or tandem walking with eyes open and closed. Tandem walking also is impaired when ataxia is caused by vestibular dysfunction. Tests of vestibular function such as an electronystagmogram, an audiogram (to differentiate lesions of the cochlea from those of the underlying nerve), visual function flicker-fusion tests, and x-rays of the skull may help to localize the lesions more precisely.

Occurrence

No reliable data exist. (*See* Occurrence section of Peripheral Neuropathy.)

Causes

Some neurotoxins in the workplace produce ataxia if exposure is significant. Ataxia can present initially as a transient symptom, which then progresses with prolonged exposure to peripheral neuropathy and/or encephalopathy. However, the functional basis and pathological target of each chemical that produces ataxia have not been identified. Sources of exposure and major occupational uses of particular chemicals that can produce ataxia are presented in Table 1 at the end of this entry. Nonoccupational causes of ataxia include dysfunction of the vestibular system, such as in Ménière's disease; infections of the middle and inner ear; and seizure disorders.

Pathophysiology

Workers exposed to mercury commonly present with ataxia due to sensory neuropathy and cerebellar dysfunction. Those exposed to lead may present with ataxia due to its effects on the peripheral nerves and the cerebellum. Workers exposed to organic solvents, such as carbon disulfide and toluene, commonly present with transient ataxia that progresses to toxic encephalopathy. Also, there is evidence that chronic toluene and trichloroethylene exposure affects the vestibular system. Chronic carbon disulfide exposure also can progress to neuropathy with signs of diminished muscle strength and distal sensory loss. Methyl-n-butyl ketone, n-hexane, acrylamide, and organophosphate insecticide exposure produce ataxia due to their primary effect on the peripheral nerves. Cerebellar lesions have been observed postmortem following long-term solvent and mercury exposure. Postmortem examination of ataxic solvent inhalers has revealed diffuse cerebral and cerebellar cortex atrophy and both central and peripheral giant axonopathy. Permanent sensory loss and/or sensory ataxia in sensory neuronopathies usually are shown by a diffuse loss of dorsal root ganglion neurons.

Prevention

Exposure to neurotoxins should be minimized. The techniques and effectiveness of industrial hygiene monitoring and control of respiratory, gastrointestinal, and dermal routes of chemical entry are discussed in part 1. Biological monitoring of workers for specific chemical exposures may be used as a guide to check on individual exposure levels, but careful attention must be paid to the interpretation of individual results. For more information, the American Conference of Governmental Industrial Hygienists (ACGIH) has issued biological exposure indices (BEIs) for a number of chemicals.

Other Issues

Ataxia may result from a nonoccupational etiology, such as diabetes, vitamin B-12 and folic acid deficiency, hereditary diseases, and alcoholism. Acute alcohol intake can inhibit the metabolism of certain solvents. Chronic alcohol consumption can induce the hepatic metabolism of some toxic chemicals. Thus, depending on whether the chemical or its metabolites is the toxic agent, exposure to alcohol may increase the individual's sensitivity to the effects of certain toxic chemicals. Of importance, exposure to certain organic solvents such as toluene and xylene may become addictive. Solvent abusers ("huffers"), who inhale rags soaked with solvents to get high, often present with an ataxic gait disturbance.

RF, SP

Further Reading

Adams RD, Victor M, eds. *Principles of Neurology.* 2nd ed. New York, NY: McGraw-Hill; 1981.

Feldman RG, Travers PH. Environmental and occupational neurology. In: Feldman RG, ed. *Neurology: The Physician's Guide.* New York, NY: Thieme-Stratton; 1984:191-212.

Friedman J. Otoneurologic disorder. In: Wilkins RW, Levinsky NG, eds. *Medicine: Essentials of a Clinical Practice.* Boston, Mass: Little, Brown; 1983:848-853.

Rosen I. Neurophysiological aspects of organic solvent toxicity. *Acta Neurol Scand.* 1984;70:101-106.

Spencer PS, Schaumburg HH, eds. *Experimental and Clinical Neurotoxicology.* Baltimore, Md: Williams & Wilkins; 1980.

Table 1.—Exposures Associated with Ataxia

Neurotoxin	Major Uses or Sources of Exposure
Metals	
Arsenic	Pesticides
	Pigments
	Antifouling paint
	Electroplating industry
	Seafood
	Smelters
	Semiconductors

Neurotoxin	Major Uses or Sources of Exposure
Lead	Solder Insecticides Lead shot Storage battery manufacturing plants Auto body shops Foundries Smelters Lead-stained glass Lead pipes Lead-based paint
Manganese	Iron, steel industry Welding operations Metal finishing operations of high manganese steel Fertilizers Manufacturers using oxidation catalysts Manufacturers of fireworks, matches Manufacturers of dry cell batteries Electrical equipment
Mercury	Scientific instruments Amalgams Electroplating industry Felt making Photography Pigments Textiles Taxidermy
Solvents	
Carbon disulfide	Electroplating industry Manufacturing of viscose rayon Paints Preservatives Rubber cement Textiles Varnishes

Neurotoxin	Major Uses or Sources of Exposure
Methyl-n-butyl ketone	Metal-cleaning compounds Paints Paint removers Lacquers Quick-drying inks Varnishes
n-Hexane	Lacquers Printing inks Stains Rubber cement Glues
Toluene	Rubber solvents Cleaning agents Manufacturers of benzene Glues Gasoline Paints Paint thinners Automobile, aviation fuels Lacquers
Monomers	
Acrylamide	Paper, pulp industry Photography Grouting material: basements, tunnels, dams, mine shafts Dyes Water, waste treatment facilities Insecticides
Organophosphates, Chlorinated hydrocarbons (DDT, Chlordecone)	Agricultural industry Fabricators

Back Pain or Back Strain—*See* Low Back Pain Syndrome

Beryllium Disease

ICD-9 503
SHE/O

Identification

Inhalation of beryllium dust can cause two lung diseases: acute beryllium disease and chronic beryllium disease (berylliosis). Exposure to beryllium salts or dust can also cause contact dermatitis, skin granulomas, and mucous membrane irritation resulting in conjunctivitis, periorbital edema, nasopharyngitis, tracheal bronchitis, and pharyngitis.

Acute beryllium disease is a form of chemical pneumonitis caused by an overexposure to high concentrations of beryllium (>100 $\mu g/m^3$). The disease has rapid onset and is characterized by dyspnea, cough and sputum, chest pain, tachycardia, crackles, and cyanosis. The chest x-ray reveals diffuse or localized infiltrates. Pulmonary function tests may show decreased lung volumes and hypoxemia at rest. The case-fatality rate for acute beryllium disease ranges from 10% to 15%. Acute beryllium disease is now uncommon in the U.S. because of engineering and industrial hygiene controls.

Chronic beryllium disease is an interstitial lung disease characterized by noncaseating granulomas in lung tissue, which are associated with cell-mediated immunity to beryllium. Granulomatous involvement of other organs, such as liver and spleen, has been described. The most common presenting symptom is dyspnea on exertion, which may be accompanied by dry cough, weight loss, chest pain, and fatigue. Basilar rales may be present on physical exam. In more advanced disease, finger clubbing and signs of cor pulmonale can occur. Pulmonary function tests may show a decrease in diffusing capacity, an increase in alveolar-arterial oxygen gradient, airflow limitation, and a decrease in compliance. The chest x-ray shows irregular or rounded interstitial opacities and, in some cases, hilar adenopathy. These abnormalities are not diagnostic for chronic beryllium disease, and similar changes can occur in sarcoidosis, other pneumoconioses, and active tuberculosis. However, bronchoalveolar lavage can distinguish beryllium disease from these other diseases when it demonstrates lymphocytic alveolitis and beryllium-specific cell-mediated immunity in the setting of granulomas or mononuclear cell infiltrates on transbronchial biopsy. The lymphocyte transformation test to beryllium salts, available in a few specialized laboratories, is used to identify this antigen-specific immune response in blood and bronchoalveolar lavage cells. Only two or three laboratories have experience in the beryllium lymphocyte transformation test, and discrepancies exist among them in the apparent sensitivity of the blood test in identifying confirmed beryllium disease cases. This is an active area of epidemiological and clinical investigation. The presumptive diagnosis of beryllium disease can no longer be confirmed without demonstrating beryllium-specific immunity.

Occurrence

Historical information about the incidence of beryllium disease in the U.S. is available from the Beryllium Case Registry (BCR), which operated from 1951 until approximately 1980, when it was transferred to NIOSH. Since then, reporting has virtually ceased, although cases continue to occur. Approximately 900 cases were reported to the registry, which used a historical case definition, perhaps leading to some misclassification. Registry cases were primarily from Ohio, Pennsylvania, and Massachusetts, reflecting both (a) the locations of beryllium extraction and fluorescent light plants and (b) the presence of physicians who considered the diagnosis.

Acute beryllium disease is now rare in the U.S. because stringent engineering controls were brought into effect in the early 1950s and the use of beryllium was discontinued in the fluorescent light industry. Of the 892 cases reported to the BCR between 1952 and 1978, 212 (23.8%) represented acute beryllium disease; only one such case was reported between 1976 and 1981.

Chronic beryllium disease occurs in approximately 1% to 3% of all workers exposed to beryllium. The disease has an average latency period of approximately 10 years from first exposure, but the latency period can range from a few months to 40 years. Chronic beryllium disease has occurred in workers judged to have had trivial, passive, or brief exposures, raising questions about the nature of the exposure-response relationship in beryllium disease. In addition to occupational exposure, cases of beryllium disease have developed in residents living near beryllium refineries and in family members of beryllium workers who have come into contact with contaminated clothing. Neighborhood exposures have decreased from 11% of all cases of beryllium disease before 1949 to 3% of all cases after 1949, with only one neighborhood case reported to the BCR between 1973 and 1977.

Workers at risk for developing beryllium disease include those engaged in all operations producing or using beryllium and its compounds, excluding beryl ore mining. Beryllium production workers have been thought to have the highest prevalence of disease, and beryllium-copper alloy production workers, a lower prevalence. Fewer than 10% of workers with exposure to beryllium in the U.S. work in primary production. Exposure also occurs in operations that involve melting, casting, grinding, machining, and drilling of beryllium-containing products, and it is in these industries that most U.S. workers exposed to beryllium are employed. Because beryllium is a neutron moderator, nuclear workers are frequently exposed to beryllium. Exposures to beryllium also occur in the aerospace, scrap metal reclaiming, specialty ceramics, and electronics industries. The beryllium hazards by industry are listed in Table 1.

Table 1.—Beryllium Hazards by Industry

Hazard	Occupational Source
Beryllium (metallic)	Nuclear reactors and weapons, inertial guidance systems, aircraft brakes, x-ray tube windows, turbine reactor blades
Beryllium oxide	Spark plugs, laser tubes, electrical components, rocket engine liners, ceramic applications, beryllium refining (intermediate), cathode-ray tubes
Beryllium oxyfluoride	Cathode-ray tubes
Beryllium-copper alloy (2% to 4% Be in copper)	Springs, bellows, gears, aircraft engines, bearings, welding electrodes, electrical contacts, ceramic products
Beryllium fluoride	Intermediate in beryllium refining
Beryllium phosphors	Chemical manufacturing
Beryllium sulfate	Electronic equipment
Ammonium beryllium fluoride	Alloy manufacturing
Zinc beryllium silicate	Metallurgical operations, tool and die manufacturing, welding and torch cutting, nonferrous foundry production, nuclear reactors, missile technology

Causes

Beryllium disease is caused by exposure to respirable dust from metallic beryllium or its compounds. It is not believed to be caused by exposure to beryllium ores.

The toxicity of beryllium compounds is a function of the bioavailability of beryllium in different compounds. Toxicological studies have shown that smaller particle size and more soluble grades of beryllium oxide are more toxic.

Pathophysiology

Chronic beryllium disease is unusual among toxic metal diseases in that its pathophysiological mechanism depends on beryllium's ability to trigger a cell-mediated immune response. People who are exposed to beryllium and who do not develop a delayed-type hypersensitivity response to it do not develop this interstitial lung disease. Whether the development of an immune response to beryllium requires genetic susceptibility or special exposure circumstances is unknown.

Prevention

Demand for beryllium is increasing worldwide, and growing numbers of workers are exposed to this metal. Control of the beryllium air concentration is the major mode of preventing acute beryllium disease. The current PEL of 2 $\mu g/m^3$ is the lowest of any metal and was established by comparison with other toxic metals, such as lead, which have greater mass. Meeting such a low standard requires elaborate engineering controls and scrupulous housekeeping efforts in the workplace. However, it is not known whether control of beryllium air concentrations will control chronic beryllium disease, because it is a hypersensitivity disease.

The utility of screening workers exposed to beryllium with the blood lymphocyte transformation test for secondary prevention of clinical beryllium disease is under active investigation. In the meantime, medical monitoring of beryllium-exposed workers should include inquiry about respiratory symptoms, physical examination for signs of lung disease, spirometry, diffusion capacity, and chest x-ray. Because workers with incidental or passive exposure to beryllium are at risk for chronic beryllium disease, medical surveillance programs should be designed to include people with potential beryllium exposure, including security guards, building trades workers, secretaries, and inspectors who may not be thought a priori to be exposed to beryllium.

Other Issues

The relative roles of exposure-response relationships and immunologic or individual host factors in the chronic disease need further clarification. If workers can become sensitized by brief, very high exposures, an 8-hour time-weighted average air concentration may not be an appropriate measure on which to base a standard. With the advent of screening for beryllium sensitization, the natural history of sensitization and subclinical beryllium disease requires study. Whether such screening can modify the course of chronic beryllium disease by removing a sensitized worker from further exposure or by treating that worker early with corticosteroids will be demonstrated only by controlled clinical trials.

Beryllium is an animal carcinogen and is associated with elevated risk of lung cancer in humans.

KK, DCC

Further Reading

Eisenbud M, Lisson J. Epidemiologic aspects of beryllium-induced nonmalignant lung disease: a 30-year update. *J Occup Med.* 1983;25:196-202.

Kreiss K, Newman LS, Mroz M, Campbell P. Screening blood test identifies subclinical beryllium disease. *J Occup Med.* 1989;31:603-608.

Kriebel D, Brain JD, Sprince NL, Kazemi H. The pulmonary toxicity of beryllium. *Am Rev Respir Dis.* 1988;137:464-474.

Newman LS, Kreiss K, King TE, Seay S, Campbell PA. Pathologic and immunologic alterations in early stages of beryllium disease: reexamination of disease definition and natural history. *Am Rev Respir Dis.* 1989;139:1479-1486.

Rossman MD, Kern JA, Elias JA, et al. Proliferative response of bronchoalveolar lymphocytes to beryllium. *Ann Intern Med.* 1988;108:687-693.

Black Lung—*See* Coal Workers' Pneumoconiosis

Bladder Cancer

ICD-9 188
SHE/O

Identification

Bladder neoplasms appear clinically with microscopic or gross hematuria, with or without pain. Other symptoms may include a change in the time and frequency of urination, and pain on urination. Diagnosis is confirmed with cystoscopy and biopsy. Because of a long preclinical period, early detection by urine cytology may be considered in high-risk groups.

Occurrence

There are about 35 000 cases of bladder cancer annually in the U.S., with about 10 000 deaths each year. Incidence is slowly rising. Mortality is declining, except among black males. Incident cases are 2.5 to 3.0 times more common in men than in women and about twice as common in whites as in nonwhites. Mortality is sharply age dependent, with rates of 10 per 100 000 among men aged 25 and older and of 75 per 100 000 among men aged 75 to 84. Workers at high risk (and their risk ratios) include rubber workers (1.5); dyers in the textile, fur, and leather industries (1.7); chemical workers (1.3); and paint manufacturing workers (1.3). Other exposed workers (with unknown risk ratios) include printing press operators, nickel and copper refiners, petroleum refiners, truck drivers, kitchen workers, and health care workers.

Causes

About 30% to 40% of bladder cancer in men is attributable to cigarette smoking, 10% to 50% is due to occupational exposure, and the balance is of unknown etiology. In women, 5% to 10%

of bladder cancer is occupationally related, and some is associated with cigarette smoking, the use of analgesics containing phenacetin, or a history of pelvic irradiation.

The best-known occupational association with bladder cancer is exposure to aromatic amines. These include beta-naphthylamine, used as an antioxidant in rubber manufacturing, and 4,4′-methylenebis (2-chloroaniline) (MBOCA), used as a curing agent in polyurethane foam manufacturing. Beta-naphthylamine is not now used in the U.S.; MBOCA is still used. The number of workers previously exposed to MBOCA is estimated to be from 1400 to 33 000. Other aromatic amines include benzidine, o-toluidine, 4-chloro-ortho-toluidine, and aromatic amines that are products of destructive distillation of carbon, hydrogen, and nitrogen, such as in tobacco smoke. Bladder cancer is also associated with exposure to polycyclic aromatic hydrocarbons. Occupational exposure occurs with exposure to nearly all products of combustion: engine exhaust, smelting, coke oven emissions, manufacture of creosote, and roofing tars. Dust and heat are risk factors for the development of bladder cancer. A synergistic effect between smoking and occupational exposure to bladder carcinogens has been reported (risk ratio of 11.7) but not confirmed. Within the U.S., the geographic distribution of age-adjusted death rates parallels the distribution of industrialization in general and of male employment in industries using aromatic amines in particular. Familial aggregation has occurred in regions where nephropathy is common, but this probably is due to analgesic abuse.

Pathophysiology

Neoplasia of the urinary bladder usually occurs in the transitional epithelium. Lesions of the bladder may occur in more than one location and recur following treatment. There are two pathologically distinguishable but overlapping types: papillary and nonpapillary. Individual low-grade papillary tumors have a relatively good prognosis, with only 5% of cases developing invasive disease, but with multiple papillary tumors, 30% develop invasive disease. Most invasive carcinomas of the bladder appear to develop from flat epithelial abnormalities, either carcinoma in situ or a markedly atypical urothelium. The natural progression of occupational bladder cancer proceeds from a period of negative cytological and histological appearance of cells, to a period where cytology is positive and cystoscopy is negative, to a period when both are positive. Occupational bladder cancer is at first clinically inapparent. On average, occupationally related bladder tumors occur 15 years earlier than nonoccupationally related bladder tumors. Latency ranges from 4 to 40 years from first exposure, with a mean of 20 years (*see* Overview of Occupational Cancer in part 3).

Prevention

Primary prevention entails substituting noncarcinogenic substances for known bladder carcinogens; implementing engineering controls such as ventilation; and avoiding other risk factors, mainly cigarette smoking. Secondary prevention, including worker education about early symptoms and medical screening of workers that entails urinalysis, urine cytology, and cystoscopy, is of uncertain value in finding early lesions. A further preventive measure is surveillance of health effects and of exposures.

Other Issues

There have been reports of an increased risk in workers exposed to aromatic amines who have the "slow" phenotypic variant for the autosomal recessive trait for the liver-bound enzyme N-acetyltransferase, but the issue is not resolved. Risk may also be increased for the phenotype reflecting the ability to hydroxylate drugs such as debrisoquine and mephenytoin. These phenotypes pertain to activities of specific isoenzymes of cytochrome P-450 and, hence, to substances metabolized by these isoenzymes.

MT

Further Reading

Cartwright RA. Screening for bladder cancer with particular reference to industrial groups. In: Prorok PC, Miller AB, eds. *Screening for Cancer.* Geneva, Switzerland: International Union Against Cancer; 1984:144-160.

Kaisary A, Smith P, Suczq E, et al. Genetic predisposition to bladder cancer: ability to hydroxylate debrisoquine and mephenytoin as risk factors. *Cancer Research.* 1987;47:5488-5493.

Matanoski GM, Elliot EA. Bladder cancer epidemiology. *Epidemiol Rev.* 1981;3:203-229.

National Cancer Institute. *1987 Annual Cancer Statistics Review.* Washington, DC: U.S. Department of Health and Human Services, Division of Cancer Prevention and Control; 1988. NIH publication 88-1789.

Schulte PA, Ringen K, Hemstreet GP. Optimal management of asymptomatic workers at high risk of bladder cancer. *J Occup Med.* 1986;28:13-17.

Brain Cancer

ICD-9 191

Identification

Brain cancers are identified clinically on the basis of symptoms, signs, and specific diagnostic tests. Symptoms vary considerably, depending on the type and location of tumor, but often include headache, fatigue, weight loss, and a variety of neurological abnormalities. Some symptoms are due to a direct effect of the tumor—for example, destruction or irritation of surrounding nerve tissue—and some are due to indirect effects—for example, increased intracranial and intraspinal pressure. Suggestive signs are primarily neurological deficits. Definitive diagnosis is based on some combination of computed tomography (CT) scanning, magnetic resonance imaging (MRI), arteriography of the cerebral blood vessels, open biopsy, and other diagnostic tests. The brain is a frequent site of the metastatic spread of cancers from other sites, most notably lung cancer in men and breast cancer in women; therefore, primary brain tumors need to be differentiated from the secondary spread of tumors from elsewhere in the body.

Occurrence

The annual incidence of brain cancer in the U.S. is approximately 5 per 100 000 males and approximately 4 per 100 000 females. The most common types of brain cancer are glioblastoma, meningioma, and astrocytoma. Meningioma is the only relatively frequent primary brain tumor with a higher incidence in females than in males in the U.S. The occurrence of occupationally related brain cancer is not known.

Causes

For the vast majority of brain cancers, a cause is never established. Studies have found increased occurrence of brain cancer among oil refinery and petrochemical production workers, synthetic rubber manufacturing workers, and polyvinyl chloride production workers. These workers share some similar exposures, including organic solvents, formaldehyde, polycyclic aromatic hydrocarbons, vinyl chloride, lubricating oils, and phenols. Some studies have suggested a relationship between employment in jobs with exposure to electromagnetic fields and brain cancer deaths, although this relationship has not been confirmed. A study of workers in a nuclear fuels fabrication plant found excess brain cancer, despite the fact that cancer and death rates overall were less than expected. An excess of glioblastoma multiforme has been demonstrated among vinyl chloride workers. Some studies have shown a relationship between brain cancer and both trauma and x-radiation to the head.

Several studies suggest an association between childhood brain tumors and parental exposure to solvents, metals, and ionizing and nonionizing radiation. In utero exposure to N-nitroso compounds, several of which are potent neurocarcinogens, may increase the risk of childhood tumors of the central nervous system. And genetic factors seem to play an important role in the etiology of some rare brain cancers, such as bilateral retinoblastoma.

Pathophysiology

Because of the various types of brain cancer, it is beyond the scope of this book to describe the pathophysiology of this condition.

Prevention

Because brain tumors are often incurable, emphasis is on primary prevention, limiting exposure to factors known to cause or suspected of causing brain tumors.

BSL

Further Reading

Jones RD. Epidemiology of brain tumors in man and their relationship with chemical agents. *Food Chem Toxicol.* 1986;24:99-103.

Kessler E, Brandt-Rauf PW. Occupational cancers of brain and bone. *State of the Art Reviews: Occupational Medicine.* 1987;2:155-163.

Thomas TL, Waxweiler RJ. Brain tumors and occupational risk factors: a review. *Scand J Work Environ Health.* 1986;12:1-15.

Wilkins JR III, Sinks T. Parental occupation and intracranial neoplasms of childhood: results of a case-control interview study. *Am J Epidemiol.* 1990;132:275-292.

Bronchitis, Acute—*See* Respiratory Tract Irritation

Bronchitis, Chronic

ICD-9 491
SHE/O

Identification

Chronic bronchitis is defined as the presence of chronic cough with sputum production for at least 3 months per year for at least 2 consecutive years. Its hallmark is chronic production of sputum (either clear or colored). In the clinical setting, chronic bronchitis is often accompanied by emphysema and may include shortness of breath and airways obstruction.

Diagnosis of chronic bronchitis depends on the presence of symptoms, often identified by questionnaire. The British Medical Research Council questionnaire on respiratory symptoms is internationally recognized as a valid and reliable method for eliciting symptoms of chronic bronchitis (*see* Lung Disease Testing).

Chronic bronchitis can occur with or without dyspnea (shortness of breath) on exertion. Although the presence of excess mucus and cough is a response to airways insult, it does not, by itself, define the seriousness of the disorder or its prognosis. Dyspnea implies more serious chronic airways disease, and affected people develop respiratory insufficiency more rapidly than do people with chronic bronchitis alone.

The chest sounds of chronic bronchitis include rhonchi or coarse rales on inspiration and, in some people, wheezing on expiration. Expiratory time is increased.

Pulmonary function tests are usually normal when dyspnea is absent. In chronic bronchitis with airflow obstruction, flow rates are decreased early in the course of disease. Later, forced expiratory function in 1 second (FEV_1) is also decreased, and total lung capacity (TLC) is generally normal or increased. Single-breath diffusion capacity is generally normal, but if emphysema is also present, the diffusion capacity will also be decreased.

Occurrence

Chronic bronchitis affected about 8 million people in the U.S. in 1981. This is probably an underestimate, because a common practice is to include only by implication a diagnosis of chronic bronchitis in chronic obstructive pulmonary disease, which typically includes both emphysema and bronchitis. Chronic bronchitis is more common among smokers, men, people over 40, and urban residents. It is usually preceded by 10 to 20 years of exposure to causative agents, with this time period declining with age. In some populations, prevalence is 10% to 40%. Prevalence has increased steadily in the past 25 years.

Chronic bronchitis (with or without emphysema) ranks fifth as a cause of mortality in the U.S., accounting for 3.3% of total deaths. In 1980, approximately 3300 deaths were attributed to chronic bronchitis. That year, there were also about 14 000 deaths attributed to emphysema and 35 000 deaths attributed to chronic airways obstruction, not otherwise classified. Because chronic airways obstruction often includes chronic bronchitis and because chronic bronchitis often occurs along with emphysema, chronic bronchitis was likely a contributing factor to these deaths.

Among those with chronic bronchitis, mortality is 2.4 times more frequent in men than in women, although in those aged 45 to 54, mortality for women approaches that for men. The risk of dying from chronic bronchitis is 10 times greater among smokers than among nonsmokers, and is even greater for male cigarette smokers.

Causes

Chronic bronchitis is caused by the sum of all harmful environmental factors, including cigarette smoking and occupational hazards. In the population as a whole, cigarette smoking is by far the most common cause. Among certain groups of workers, however, occupational exposures are also important causes. Occupational causes have been identified primarily on the basis of clinical and epidemiological data. Many gases, fumes, and aerosols are acutely irritating to the respiratory tract, causing inflammation of the airways lining and other parts of the lung. Repetitive or continuous exposure to such irritants can result in chronic disease. Occupational exposures leading to chronic bronchitis can be grouped into two broad categories: (a) specific chemicals, such as ammonia, arsenic, chlorine, cadmium oxide, chromium, oxides of nitrogen, osmium tetroxide, phosgene, tungsten carbide, vanadium, sulfur dioxide, isocyanates, and possibly chlorinated hydrocarbons; and (b) complex dusts and aerosols, such as smoke; those dusts and aerosols found in brickworks, cement plants, coke ovens, foundries, coal and other mines, rubber plants, quarries, rock-crushing operations, smelters, and welding operations; and dusts of cotton, flax, soft hemp, mixed potash, and phosphates.

Pathophysiology

The pathology of chronic bronchitis due to occupational exposures is the same as that due to inhalation of tobacco smoke. It is an inflammation of the large airways in response to repeated physical and chemical irritation. It is characterized by enlargement of goblet cells and mucus glands and by mucus obstruction of the small airways. The volume density of mucus glands and the ratio of the thickness of mucus glands to that of the airways lining, known as the Reid Index, is sometimes used to assess bronchitis at autopsy.

The pathogenesis of chronic bronchitis is not well established. In addition to increases in sputum production and mucus obstruction of small airways, there may also be airway wall scarring produced by a variety of agents.

Prevention

Chronic bronchitis can be controlled by reducing exposure to respiratory irritants using standard means of industrial hygiene, such as positive engineering to eliminate hazards, ventilation to remove hazards from the workplace, changes in work processes to reduce exposures, and use of respirators in some circumstances.

Medical surveillance by periodic questionnaires is recommended for workers at risk of developing occupational chronic bronchitis. If exposed individuals develop symptoms, similarly exposed workers should be examined and exposure should be monitored. Pulmonary function testing may also be helpful.

Other Issues

Efforts are needed to reduce pollution of the personal environment by tobacco smoke. Cigarette smoke usually is additive to other risks for chronic bronchitis, but it may also be synergistic, as it is with cotton dust exposure (ie, the combination of cigarette smoke and cotton dust is more than additive).

DCC

Further Reading

Becklake MR. Chronic airflow limitation: its relation to work in dusty occupations. *Chest.* 1985;88:608-617.

Jamal K, Cooney TP, Fleetham JA, Thurbleck WM. Chronic bronchitis: correlation of morphologic findings to sputum production and flow rates. *Am Rev Respir Dis.* 1984;128(5):719-722.

Kilburn KH. Chronic bronchitis and emphysema. In: Merchant JA, ed. *Occupational Respiratory Diseases.* Washington, DC: U.S. Government Printing Office; 1986:503-522. DHHS (NIOSH) publication 86-102.

Korn RJ, Dockery DW, Speizer FE, Ware JH, Ferris BG. Occupational exposures and chronic respiratory symptoms. *Am Rev Respir Dis.* 1987;136:298-304.

Medical Research Council. Standardized questionnaire on respiratory symptoms. *BMJ.* 1960;2:1665.

Morgan WKC. Industrial bronchitis. *Br J Ind Med.* 1978;35:285-291.

Mullen JR, Wright JL, Wiggs BR, Pare PD, Hogg JC. Reassessment of inflammation of airways in chronic bronchitis. *BMJ.* 1985;281(6504):1235-1239.

Oberholzer M, Dalquen P, Wyss M, Rohr HP. The applicability of the gland/wall ratio (Reid Index) to clinicopathological correlation studies. *Thorax.* 1978;33(6):779-784.

Samet JM. A historical and epidemiologic perspective on respiratory symptoms questionnaires. *Am J Epidemiol.* 1978;108(6):435-446.

U.S. Public Health Service. *The Health Consequences of Smoking: Cancer and Chronic Lung Disease in the Workplace. A Report of the Surgeon General.* Washington, DC: U.S. Government Printing Office, 1985. DHHS (PHS) publication 85-50207.

Brown Lung—*See* Byssinosis

Building-Related Illness

Identification

Building-related health problems can be divided into two groups: those characterized by a relatively specific clinical picture for which a specific etiology can often be identified, and those with reports of a wide spectrum of symptoms, temporally related to work but usually of unclear etiology. The former set

Table 1.—Specific Building-Related Illnesses

Illness	Etiologic Agent(s)	Basis of Diagnosis
Respiratory infection	*Legionella* species, *Mycobacterium tuberculosis*, Respiratory viruses	Culture, appropriate stains, serology
Hypersensitivity pneumonitis, including humidifier fever	Avian proteins, thermophilic actinomycetes, protozoans, fungi, contaminated humidifiers and HVAC systems	Serum precipitins, clinical responses to removing exposure, challenge
Asthma	Organic antigens, toluene diisocyanate, formaldehyde	Specific challenge, skin testing, clinical picture
Carbon monoxide intoxication	Carbon monoxide from combustion sources, vehicle exhaust	Elevated carboxyhemoglobin level

of problems is often referred to as specific building-related illness and the latter as sick (or tight) building syndrome.

Specific Building-Related Illnesses. The principal specific building-related illnesses result from exposure to biological and chemical agents (Table 1). The diagnosis of specific respiratory infections, such as pneumonia caused by *Legionella* species, is usually based on the results of cultures, stains of respiratory secretions or other materials, or diagnostic increases in antibody titers in serum. Hypersensitivity pneumonitis has a variable clinical picture (*see* Hypersensitivity Pneumonitis). Humidifier fever has the systemic components of hypersensitivity pneumonitis but without evidence of pulmonary involvement. The symptoms of carbon monoxide intoxication are varied, ranging from subtle neuropsychological effects to coma and death (*see* Asphyxiation). In identifying these illnesses, the conventional clinical criteria should be applied in relation to the building environment. The temporal relationship of the illness to exposure in the building must be carefully assessed, as should the occurrence of disease in co-workers in the same area or building.

Sick Building Syndrome. Although a standard, accepted definition is lacking, it has been suggested that SBS should be considered as the occurrence of a specific constellation of symptoms in over 20% of the work force in a building or in a particular area within a building. The constellation of symptoms includes primarily headache, fatigue, and eye and mucous membrane irritation; nausea, and chest symptoms such as cough and chest tightness, may also occur. Employees typically find that these symptoms disappear shortly after they leave the building. Evidence of illness on physical examination or laboratory tests is nearly always lacking.

The diagnosis of SBS in an individual patient must depend on epidemiological as well as clinical characteristics. A careful history and physical examination, and any indicated diagnostic tests, are necessary to rule out other problems. Key items in the history include any recent changes in the home or work environment, including the use of new equipment, materials, or processes; similar symptoms of recent onset among family members or co-workers; and the temporal relationship between symptom occurrence and participation in activities or environmental exposures.

Occurrence

Information on the incidence of specific building-related illnesses is lacking. In 1987 and 1988, respectively, 1085 and 1038 cases of *Legionella* pneumonia in the U.S. were reported to CDC.

Valid estimates of the incidence or prevalence of SBS are not available because of (a) the lack of uniform terminology to describe health problems related to buildings; (b) the diversity of agencies and organizations that investigate problem buildings; and (c) the paucity of systematic large-scale epidemiological studies of SBS. In a 1986 nationwide survey of 600 workers, about 20% reported dissatisfaction with the air quality at work.

Causes

By definition, specific building-related illnesses are associated with etiologic agents (Table 1). In some outbreaks of hypersensitivity pneumonitis (or humidifier fever), the illness cannot be linked to a specific biological agent, but the affected people generally have precipitating antibodies to extracts of the biological mixture contaminating the responsible humidifier or heating, ventilation, and air conditioning (HVAC) system.

In the 1960s, building construction practices changed significantly. New construction emphasized the use of prefabricated exterior sections mounted on a steel frame, creating a much tighter building envelope. Windows were made inoperable, and mechanical ventilation, often under centralized control, replaced natural ventilation under individual control. The rapid rise in SBS complaints followed the strong push toward energy conservation that resulted from the dramatic increase in oil prices in the mid-1970s. Since heating and cooling outside air constitutes a major part of a building's energy costs, the proportion of outside air used in the ventilation system was often drastically reduced.

Typically, air movement is controlled entirely by an HVAC system that cannot be controlled directly by the occupants of a particular space. Often, the use of a space is not consistent with the original design, so ventilation may be inadequate. Inadequate maintenance of the HVAC system may also threaten indoor air quality in commercial buildings.

At the same time that ventilation has decreased, the number and variety of potentially toxic agents in the office environment have increased. The sources are diverse (Table 2). Outdoor pollutants may enter through improperly placed ventilation intakes. Bacteria, fungi, and other biological agents may grow on moist surfaces in the building and on air conditioners, ducts, filters, and humidifiers. Building materials and furnishings may release formaldehyde and other volatile organic compounds; sources include building insulation, adhesives, tiles, vinyl wall coverings, rugs, carpets, copying machines, and furniture. Cleaning products and pesticides may also contaminate the air. Cigarette smoke is a significant source of noxious gases and respirable particles. Human bioeffluents, such as butyric acid and body odor, may also contribute to an unsatisfactory work environment.

Table 2.—Indoor Pollutants and Their Sources in Commercial Buildings

Source	Pollutants
Tobacco smoke	Carbon monoxide, particles, organics
Gas boilers, furnaces, cookers	Carbon monoxide, nitrogen dioxide, particles, organics
People	Carbon dioxide, odors (bioeffluents), bacteria, viruses
Standing water, water damage	Biological agents
Furnishings, building materials	Formaldehyde and other organics, fibers
Computers, copiers, correction fluids, typesetting equipment	Organics, particles
Air from garages and loading docks	Carbon monoxide, particles, organics
Outdoor air	Carbon monoxide, nitrogen dioxide, ozone, sulfur dioxide, particles, organics, pollen, fungi
Soil gas	Radon, biocides, organics

The perception of comfort depends not only on the concentrations of specific pollutants, but also on the movement of air, temperature, lighting, and humidity. Workers' responses to indoor air-quality problems may also be affected by the quality of labor-management relations. Thus, building-related problems often have a multifactorial etiology.

There are at least three separate, but overlapping, perspectives on the causes of SBS: (a) SBS symptoms represent a nonspecific response to an extensive array of low-level stressors, including air contaminants as well as thermal, lighting, and acoustic properties of the office environment. (b) SBS is caused by a specific agent or agents. Both of these two perspectives include the belief that the HVAC system may play a role if inadequate design or operation permits the buildup of contaminants to objectionable levels. (c) SBS does not have a physical basis but occurs because of poor labor-management relationships, stress, or other psychosocial factors. This view is not widely held.

Pathophysiology

The pathophysiology of the specific building-related illnesses is varied. Hypersensitivity pneumonitis, including humidifier fever, results from systemic immune responses (*see* Hypersensitivity Pneumonitis). Pollutants in indoor air may both induce a state of airway hyperresponsiveness and exacerbate already established asthma. The pathophysiology of carbon monoxide intoxication is described in the entry on asphyxiation.

The pathophysiology of SBS is not known. The symptoms associated with SBS are mostiy those of irritation and, perhaps, central nervous system effects. The symptoms appear to be transient and noncumulative, and they are not thought to result in chronic health impairment.

Prevention

Preventing specific building-related illnesses requires identifying the source of the specific etiologic agent(s) and applying standard industrial hygiene and infectious disease control strategies. Positive engineering to control the sources of toxic agents is preferable. Dilution by increasing the supply of clean air may be effective in some situations.

Specific illnesses may be prevented by careful planning as buildings are designed, changed, and used. New uses for space and introduction of new processes should be evaluated for their potential to cause adverse health effects. Ongoing maintenance of HVAC systems is essential. Proper cleaning and disinfection of cooling towers prevents the growth of *Legionella pneumophila*.

Once the diagnosis of a specific building-related illness has been established, evaluation of the building is indicated; immediate evaluation may be necessary for carbon monoxide inhalation or outbreaks of infectious respiratory diseases. Preventing SBS necessitates a more complete understanding of its causes, particularly if certain substances that could be eliminated play a major role.

The HVAC system is important in controlling both specific illnesses and SBS. Problems with HVAC systems often include inadequate outdoor air because of either fixed minimum amounts of outdoor air or no outdoor air intakes. Outdoor air intakes may be contaminated either at ground level by soil or vehicular exhaust, or on the roof by cross-contamination from other towers or exhaust vents. Other problems include poorly placed air supply and return devices, and airflow imbalances from short circuiting of the air supply. Common operational problems consist of inadequate maintenance, excessive thermal or contaminant loads, and changes in control strategies. Reduction of the amount of outside air to conserve energy is a common problem. These findings have important implications for both primary and secondary prevention.

Problems can occur during the life of a building, even when the original design is adequate. Very commonly, the thermal load within an office is increased by the introduction of lights, computers, and other office machinery, or the occupant load increases more than was planned. Another increasingly common change is the use of modular equipment to create separate cubicles in an open bay. Besides increasing noise levels and occupant density, this modification often dramatically decreases the ventilation efficiency, particularly within the cubicles. Changes in the interior of the building can thus have a dramatic effect on air quality, and the potential consequences should be considered routinely in any planning process before changes are implemented.

For investigating buildings and developing controls, the systems approach known as "building diagnostics" is recommended. The investigative team should be interdisciplinary, including a physician, an industrial hygienist, and an HVAC engineer. Conventional industrial hygiene methods are often not sensitive enough to detect the causes of most building-related problems. The concentration of air contaminants may be one or two orders of magnitude below the sensitivity of the most commonly used industrial hygiene instruments.

Building diagnostics begins with a walk-through inspection of the workplace, during which time complaints are gathered from occupants and the HVAC system and its controls are examined. Attention is paid to thermal, lighting, and acoustic conditions as well as to air quality. If needed, more detailed investigation is conducted, including simulation, testing, and measurement of the HVAC system; air sampling; and use of a questionnaire.

Other Issues

Although most investigators have largely discounted the notion that SBS is generally a form of collective stress disorder (mass psychogenic illness), there have been several reports of illness outbreaks due to collective stress disorder. These outbreaks are of illness that usually does not resemble SBS: The onset is much more acute than with SBS, and the symptoms are more varied. Distinguishing these incidents from an acute chemical exposure may be difficult. Characteristics that may help to distinguish an episode of collective stress disorder from chemical exposure include a pattern of spread that is not consistent with ventilation patterns; a time sequence not consistent with ventilation flow rates; an absence of medical findings compatible with exposure; and an epidemic curve consistent with "person-to-person" rather than "common source" exposure. Sometimes, poor air quality may trigger an episode of collective stress disorder, suggesting the importance of considering air quality even when the features of the outbreak suggest that anxiety is the etiologic agent (*see* Collective Stress Disorder).

The home is a predominant source of personal exposure to many air pollutants. Indoor air pollution at home may also cause disease, including hypersensitivity pneumonitis, respiratory infection, asthma, and carbon monoxide poisoning. Comprehensive inventories of environmental exposure should cover not only the workplace but also the home.

JS, MM

Further Reading

Finnegan MJ, Pickering CA, Burge PS. The sick building syndrome: prevalence studies. *Brit Med J.* 1984;289:1573-1575.

Guidotti TL, Alexander RW, Fedoruk MJ. Epidemiologic features that may distinguish between building-associated illness outbreaks due to chemical exposure or psychogenic origin. *J Occup Med.* 1987;29:148-150.

Molhave L, Bach B, Pederson OF. Human relations during controlled exposures to low concentrations of organic gases and vapours known as normal indoor air pollutants. In: Berglund B, Lindvall T, Sundell J, eds. *Proceedings of the Third International Conference on Indoor Air Quality and Climate.* Vol 3. Stockholm, Sweden: Swedish Council for Building Research; 1984:431-436.

National Research Council, Committee on Indoor Air Quality. *Policies and Procedures for Control of Indoor Air Quality.* Washington, DC: National Academy Press; 1987.

Samet JM, Marbury MC, Spengler JD. Health effects and sources of indoor air pollution: part 2. *Am Rev Respir Dis.* 1988;137:221-242.

Woods JE, Drewry GM, Morey PR. Office worker perceptions of indoor air quality effects on discomfort and performance. In: Seifert B, et al., eds. *Proceedings of the Fourth International Conference on Indoor Air Quality and Climate.* Vol 2. Berlin: Institutes of Water, Soil and Air Hygiene; 1987:464-468.

Burn Injury

ICD-9 940-949

Identification

Thermal burns are classified according to the depth of the wounds and the percent of the body surface area affected. First-degree burns, the least severe category, affect only the epidermis and are characterized by redness and swelling of the affected areas. Sunburns exemplify this depth of burning. Second-degree burns affect not only the epidermis but also the dermis and are characterized by blistering of the affected areas. If uninfected, second-degree burns usually heal by themselves in 7 to 10 days. Third-degree ("full-thickness") burns affect both the epidermis and the dermis, but, unlike first- or second-degree burns, third-degree burns larger than approximately 1 inch in diameter will not heal by themselves. These burns require skin grafting, because all the dermis in the affected area has been destroyed. Fourth-degree burns, affecting the muscle, constitute the most severe category and, like third-degree burns, require skin grafting to close the wounds.

Occurrence

Burns are the fourth leading cause of injury death in the U.S., accounting for approximately 6000 deaths each year, about 400 of which are occupational. An additional 90 000 persons are hospitalized annually for the treatment of burns. Deep burns can cause severe, permanent disability and disfigurement; for this reason, the social and economic costs of burns exceed their expected costs based on incidence alone.

Known risk factors for burn injury include youth and old age, alcohol use, and functional or mental impairment. In the U.S., the burn incidence rate is higher for males than for females and higher for blacks than for whites. Economic well-being is a strong predictor of the risk of burning, with poorer economic status associated with higher rates for most types of burns.

The literature on burn epidemiology and control focuses primarily on domestic burns, such as scalds to young children and flame burns involving ignited clothing, flammable liquids, or structural fires. Work-related burns are less well investigated, with the literature typically limited to the evaluation of burn victims treated at one or several hospitals or in a small cluster of communities. However, the literature has not specifically focused on this type of burn.

Overall, in the U.S., an estimated 45% of medically treated burns to males are work related, as are 10% of such burns to females. Occupational burn rates are approximately two to three times higher for blacks than for whites, eight times higher for males than for females, and, among males, four to five times higher for workers aged 16 to 24 years than for workers aged 65 years or older, with intermediate rates for other adult males. In New England, burn rates for people treated in hospitals are 62 burns per 100 000 person-years for males versus 6 burns for females; 79 burns per 100 000 person-years for blacks versus 31 burns for whites; and 100 burns per 100 000 person-years for males aged 16 to 24 years versus 22 burns for males aged 65 years or older.

The total number of work-related burn injuries treated in hospital emergency departments each year in the U.S. is estimated to be 150 000, based on data derived from the NIOSH-CPSC National Electronic Injury Surveillance System. Of these burns, approximately 30% affect the face and 30% affect the hands.

Substantial variation in burn rates occurs by occupation, with rates for operators and laborers 5 times higher than those for managers, professionals, and administrators and 10 times higher than those for workers in clerical and sales occupations.

Considerable geographic variability in occupational burn incidence and mortality probably exists. In New England, for example, incidence rates of hospitalized occupational burns for individuals aged 20 years or older are highest in Maine for both males and females (88 burns per 100 000 person-years for males versus 20 burns for females); the lowest rate for males occurs in Vermont (40 burns per 100 000 person-years), and the lowest rate for females occurs in Rhode Island (2 burns per 100 000 person-years).

Causes

There are six broad types of burn injuries. These types and their approximate proportions among people who are hospitalized for work-related burn injuries are scald (40%); flame/flash/smoke inhalation (30%); chemical (10% to 15%); contact (10%); electrical (5% to 10%); and, only rarely, radiation burns. Radiation burns (eg, flash burns to the cornea and conjunctiva associated with arc welding) are more highly represented among work-related burns that do not require hospital care (*see* Eye Injury).

Overall, the job tasks most commonly associated with work-related burn injury are similar to the activities that contribute to nonoccupational burn injury. Activities associated with food preparation and consumption account for 40% of all hospitalized occupational burns (35% for males and 70% for females). Products such as stoves, ovens, grills, cups, and dishwashers feature prominently among the environmental hazards contributing to these injuries. Job tasks associated with motor vehicle maintenance and use, the second most frequent activity associated with burns at work, account for 10% to 15% of all hospitalized work-related burn injuries (approximately 15% for males and 3% to 5% for females). Usual types of motor vehicle-related burns include scalds from radiator fluids, flame burns from gasoline (eg, from "priming" a carburetor) or from postcrash fires, contact burns from tailpipes, and chemical burns from battery acid.

The epidemiology of people hospitalized with occupational burns reveals the following: (a) The large majority of scald and contact burns involve either food preparation and consumption, or motor vehicles. (b) Approximately 75% of all flame/flash burns or cases of smoke inhalation involve flammable liquids, particularly gasoline. Only 10% to 15% of these burns occur in structural fires, such as building and motor vehicle fires. The low frequency of structural fire-related burns contrasts with the typical pattern for nonoccupational flame/flash/smoke inhalation injuries, in which 20% to 30% of such burns are caused by structural fires. (c) At least half the chemical burns (in contrast to occupational chemical toxicity) are caused by contact with motor vehicle battery acid. Another common cause of chemical burns is contact with household cleaning products. (d) Most of the electrical burns occur while people are working with outdoor electrical supply lines. (e) In contrast to nonoccupational burns, few work-related burns are known to be associated with alcohol use.

Pathophysiology

The pathophysiology of burns is very complex. Large burns affect the functioning of every body system and place enormous demands on the body's ability to regulate normal body temperature, protect against infection, and repair damaged tissue.

Prevention

Primary prevention of occupational burn injury entails adherence to various engineering and administrative control strategies, including the following:

1. In-depth investigations of burn injury occurrence should be conducted to identify the subtle factors that alter the balance between safety and tragedy in a given type of situation. Promptness in conducting an investigation is imperative if accurate information about the events leading to injury is to be obtained.

2. Unambiguous warning labels should be placed on containers of flammable liquids, motor vehicle carburetors, and industrial equipment and processes that use flammable liquids. Adequate pictorial warnings to convey the degree of flammability of the material should be used. Flammable liquids and chemicals that burn, such as strong acids and bases, should be stored in safety containers and clearly labeled (in languages understood by the work force) with their contents. Substances that react with one another (eg, acids and bases) should have separately vented storage areas. A review of recommendations to prevent burn injury from flammable liquids and other flammable materials has been published.[1]

3. Ignition sources, such as lighted cigarettes and pilot lights, should be eliminated from areas where flammable liquids are being used. "Intrinsically safe" equipment, such as sparkless electrical devices, should be substituted for less safe equipment when-

ever the possibility of exposure to explosive substances exists.

4. Products and work areas associated with food preparation and consumption should be modified to incorporate sound ergonomic design criteria. Cups should be designed to prevent easy spillage; stoves, ovens, and grills should be designed to discourage users from reaching across hot surfaces and burners to reach control knobs. Control knobs should be positioned either to the side or in front of hot surfaces and burners, not behind them. Barriers (guards and enclosures) can be used to prevent contact with hot products and surfaces.
5. Supervisors and other individuals who work in occupations associated with increased risk of burn injury should receive focused, evaluated training programs that cover the fundamental principles and practice of burn injury control.

Secondary prevention of burn injuries requires prompt treatment in the form of fluid replacement therapy in the early postburn period to prevent complications arising from the burn shock phase of the injury. This therapy should be initiated immediately (in the ambulance) for any burn affecting more than 20% to 30% of the body surface area. For all burns, infection prevention and control is of paramount concern to minimize the physiological load from this source and to promote healing. Prevention of infection by applying dry sterile dressings to the burn wounds should be accomplished immediately after the burned areas have been cooled. In addition, burns to the face and head, or knowledge that inhalation of smoke and toxic gases has occurred, may suggest the need to provide respiratory support to the victim.

In general, because even a 5% full-thickness burn to an older person can be fatal, people aged 50 years or older who have sustained burns should be evaluated in a hospital. In addition, people of any age who have sustained second-degree or full-thickness burns to the hands or face should receive prompt in-hospital care to minimize disfigurement and disability.

Third- and fourth-degree burns require skin grafting to close the burn wounds. The benefits of prompt skin grafting are many and include control of infection, prevention of the increased metabolic and pathophysiologic responses to open wounds, and reduced lengths of hospital stay.

Counseling for people disfigured by burns is indicated during the initial hospital stay, especially as the date of discharge approaches. Additional counseling may be necessary over the next several years as part of the patient's follow-up care.

Other Issues

Treatment of large burns and treatment of smoke inhalation are among the most difficult tasks in medicine. Specialized burn treatment units and centers have been established to treat large or difficult burn injuries. The cost for care in these specialized facilities is often high, because long hospital stays may be required to treat the acute burn injury and because the staff-to-patient ratio is high. In addition, reconstructive procedures, if necessary, may require additional hospital admissions over an extended period of time. The American Burn Association has established guidelines that address the need to transfer a burn victim to a specialized burn treatment unit or center. The guidelines are based on the extent (size and depth) of the burn injury; the age of the victim; and the presence of any pulmonary involvement, other traumatic injury in addition to the burn injury, and preexisting disease. In addition, the decision of whether to transfer takes into account the range of treatment options available to the patient at the initial treating facility.

High-risk groups for burn injury include young workers, black workers, males, and operators and laborers. One obvious explanation for the higher risks for these workers is increased exposure to hazardous situations and environmental factors.

Research and control programs directed toward jobs held by these workers might help to identify and reduce the environmental and behavioral causes of burn injury to these groups.

AMR

Reference

1. Marshall G. *Safety Engineering*. Monterey, Calif: Brooks/Cole, Engineering Division; 1982.

Further Reading

Anastakis DND, Douglas LG, Peters WJ. Work-related burns: a six-year retrospective study. *Burns.* 1991. In press.

Baker SP, O'Neill B, Karpf RS. *The Injury Fact Book.* Lexington, Mass: DC Heath; 1984.

Barancik JI, Shapiro MA. Pittsburgh burn study. Springfield, Va: National Technical Information Service; 1975. Report PB250-737.

Burke JF, Bondoc CC, Quinby WC. Primary burn excision and immediate grafting: a method of shortening illness. *J Trauma.* 1974;14:389-394.

Chatterjee BF, Barancik JI, Fratianne RB, et al. Northeastern Ohio trauma study, V: burn injury. *J Trauma.* 1986;26:844-847.

Feck G, Baptiste MS. The epidemiology of burn injury in New York. *Public Health Rep.* 1979;94:312-318.

Feck G, Baptiste M, Greenwald P. The incidence of hospitalized burn injury in upstate New York. *Am J Public Health.* 1977;67:966-967.

Inancsi W, Guidotti TL. Occupation-related burns: five-year experience of an urban burn centre. *J Occup Med.* 1987;29:730-733.

Iskrant AP. Statistics and epidemiology of burns. *Bull NY Acad Med.* 1967;43:636-645.

Jelenko C. Chemicals that "burn." *J Trauma.* 1974;14:65-72.

Occupational injury surveillance. *MMWR.* November 1981; 30:578-580.

Rossignol AM, Boyle CM, Locke JA, et al. Hospitalized burn injuries in Massachusetts: an assessment of incidence and product involvement. *Am J Public Health.* 1986;76:1341-1343.

Rossignol AM, Locke JA, Boyle CM, et al. Motor vehicle-related burn injuries: Massachusetts. *MMWR.* October 1985;39:597-600.

Rossignol AM, Locke JA, Boyle CM, et al. Epidemiology of hospitalized, work-related burn injuries in Massachusetts. *J Trauma.* 1986;265:1097-1101.

Rossignol AM, Locke JA, Burke JF. Employment status and the frequency and causes of burn injuries in New England. *J Occup Med.* 1989;31:751-757.

Wilson GA, Sanger RG, Boswick JA. Accidental hydrofluoric acid burns of the hand. *J Am Derm Assn.* 1979;99:57-58.

Byssinosis

ICD-9 504
SHE/O

Identification

Byssinosis is an acute and chronic airways disease caused by exposure to the dust of cotton, flax, hemp, or sisal. The acute response is reversible and is characterized by a sensation of chest tightness and/or shortness of breath upon return to exposure following a weekend or holiday break. In societies in which Monday is the first workday of the week, it is sometimes referred to as "Monday morning syndrome." Cough may be present and may become productive. Ventilatory function (forced expiratory volume in 1 second, or FEV_1) on return to work often decreases from beginning to end of the work shift. For most affected people, these symptoms will decrease or disappear on the second day of work. However, if workers are exposed virtually every day of the workweek, as is the case in many developing countries, or if they work on rotating shifts, the periodicity may be altered.

Onset of chest tightness usually occurs 2 to 3 hours after start of exposure, distinguishing byssinosis from asthma, which usually has either immediate or delayed (6 or more hours) onset following exposure to allergens.

With prolonged exposure, both chronic cough (productive or nonproductive) and decline in ventilatory function become more severe. Dyspnea becomes a prominent complaint while acute declines in expiratory flow rates or FEV_1 over the work shift are marked. In advanced stages, clear clinical and physiological evidence of irreversible chronic obstructive lung disease emerges, and cross-shift declines are more pronounced.

A standard definition for use in epidemiology was developed by Schilling and adopted by the British Medical Research Council:[1]

Grade 0: No symptoms of chest tightness or breathlessness on Monday (or on return to exposure).

Grade 1/2: Occasional chest tightness or mild symptoms, such as respiratory tract irritation, on return to exposure.

Grade 1: Chest tightness and/or breathlessness on return to exposure only.

Grade 2: Chest tightness and/or breathlessness on return to exposure and on other days.

There are no characteristic signs on physical examination unless severe bronchospasm is present. In this case, wheezing on expiration may be heard. Among those with advanced disease, decreased breath sounds and an increased expiratory phase will be noted. These findings are characteristic of chronic obstructive lung disease and are not specific to byssinosis. Therefore, an accurate diagnosis depends on well-characterized exposure, a compatible history, and documentation of cross-shift declines in FEV_1 and expiratory flow rates.

Although the chest x-ray can produce no specific information to associate impairment with occupational exposure, it is important to eliminate other pulmonary pathologies, such as tuberculosis, lung cancer, and pulmonary fibrosis. In cases in which symptoms and functional changes are inconsistent, a complete set of pulmonary function tests, including static lung volumes and diffusion capacity, may be useful.

Occurrence

Byssinosis occurs worldwide, with several million workers occupationally exposed. Prevalence averages about 20% among exposed workers and has been reported to be as high as 83% among active workers with the highest exposure. It is most often associated with dust in cotton textile mills, but it also occurs with exposure to the dust of flax, hemp, and sisal and among workers in the cotton waste-processing industries. In the U.S., approximately 500 000 workers are at potential risk, with perhaps half this number in high-exposure jobs (eg, yarn manufacture and agricultural processing of cotton).

In the cotton textile industry, the prevalence of byssinosis is highest among workers employed in the initial stages of cotton processing (opening, picking, carding, stripping, and grinding) in mills that process a coarse grade of cotton and produce higher concentrations of dust. Prevalence is lower among workers in yarn processing and lowest among workers in slashing, weaving, and cloth handling.

Byssinosis also occurs among cotton gin workers and among workers in the garneting or bedding industry. It has been reported among workers in the delinting process of cottonseed oil extraction plants, but it has not been reported among workers in cottonseed mills. Nor has it been reported in the cotton fiber reuse or medical cotton industries or when cotton has been cleaned before processing.

Prevalence increases with the duration of exposure up to about 20 years, and then it decreases. This decline is attributed to workers who retire early because of impairment, leaving a survivor population. Prevalence is significantly lower among workers in the U.S. who were first exposed after the OSHA PEL was enforced (1978).

There is no evidence that race or ethnic background affects the risk of byssinosis. After controlling for smoking and dust exposure, there is no difference in risk between males and females.

Causes

The actual etiologic agent in cotton dust that causes the syndrome has not been identified, but current research has focused much attention on the role of gram-negative bacterial endotoxins. In experimental models, cotton dust extracts combined with purified *Escherichia coli* endotoxins in rabbits over a 20-week period resulted in severe bronchitis and bronchiolitis. When humans selected for their sensitivity to cotton dust were exposed to cotton dust in a model card room, acute changes in lung function tests related well with the gram-negative endotoxin content of the air. Although gram-negative endotoxin exposure clearly results in acute symptoms, a single causal agent to explain both acute and chronic reactions has not yet been identified.

Pathophysiology

Byssinosis appears to be pathologically similar to chronic bronchitis. It remains unclear whether emphysema is a significant part of the pathological picture in late stages of byssinosis. Pathological studies that include good occupational and smoking histories are needed. Potential pathophysiological mediators and mechanisms include histamine and histaminelike substances, immediate hypersensitivity, and simple irritation.

Prevention

Reduction of risk depends on reduction of cotton dust concentrations with the standard industrial hygiene practice of local exhaust ventilation. A second control technology is cotton washing, which, although effective, may not be technically feasible on a large scale.

The OSHA cotton dust regulation (29 CFR 1910.1043) is complex. It requires exposure monitoring, reduction of exposure, and medical surveillance, and it prescribes conditions under which respirators, as opposed to engineering controls, are permitted. The current PEL for cotton dust ranges from 200 to 750 $\mu g/m^3$ and is dependent on work location. In early stages of processing (from opening, picking, and carding to yarn preparation), the PEL is a time-weighted average of 200 $\mu g/m^3$. From slashing and weaving, the PEL is 750 $\mu g/m^3$. In the nontextile cotton-processing industry—involving cottonseed oil, batting, and bedding—and in waste processing in textile mills, the PEL is 500 $\mu g/m^3$. Employers are required to conduct medical surveillance and to monitor exposure regularly. If exposure reaches an "action level" of half the PEL, more surveillance and monitoring are required.

Medical surveillance should include use of questionnaires for symptoms of byssinosis following return to work, physical examination, and spirometry before and after the work shift on Monday or on return to work. These examinations should identify many of those people who are acutely affected and probably all of those with significant impairment. Because smoking contributes significantly to the risk of chronic bronchitis and airways obstruction, exposed workers should be encouraged to stop smoking.

Acute responsiveness should be treated as a sentinel health event. Workers in the same jobs or with the same exposure as those with stage 1/2 or higher byssinosis should be examined, and exposure on these jobs should be evaluated and, if indicated, reduced. If these primary preventive measures are ineffective, affected workers should be allowed to transfer, with wage retention, to jobs with less exposure. They should then be reevaluated. Respirators are useful as temporary protective measures or when there is no alternative, but they are not practical for permanent protection. If workers are symptomatic, employers in the U.S. are required by OSHA to monitor both exposure and workers' health more frequently.

If workers are transferred to jobs with lower exposure, the OSHA regulation legally protects their wages if (a) the workers show signs of byssinosis grade 1/2 or higher; (b) the workers are unable, in their physician's opinion, to wear respirators; and (c) exposure at their current job exceeds the PEL (*see* 29 CFR 1910.1043(f)).

Other Issues

Smokers are at increased risk for chronic obstructive lung disease, which adds to the risk of byssinosis. There is no evidence that asthma is an additional risk factor for development of byssinosis.

Current knowledge is limited by the absence of prospective studies with good exposure measurements that are not greatly affected by high rates of migration out of the cohort, although several such studies are now under way in various parts of the world. Out-migration leaves a survivor population and results in an underestimate of risk. Available data are largely from cross-sectional studies. It is not clear whether chronic byssinosis is always preceded by the acute response, whether the acute response with uncontrolled exposure inevitably results in chronic disease, or whether the natural history of byssinosis is a steady progression from grade 1/2 to higher grades.

DCC

Reference

1. Medical Research Council. Standardized questionnaire on respiratory symptoms. *Brit Med J.* 1960;2:1665.

Further Reading

Castellan RM, Loenchock SA, Kinsley KB, Hankinson JL. Inhaled endotoxin and decreased spirometric values: an exposure-response relation for cotton dust. *N Engl J Med.* 1987;317:605-609.

Kennedy SM, Christiani DC, Eisen EA, et al. Cotton dust and endotoxin exposure-response relationships in cotton textile workers. *Am Rev Respir Dis.* 1987;135:194-200.

Merchant JA. Byssinosis. In: Merchant JA, ed. *Occupational Respiratory Diseases.* Washington, DC: U.S. Government Printing Office; 1986. DHHS (NIOSH) publication 86-102.

Cardiomyopathy—*See* Congestive Heart Failure

Carpal Tunnel Syndrome

ICD-9 354
SHE/O

Identification

Carpal tunnel syndrome (CTS) is a nerve compression disorder affecting the median nerve, one of the three nerves that supply the hand with sensory and motor capabilities. The "carpal tunnel" in the wrist is a tunnel formed by the carpal bones on three sides and a ligament across the front. The median nerve, a large blood vessel, and several tendons pass from the forearm through this tunnel into the hand. The syndrome develops when there is compression or entrapment of the nerve in the wrist area.

Carpal tunnel syndrome usually begins with gradual onset of symptoms—specifically, intermittent tingling and numbness in the fingers of one or both hands, initially at night. The area of median nerve distribution of the hand is affected: usually part of the thumb; the index, third, and part of the fourth fingers; and the corresponding area of the palm and sometimes of the wrist, although many variations in nerve supply to the hand have been known to occur. The symptoms may progress to burning pain, severe and painful numbness, a sensation of swelling without objective signs, or a sensation of grip weakness (due to loss of sensory nerve function). Occasionally, the pain spreads to the forearm or higher. The loss of manual dexterity and motor control often creates difficulty in performing certain hand movements, such as opening bottles, turning doorknobs or keys, or grasping small items.

The symptoms may be provoked or worsened by Phalen's test (sustained wrist flexion for 1 minute) or Tinel's test (tapping over the median nerve at the carpal crease). There may be objective loss of strength in the thumb or in any of the affected fingers (see above), or loss of sensation noted on palpation or two-point discrimination, especially in the affected fingertips. Wasting away of the muscles at the base of the thumb is observed in severe cases. However, all of these tests have relatively low sensitivity, even when performed in a carefully standardized manner. Hand pain diagrams, on which affected individuals mark the location and quality of symptoms (to be compared with the usual CTS pattern), have better predictive value than most other clinical findings.[1,2]

Various laboratory tests may be used to attempt to confirm the diagnosis. The most common is nerve conduction testing, which measures whether the nerve impulse is conducted more slowly (decreased nerve conduction velocity) or less efficiently (decreased nerve conduction amplitude) through the wrist than along other segments of the nerve or in the corresponding area of the other hand (if only one hand is symptomatic). A relatively new approach to early detection is vibrometry, or measurement of the intensity of vibration that can be detected under the fingertips (similar to the method for testing hearing loss). However, its use in CTS diagnosis assumes that the vibration-sensitive nerve fibers are those most affected by the disorder, which has not been firmly established. Other tests that have been proposed include magnetic resonance imaging and computed tomography, both of which are expensive and unlikely to be used routinely in most workplaces. The validity of these tests remains to be demonstrated; the tests tend to differentiate groups of affected workers from unaffected workers but demonstrate poor validity on an individual basis.

The differential diagnosis includes ruling out compression of the median nerve at the forearm near the elbow (pronator teres syndrome) or at the shoulder (*see* Peripheral Nerve Entrapment Syndromes).

Occurrence

In the general population, the rate of CTS increases slightly among women with age.[3] However, in most occupational studies, there has generally been no trend or only a weak one with age. This may be an artifact due to a survivor effect ("healthy worker selection") operating in cross-sectional studies. In addition, in occupational cases, it is difficult to distinguish age from the duration of cumulative exposure to ergonomic stressors. Carpal tunnel syndrome is reported to be more common among women than men, especially in middle age, although this may be true only for nonoccupational cases. Among men and women performing the same or nearly identical work, little difference in risk has been observed;[4] the effect of ergonomic factors, especially repetitive work, appears to be far more important.

The prevalence of CTS in various workplaces has been estimated to range from 2% to 53% of the workers in ergonomically stressful jobs, and the annual incidence has been estimated to range from less than 1% to 26% of the work force per year. The rates depend on both the criteria used to define a case of CTS (self-reported symptoms, signs, or test results) and the intensity of exposure to repetitive or forceful exertions. Recent work in jobs requiring a flexed wrist posture increased the risk by about three times; there was also a strong trend with more hours per week when either wrist flexion or extension was required.[5] Highly repetitive manual work has been estimated to increase risk of CTS by three to six times, and the risk associated with highly forceful exertions has been estimated at two to three times that associated with low-force jobs.[4,6] The interaction of forceful and repetitive work resulted in an increase of 15 times the risk. The use of vibrating hand tools has been reported to elevate the risk by 6 to 14.[6,7] The latter estimate may have been partially confounded by the use of tools in nonneutral wrist postures. The risk due to repetitive motion, forceful loading, *and* segmental vibration was tripled again when occupational exposure lasted for more than 20 years.[6]

Causes

Occupational activities linked to CTS include typing and other data input operations; many manufacturing, assembly, and packing jobs; work with vibrating hand tools; and a wide variety of other occupations that involve highly repetitive or forceful manual tasks. (*See* Overview of Musculoskeletal Disorders in part 3 for a list of industries and occupations.) Specific risk factors include frequent, prolonged, or forceful use of a pinch grip or other finger activity, and bending of the wrist in the direction of either the palm or the back of the hand (palmar or dorsal flexion). The hand has a much greater biomechanical advantage in a power grasp (when the object is in contact with the full hand, not just the fingertips) than in a pinch grip (when greater forces are generated internal to the carpal tunnel). Repetitive or forceful pinch grips or other finger activities while the wrist is flexed are particularly likely to result in CTS. Exposure to local or environmental cold may intensify the effects of posture, force, and repetitiveness by decreasing finger sensitivity and requiring a worker to increase grip force to be sure that an object is held securely. Vibration has a similar effect through loss of sensation as well as direct damage to the nerve tissues. In addition, prolonged contact of the wrist with hard or sharp edges or other sources of mechanical compression may directly damage the median nerve, along with the blood vessel that supplies it.

In addition to ergonomic factors, other causes that may be either occupational or nonoccupational are exposure to solvents or heavy metals (eg, lead, mercury) and acute trauma to the wrist (eg, Colles' fracture) (*see* Peripheral Nerve Entrapment Syndromes).

Common nonoccupational associations include diabetes mellitus, rheumatoid arthritis, thyroid disease, pregnancy, obesity, alcoholism, and possibly osteoarthritis, estrogenic agents, and gynecologic surgery (eg, bilateral oophorectomy).[3,8] It should be emphasized that although these conditions are so prevalent in the general population, they still do not explain many occupational cases, and they should not be permitted to deflect attention from the workplace and easily preventable causes. Uncommon nonoccupational associations include Paget's disease of bone, gout, myxedema, acromegaly, possibly severe vitamin B_6 (pyridoxine) deficiency or excess, renal disease, toxic shock syndrome, and neoplasms including multiple myeloma.

Pathophysiology

The mechanism of damage is compression or irritation of the median nerve as it passes into the hand between the carpal (wrist) bones and the transverse carpal ligament across the front of the wrist. This causes slowed or incomplete transmission of sensory and motor nerve signals through the wrist to the part of the hand supplied by the median nerve, resulting in discomfort and impaired neuromuscular function. Compression may result directly from using the hand with the wrist flexed, so that the space within the carpal tunnel is narrowed and the nerve is squeezed against the other tissues. In a pinch grip, high forces are generated internal to the carpal tunnel, increasing the pressure on the median nerve; also, the tendons within the tunnel may become irritated and swollen as a result of repetitive or forceful finger activity, so that they crowd the carpal tunnel and compress the nerve. There is evidence that the mechanism is vascular, involving ischemia (loss of blood supply) to the nerve itself. Segmental vibration and low temperatures may also interfere with the microcirculation of the nerve and exacerbate the effects of repetitive motion and postural stressors (*see* Hand-Arm Vibration Syndrome). When vibration is the predominant occupational exposure, mechanical compression of the nerve in the carpal tunnel may not also be necessary for development of the syndrome.

Prevention

Primary prevention involves the use of engineering controls (appropriate ergonomic design and selection of tools, tasks, and workstations) to reduce the exposure to occupational ergonomic stressors (*see* Overview of Musculoskeletal Disorders in part 3). Particular attention should be paid to eliminating the need for pinch grips or wrist flexion by redesigning tools, equipment, and work methods. A keyboard, workbench, or other work surface may need to be raised or lowered to permit the wrist to be kept straight. A tool with a bent handle may eliminate the need to bend the wrist, but there is no such thing as an "ergonomic tool" per se; every tool must be selected and installed appropriately for the particular application. Reducing the force required to hold and operate tools or equipment will also reduce the forces generated within the carpal tunnel. Other factors that contribute to force requirements include low friction between hand and parts or equipment, mechanical resistance, infrequently sharpened knives or scissors, the torque produced by power hand tools, and poorly fitting gloves.

In general, attention should be paid to reducing work pace and repetitiveness, minimizing the force of manual exertions, and avoiding local or environmental exposure to cold temperatures. Piece-rate wages and machine-paced work should be avoided whenever possible because of the pressure on workers to maintain a constantly high speed, even when they are in pain. Sources of external mechanical compression of the wrist should be eliminated by measures such as padding the edges of a desk or bench where the wrists rest.

Where vibrating hand tools are used or where there is other exposure to segmental vibration, mechanical isolation and damping should be used to reduce the amplitude of the vibration transmitted to the hand and arm. Installation of a tool on an articulating arm or overhead suspension will also help to reduce vibration transmission. Tools should be selected for low vibration intensity and possibly for vibration frequencies above the natural resonances of the upper extremity (30-300Hz) (*see* Hand-Arm Vibration Syndrome).

As a secondary preventive measure, health care providers should be trained in the appropriate interview and clinical examination procedures to identify occupational CTS.[2] Workers who report symptoms typical of CTS (eg, numbness and tingling in the relevant part of the hand, especially at night) should receive immediate attention, even if physical findings and laboratory tests are negative or inconclusive. Once reported, cases should be treated conservatively, and jobs should be analyzed for ergonomic features that can be modified. Splint use on the job should be considered only if it does not interfere with work or require the worker to exert more force or strain another joint in order to perform the task. Removal from sources of exposure is essential to prevent the nerve damage from becoming irreversible. Follow-up is important to ensure that job modifications have been effective, that "light-duty" jobs have been correctly selected to avoid continuing stress to the wrist, and that symptoms and signs do not progress. There is no evidence that administration of pyridoxine is effective in treating the occupationally induced syndrome.

Concerning tertiary measures, surgery may be only temporarily effective if the worker is returned to an ergonomically stressful job that has not been modified; possible loss of grip strength, buildup of scar tissue, and increased vulnerability of the carpal tunnel to mechanical insult following surgery make it imperative that job assignments be selected carefully to avoid recurrence. Surgery is particularly questionable when there has been occupational exposure to segmental vibration, because mechanical compression at the carpal tunnel may not be the primary etiologic mechanism.

Other Issues

In high-exposure occupations, the contribution of nonoccupational factors such as age, gender, and medical history to overall risk appears to be relatively small compared with the effect of high exposure.

Many manual-intensive jobs involve simultaneous exposure to more than one of the causative factors listed above, and such combined exposures appear to have greater-than-additive effects on risk. For example, vibration or cold, by interfering with sensory nerve function and neuromuscular feedback, may result in increased grip forces compared with objective requirements. Work that is both forceful and repetitive increases the risk of CTS by 15 times, compared with 2 to 6 times for one factor or the other alone.

No data are available concerning the effects of combined exposures to ergonomic stressors and other peripheral neurotoxins, such as heavy metals or organic solvents.

References

1. Bleecker ML. Recent developments in the diagnosis of carpal tunnel syndrome and other common nerve entrapment disorders. *Semin Occup Med.* 1986;1:205-211.

2. Katz JN, Larson MG, Sabra A, et al. The carpal tunnel syndrome: diagnostic utility of the history and physical examination findings. *Ann Intern Med.* 1990;112:321-327.

3. Vessey MP, Villard-Mackintosh L, Yeates D. Epidemiology of carpal tunnel syndrome in women of childbearing age. Findings in a large cohort study. *Internat J Epidemiol.* 1990;19:655-659.

4. Silverstein BA, Fine LJ, Armstrong TJ. Occupational factors and carpal tunnel syndrome. *Am J Ind Med.* 1987;11:343-358.

5. de Krom M, Kaster ADM, Knipschild PG, Spaans F. Risk factors for carpal tunnel syndrome. *Am J Epidemiol.* 1990;132:1102-1110.

6. Wieslander G, Norback D, Gothe C-J, Juhlin L. Carpal tunnel syndrome (CTS) and exposure to vibration, repetitive wrist movements, and heavy manual work: a case-referent study. *Br J Ind Med.* 1989;46:43-47.

7. Cannon LJ, Bernacki EJ, Walter SD. Personal and occupational factors associated with carpal tunnel syndrome. *J Occup Med.* 1981;23:255-258.

8. Dieck GS, Kelsey JL. An epidemiologic study of the carpal tunnel syndrome in an adult femal population. *Prev Med.* 1985;14:63-69.

Further Reading

Cailliet R. *Hand Pain and Impairment.* 3rd ed. Philadelphia, Pa: FA Davis; 1982.

Occupational disease surveillance: carpal tunnel syndrome. *MMWR.* 1989;38:485-489.

Silverstein BA, Fine LJ, Armstrong TJ. Carpal tunnel syndrome: causes and a preventive strategy. *Semin Occup Med.* 1986;1:213-221.

Chronic Obstructive Pulmonary Disease—*See* Bronchitis, Chronic, and Emphysema

Cirrhosis—*See* Fatty Liver Disease and Cirrhosis

Coal Workers' Pneumoconiosis

ICD-9 500
SHE/O

Identification

Coal workers' pneumoconiosis (CWP) is one of the lung diseases arising from inhalation and deposition of coal mine dust in the lungs and from the reaction of the lungs to the dust. It is a chronic, irreversible disease of insidious onset, usually—but not always—requiring 10 or more years of dust exposure before appearing on chest x-ray. It is characterized by abnormalities visible as small or large opacities (spots) on chest x-ray. When only small opacities are present, the condition is called simple CWP. Complicated CWP or progressive massive fibrosis (PMF) are the terms used when opacities greater than 1 centimeter and attributable to coal dust exposure are present on x-ray. Obsolete terms applied to the same conditions include anthracosis, anthracosilicosis, miners' phthisis, and miners' asthma.

For purposes of compensation, various jurisdictions define pneumoconiosis differently. For example, in the U.S. federal compensation system for miners, pneumoconiosis is defined as "a chronic dust disease of the lung arising out of employment in an underground coal mine" (Federal Mine Safety and Health Act, Sec 402[b]). Thus, eligibility for benefits under the federal act is not limited to dust effects visible on chest x-ray. However, disease definitions delineating eligibility for state workers' compensation benefits also vary from state to state.

The International Labour Organization (ILO) disseminates a conventional method of x-ray interpretation that is often used to diagnose and classify CWP. This method classifies opacities according to their shape, size, location, and profusion. Profusion is determined by comparing the miner's film with "standard" ILO films. There are four major categories of increasing profusion of opacities: 0, 1, 2, and 3. Whichever standard film most closely matches that of the miner determines the "major category" of profusion. If the film is in a border area between two major categories, both categories are noted, with the category most like the film noted first. For example, a film that shows a higher profusion of opacities than the category 1 standard film, but that is more like the 1/1 standard than like the 2/2 standard, is classified 1/2. Complicated pneumoconiosis (PMF) is classified as category A, B, or C, depending on the size of the large opacities.

The ILO system was originally established to achieve consistency in film interpretation during the conduct of health surveillance or epidemiological investigations. In the U.S., readers trained in this method of interpretation who pass a competency test administered by the National Institute for Occupational Safety and Health (NIOSH) are designated as "B-readers." However, despite efforts to achieve standardized interpretations of chest x-rays through use of the ILO system, readers still disagree significantly among themselves or among different cases on the presence or absence of CWP.

Coal workers' pneumoconiosis has characteristic pathological features, which can be seen on autopsy or biopsy and are described in standard reference works. Tissue examination is not necessary for the diagnosis of the disease.

Occurrence

The prevalence of CWP increases with increasing dust exposure. A cross-sectional epidemiological study of coal miners conducted from 1985 to 1988 showed that approximately 4% of working underground miners exposed to coal mine dust under current exposure limits had developed CWP after 15 years and that at least 19% of miners with 30 or more years' experience showed category 1/0 or greater CWP. Projections based on British data suggest that 6.2% of miners will have simple CWP and that 0.4% will develop PMF after 35 years of exposure at the current U.S. limit of 2 mg/m^3. In the U.S., more than 2000 miners die each year with CWP noted on their death certificates. Surface coal workers are also at risk of developing lung diseases from dust exposure. This risk is increased for certain workers such as drillers, as well as for surface miners with prior underground dust exposure.

Causes

Respirable coal mine dust causes CWP. Respirable dust is any dust that is small enough to be deposited in the terminal bronchioles or alveolar air spaces (generally less than 5 μm in diameter).

Coal mine dust is a mixed dust consisting mostly of coal but also including other minerals found in mines. Silica dust, or silica mixed with coal dust (as is often found in the coal mine environment), can cause an x-ray picture no different from that of CWP, although, at times, there are distinguishing characteristics. On autopsy or biopsy examination of lungs, findings of CWP and silicosis may coexist. Other carbonaceous dusts such as carbon black can cause a lung disease identical to CWP.

Pathophysiology

Inhaled fine particles of coal dust are scavenged by specialized lung cells (macrophages). These cells and dust particles accumulate deep in the respiratory tree near or in the air exchange units (alveoli). Abnormal fibrotic material (reticulin and collagen) may be present. Localized areas of tissue destruction—focal emphysema—may also occur. The areas demonstrating a combination of these abnormalities are called coal macules. In some cases, nodules consisting of macrophages and a greater quantity of abnormal fibrotic tissue are present. These can lump together (coalesce) to form the large lesions of PMF. Generally, there is significant destruction of lung tissue as the nodules coalesce.

The definition of the exact physiological abnormalities resulting from the development of CWP has been complicated, in part, by the diversity of pulmonary responses to coal mine dust. Increasing dust exposure has been associated with progressive loss of lung function, resulting in the development of obstructive lung disease. This loss appears similar in magnitude to that caused by regular cigarette smoking. Miners have increased rates of emphysema and chronic bronchitis.

Coal workers' pneumoconiosis is itself an effect of exposure to dust. When PMF destroys and distorts lung tissue, it is usually associated with loss of lung function on a variety of measures. Simple CWP alone may or may not be associated with diminished function in any particular individual. Nevertheless, individuals with simple CWP may experience significant loss of lung function as a result of the same exposure (coal mine dust) that caused the development of CWP (*see* Emphysema and Bronchitis, Chronic).

Prevention

Primary prevention of CWP is achieved by reducing exposure to coal dust through improved ventilation and dust suppression supported by enforcement of strict dust control standards. Current preventive efforts focus on suppressing the "respirable" fraction of coal mine dust (less than 10 μm in diameter) that appears to cause CWP. In the U.S., the permissible exposure level is a time-weighted average of 2.0 mg/m^3 measured as a personal sample. Control of other lung diseases associated with dust exposure in mining, which may be caused by larger dust particles, may or may not be achieved by suppressing the respirable dust fraction.

Prevention of pulmonary impairment associated with PMF has been based on an assumption that PMF almost invariably develops on a background of advanced simple CWP; therefore, secondary prevention efforts have been directed at identifying miners with early simple CWP (category 1) and at further reducing the dust exposure for these miners. Recent reports have brought the underlying assumption into question, however, as an increasing percentage of PMF diagnoses are made in miners with a background of early CWP, although the overall incidence of PMF is falling.

Coal workers' pneumoconiosis does not resolve with elimination of exposure to coal dust and may, in some cases, progress from simple CWP to PMF in the absence of additional exposure. Rehabilitation efforts (tertiary prevention) are the same as those employed for anyone with disabling lung disease: elimination of adverse environmental exposures; immunization against influenza and pneumococcal infection; early recognition and treatment of infection; education directed at improved levels of self-care; graded exercise; and consideration of medications such as bronchodilators.

Other Issues

1. Efforts to control CWP through the monitoring and reduction of respirable dust exposure may not be adequate to control the development of pulmonary impairment in miners. The absence of a positive finding for CWP on x-ray does not ensure the absence of significant disease from coal mine dust exposure.

2. Preventive strategies directed toward eliminating PMF through early identification of miners with simple CWP may be inadequate.

3. Some miners develop disabling lung disease in the absence of PMF. Periodic pulmonary function testing might be useful in early disease recognition and secondary prevention.

4. Coal mining communities are often isolated, with few local alternative sources of employment. Miners developing lung disease are reluctant to eliminate exposure to dust by leaving the industry when unemployment is the only alternative.

5. Cigarette smoking has no apparent effect on the formation of the coal macule but adds to the miners' risk of developing chronic bronchitis and emphysema.

6. Modern underground mining techniques such as longwall mining result in high levels of coal mine dust being generated and thereby requiring careful identification of sources of dust and attention to control methods.

7. Dust regulations in the U.S. assume a 5-day, 8-hour-per-shift workweek; however, other considerations have resulted in increasing numbers of shifts per worker in some areas. The impact of these changes on health has not been evaluated.

GRW

Further Reading

Attfield MA. A review of epidemiologic information on the prevalence of CWP in U.S. miners 1960-1988. Unpublished data.

International Labour Organization. *1980 International Classification of Radiographs of the Pneumoconioses.* Geneva, Switzerland: International Labour Office Technical Publication.

Kleinerman J, et al. Standards for the pathology of coal workers' pneumoconiosis. *Arch Path.* 1979;103(8):375-432.

Merchant JA. Coal workers' pneumoconiosis. In: Last JM, ed. *Maxcy-Rosenau: Public Health and Preventive Medicine.* 12th ed. Norwalk, Conn: Appleton-Century-Crofts; 1986:545-552.

Weeks JL, Wagner GR. Compensation for occupational disease with multiple causes: the case of coal miners' respiratory diseases. *Am J Public Health.* 1986;76:58-61.

Cobalt-Induced Interstitial Lung Disease

(Hard Metal Disease)

Identification

This disease is a diffuse, interstitial pulmonary fibrosis that results from overexposure to cobalt-containing dusts present in the manufacture of hard metals or cemented tungsten carbide. It is distinguishable from similar conditions only by occupational history. Progressive dyspnea can be either insidious, developing after protracted cobalt exposure, or relatively abrupt, developing after brief and intense exposure. Chest x-ray shows diffuse or patchy infiltrates. Pulmonary function testing shows restrictive changes. Asthma and other obstructive diseases are also associated with cobalt exposure, so one cannot rely on pulmonary function testing alone to determine whether a cobalt-induced disease is present.

Occurrence

There is little information on the occurrence of cobalt-induced interstitial lung disease. Approximately 30 000 workers in the U.S. may be exposed to tungsten and its compounds. Between 1% and 10% of exposed workers may develop disease. Those at risk are involved in the manufacture or grinding of tungsten carbide tools. Because the condition bears the initial features of a hypersensitivity pneumonitis (extrinsic allergic alveolitis), it is possible that cases often are diagnosed without an assignment of etiology; therefore, unrecognized cases may occur.

Causes

The cause of this condition is cobalt, a binder used in combination with tungsten carbide in the manufacture of cemented carbide metals. These metals are used as abrasives or as cutting tips for tools used in the high-speed cutting of metals, very hard woods, cement, or other hard materials. Tungsten carbide, the major component of these products, is considered to be biologically inert. Cobalt content is usually less than 10% but may be as high as 25%. For workers to be at risk, it has been suggested that the content of cobalt in the alloy needs to be at least 2%.

Pathophysiology

Little formal study of the pathogenesis of this disorder exists. Cobalt may be both allergenic and cytotoxic, capable of provoking release of a fibrogenic agent from macrophages. There is little correlation between the amount of cobalt recovered from the lung and the amount of disease in cases that have undergone detailed study. Cobalt is highly soluble in biologic fluids and combines readily with protein, which suggests that cobalt may act as a hapten and promote immunologic reactions. A second condition often seen in workers with hard metal disease is occupational asthma associated with cobalt exposure. This is further evidence that the interstitial disease may result from an immunologic process.

Prevention

Dust suppression or elimination is essential in cemented tungsten carbide manufacturing facilities and in locations where tungsten carbide tools are ground. Because cobalt is soluble in certain cutting fluids (eg, water-soluble oils), those fluids with the lowest capacity for dissolving cobalt should be selected when machining or grinding tools or other materials made from cemented tungsten carbide. Annual chest x-ray and pulmonary function tests are indicated for workers who may be exposed.

Other Issues

Although there is a clear immunologic component to this disease, the associated risk factors are not understood. Therefore, excluding certain individuals (eg, those who are atopic) from exposure is not justifiable on the basis of current scientific evidence. Certain exposures to cobalt have also been associated with increased risk of cardiomyopathy.

DHW

Further Reading

Balmes JR. Respiratory effects of hard-metal dust exposure. *State of the Art Reviews in Occup Med.* 1987;2:327-344.

Merchant JA, ed. *Occupational Respiratory Diseases.* Washington, DC: U.S. Government Printing Office; 1986. DHHS (NIOSH) publication 86-102.

Parkes WR. *Occupational Lung Disorders.* 2nd ed. London, England: Butterworths; 1982.

Cold-Related Disorders

ICD-9 991, E901

Identification

Cold-related disorders include hypothermia, frostbite, trenchfoot, and immersion foot. These disorders are summarized below:

Hypothermia. Hypothermia is a progressive decrease in the body temperature to 80°F or below, at which point unconsciousness and death generally occur. Most cases occur between 30°F and 50°F, but cases can occur as high as 65°F in air or 72°F in water. Body heat is lost quickly when an individual becomes wet and even more quickly when that individual is also exhausted; when the core temperature drops below 86°F, physiological mechanisms to reduce heat loss become ineffective.

Frostbite. This condition results from the freezing of tissues in the affected part of the body. Damage can range from mild, superficial, and reversible to severe damage with gangrene.

Trenchfoot. Trenchfoot involves numbness, extreme edema, and other symptoms, sometimes including gangrene, and is due to exposure to moisture at or near the freezing point for anywhere from one to several days.

Immersion foot. Similar symptoms occur in immersion foot as in trenchfoot due to prolonged exposure (days to weeks) to cool or cold water.

In addition, extreme cold can reduce tolerance to toxic chemicals and resistance to disease and can decrease speed and accuracy of work as well as manual dexterity and sensitivity.

Occurrence

There are no reliable data on the incidence of the above conditions. Workers at risk include those working in cold climates, in cold and wet environments, and in refrigerators or freezers.

Causes

Relevant disorders are caused by unprotected or underprotected exposure to extreme cold, with or without exposure to moisture or water.

Pathophysiology

Functional changes vary with the condition.

Prevention

Neither federal standards nor threshold limit values exist. Recommended procedures and work practices include the following:

1. For work in cold outdoor environments, sufficient rest periods in heated shelters.
2. Working at a rate to prevent profuse sweating.
3. Education and close monitoring of new employees in jobs at risk for cold-related disorders.
4. Protection of the extremities, as with gloves or boots.
5. Using protective clothing for cold environments sufficient to maintain necessary warmth.
6. Job design to prevent, whenever possible, extreme coldness of the hands.

Other Issues

Workers new to an at-risk job can be at high risk.

BSL

Further Reading

Heins AP. Hot and cold environments. In: Rom WN, ed. *Environmental and Occupational Medicine.* Boston, Mass: Little, Brown; 1983;733-741.

Collective Stress Disorder ICD-9 308
(Mass Psychogenic Illness)

Identification

Collective stress disorder (commonly referred to as mass psychogenic illness) is an acute outbreak of a variety of signs and symptoms usually affecting several people at the same time and place. Symptoms are diverse and nonspecific, including headache, fatigue, dizziness, hyperventilation, nausea, vomiting, and, occasionally, fainting. An outbreak is often distressing and dramatic.

Occurrence

There are no estimates of incidence rates, and knowledge is limited to published case reports. Women are significantly more often affected than men. Episodes have been reported worldwide, often affecting workers from lower socioeconomic sectors who recently moved from rural settings to an industrial workplace. Episodes appear to be more common in light manufacturing industries and offices, but occurrences have been reported in agricultural workplaces also. Not all workers at a workplace are affected.

Causes

Almost without exception, incidents of collective stress disorder occur in a work force undergoing stress in the form of job insecurity; rigidly paced, repetitive, or boring work; pressure to work faster; or poor labor-management relations. The physical work environment may be a contributing factor with excess noise or heat or inadequate ventilation. Typically, opportunities to address these problems are nonexistent or frustrating, and affected individuals lack such coping methods as personal support networks and control. Episodes are often precipitated by a discrete event, such as an accident, injury, adverse personnel action, odor, or manifest illness.

Specific physical etiologic agents typically cannot be identified, giving rise to the term *psychogenic*. However, personality characteristics, such as neuroticism, extroversion, introversion, hysteria, or conversion reactions, do not explain which people are affected and which are not. Thus, this term is somewhat misleading, because it suggests that the individuals' psychological makeup is the problem rather than some characteristic of work.

Pathophysiology

The pathophysiology is unknown but can be characterized as an acute stress reaction brought on by anxiety, fear, and uncertainty.

Prevention

Specific strategies for preventing collective stress disorders must be tailored for each workplace. If work is boring, rigidly paced, and repetitive, it should be reorganized to provide for more worker control, work breaks, or slower pacing. Workers should be involved in planning and implementing such strategies. Contributing physical factors should be controlled (*see* Building-Related Illness).

Specific episodes should be treated by identifying and addressing both precipitating incidents and underlying problems, whether social or physical. There should be a diligent search for physical causes and appropriate controls. Use of the term *psychogenic* may impede efforts to find a solution.

Other Issues

That women are more affected than men probably does not arise from any increased susceptibility, but rather from women being assigned to certain jobs at risk for the disorder or being recent entrants into the work force.

JLW

Further Reading

Boxer P. Occupational mass psychogenic illness: history, prevention and management. *J Occup Med.* 1985;27:867-872.

Brodsky CM. The psychiatric epidemic in the American workplace. *State of the Art Reviews in Occup Med.* 1988;3:653-662.

Colligan M. Mass psychogenic illness: some clarification and perspectives. *J Occup Med.* 1981;23:635-638.

Hall EM, Johnson JV. A case study of stress and mass psychogenic illness in industrial workers. *J Occup Med.* 1989;31:243-250.

Colorectal Cancer

ICD-9 153, 154

Identification

Abdominal pain, weight loss, symptoms related to anemia, and stool changes are the leading symptoms of colorectal cancer, varying in part because of the differing locations of the cancer and the stage of disease. Early physical findings are uncommon, although approximately 50% of affected individuals will test positive for occult blood in the stool. Sigmoidoscopy, colonoscopy, and double-air-contrast barium enema are possible diagnostic procedures. The diagnosis requires a histological demonstration of malignant lesions in tissue. Cost considerations, together with reviews of successful diagnostic and treatment protocols, suggest that, in the long run, initial evaluation with colonoscopy of those at high risk or those who test positive for occult blood is the most efficient and useful strategy.

Occurrence

An estimated 151 000 new cases of colorectal cancer and 61 300 deaths occurred in 1989 in the U.S. One in seven new cancer cases are colon and rectal cancers, ranking colorectal cancer second to lung cancer (excluding common, often curable, skin cancers). Because of the large variations in incidence around the world, colon cancer is thought to have major environmental determinants, with diet considered an important factor.

Causes

It appears that the best-defined occupational exposure conferring an increased risk for colon cancer is asbestos exposure, although other factors are generally considered more important. Despite the small sample size in studies regarding the association between asbestos and colorectal cancers and despite the difficulty in identifying the specific contribution of asbestos for workers who are exposed to multiple carcinogens, the literature overall has consistently reported a mild excess of colorectal cancers with exposure to asbestos and has indicated a dose-response relationship. Increased risk of colorectal cancer has been demonstrated among pattern- and model-makers and among workers exposed to acrylonitrile, ethylacrylate, and methyl methylacrylate. Increased risk may also be found among synthetic-rubber workers, brewery workers, workers (including professional artists) exposed to paint, metalworkers, transportation workers, leather workers, linemen, fire fighters, paper industry workers, and laundry and dry-cleaning workers. Some studies suggest that synthetic-fabric workers; workers exposed to polybrominated biphenyls, chlordane, heptachlor, and endrin; printing workers; radium dial painters; workers in copper smelting and oil refining; embalmers; and workers exposed to vinyl chloride may be at greater risk for colon cancer. Workers in sedentary jobs may also be at greater risk, suggesting an inverse relationship between occupational activity and risk of colon cancer. Environmental exposures to trihalomethanes and ionizing radiation also have been associated with increased risk of colorectal cancer.

Genetic factors clearly predispose to the subsequent development of colorectal cancer. There is a well-defined increased incidence in first-degree blood relatives of patients. The presence of familial polyposis increases the risk. The association between ulcerative colitis and higher frequency of colorectal cancer is well-known. There is some scientific support for the theory that greater amounts of dietary fiber may reduce the risk of colon cancer and that high levels of dietary fat may increase it.

Pathophysiology

The true mechanism is unknown. Some theories suggest that delayed transit time permits extended contact between ingested toxins and the bowel wall, increasing the probability of carcinogenesis.

Prevention

Primary prevention includes reducing or eliminating exposure to known or suspected carcinogens such as asbestos. On the basis of current hypotheses, increasing dietary fiber and decreasing dietary fat intake may reduce risk.

Screening by testing feces for occult blood and by using flexible sigmoidoscopy is a useful secondary prevention strategy. Early identification and removal of adenomatous polyps prevent their maturation to adenocarcinoma. Measurement of carcinoembryonic antigen is not thought to be of use in early detection of the disease, although it may be used in following individuals at risk for recurrence.

The American Cancer Society recommends annual stool exams for occult blood and flexible sigmoidoscopy every 3 years, each beginning at age 50. For occupationally exposed workers, it is recommended that these tests begin at age 40. Concerning treatment, surgery in early stages may be curative. There is some experimental evidence for the usefulness of adjuvant chemotherapy.

HF

Further Reading

Brownson RC, Hoar S, Chang JC, et al. Occupational risk of colon cancer. *Am J Epidemiol.* 1989;130:675-687.

Frumkin H, Berlin J. Asbestos exposure and gastrointestinal malignancy: review and meta-analysis. *Am J Ind Med.* 1988;14:79-85.

Lashner BA, Epstein SS. Industrial risk factors for colorectal cancer. *Int J Health Serv.* 1990;20:459-483.

Neugut AI, Wylie P. Occupational cancers of the gastrointestinal tract. In: Brandt-Rauf PW, ed. Occupational Cancer and Carcinogenesis. *State of the Art Reviews: Occupational Medicine.* 1987;2:109-135.

Sleisenger MH, Fordtran JS. *Gastrointestinal Disease.* 2nd ed. Philadelphia, Pa: WB Saunders; 1983.

Congestive Heart Failure

ICD-9 428.0

Identification

Congestive heart failure is a clinical syndrome due to heart disease and is characterized by breathlessness and abnormal sodium and water retention, resulting in edema. The congestion may occur in the lungs or in the peripheral circulation, or in both, depending on whether the heart failure is left-sided, right-sided, or general.

Symptoms include dyspnea (difficulty breathing), fatigue, weakness, nocturia (urination at night), and, with advanced impairment, confusion, memory loss, and other symptoms that reflect decreased brain perfusion. The most common physical findings are rales, a gallop heart sound (S_3), and edema. The diagnosis is confirmed by enlargement of the heart and a change in bronchovascular markings on the chest x-ray.

Electrocardiographic (ECG) findings may reflect the underlying cause of the congestive heart failure, but there are no ECG findings specific to heart failure. Abnormalities in kidney and liver function may be detected in severe late-stage heart failure. Tests for the cause of heart failure may include echocardiography, cardiac catheterization, and endomyocardial biopsy.

Occurrence

Two million people in the U.S. are reported to have heart failure, with 250 000 new cases diagnosed annually. There are 500 000 to 1 million hospitalizations for congestive heart failure annually. The incidence increases with age; it is five times higher for men in their 60s than for men in their 40s.

Causes

Coronary artery disease and hypertension are by far the two most common causes of congestive heart failure, accounting for at least 75% of all cases. There are many less common causes, including infections, poor nutrition (both vitamin deficiency and excess alcohol), connective tissue problems, and genetic diseases. Congestive heart failure has been associated with chemical exposure in two episodes of beer contamination (with arsenic in 1900 and cobalt in the 1960s) and, very rarely, with occupational exposure (cobalt). Congestive heart failure may be a secondary occupational disease if occupational exposures have caused ischemic heart disease or hypertension.

Pathophysiology

Congestive heart failure is the failure of the heart to pump blood at a sufficient rate to meet the physiological requirements of the organs. Although it is usually caused by failure of the heart muscles (most commonly due to ischemic heart disease and hypertension), similar symptoms can be engendered by conditions that cause increased blood volume (such as renal failure) or inadequate filling of the ventricular chambers of the heart (such as constrictive pericarditis). Direct damage to myocardial muscle can occur from infectious agents, nutritional excesses or deficiencies, and toxins (such as arsenic and cobalt). Cobalt myocardiopathy causes unique pathological findings. The synergistic effect of alcohol, cobalt, and a protein-poor diet on enzyme metabolism has been suggested as the mechanism responsible for this condition.

Prevention

Prevention of the disease underlying congestive heart failure is necessary for primary prevention. This means reduction in exposures to substances that cause ischemic heart disease, hypertension, or direct myocardial toxicity. Although alcohol ingestion and a protein-poor diet are thought to play a role in cobalt cardiomyopathy, not enough is known about actual amounts to recommend specific dietary guidelines other than alcohol in moderation and a well-balanced diet for industrial workers exposed to cobalt.

Other Issues

Arsenic cardiomyopathy has been described only with beer contamination. Ischemic heart and peripheral vascular disease has been described with industrial and environmental exposure. Carbon monoxide and solvent exposure may precipitate an acute exacerbation of congestive heart failure. Individuals with underlying ischemic heart disease may be particularly susceptible.

KR

Further Reading

Morin YL, et al. Quebec beer-drinkers' cardiomyopathy. *Can Med Assoc J.* 1967;97:881-931.

National Academy of Sciences. *Arsenic.* Washington, DC: National Academy of Sciences; 1977.

Conjunctivitis

ICD-9 372.0–372.3

Identification

Irritant conjunctivitis is an acute irritation of the conjunctiva in response to a dusty or vapor-laden atmosphere. Atopic conjunctivitis is an allergic reaction to any of a large number of allergens; symptoms include tearing and itching eyes and possibly edema and may be accompanied by nasal discharge. Reaction varies among individuals.

Occurrence

Workers in agriculture and related industries may suffer allergic reactions while exposed to various plant, animal, and other allergens, including chemicals and pesticides. For example, workers exposed to pollens during artificial pollination and workers handling green plants (eg, weeping fig, or *Ficus benjamina*) have been reported to suffer from conjunctivitis and/or rhinitis and asthma. There are no good data on occupationally induced conjunctivitis.

Causes

Irritant conjunctivitis can be caused by any simple irritant, such as acids or alkalis, aerosols, solvent vapors, and airborne dusts. Atopic conjunctivitis may be caused by exposure of atopic individuals to one (or more) of many allergens, most of which are airborne. Pollens are most common; other allergens include animal epidermal products, nonpathogenic fungi, vegetable and animal proteins, hair, wool, feathers, industrial chemicals, and pesticides. Workers with occupational exposure to polychlorinated biphenyls have been reported to suffer from hypersecretion of the meibomian glands, swelling of the upper eyelids, and hyperpigmentation of the conjunctiva.

Pathophysiology

Irritant conjunctivitis is a simple irritant reaction that varies in intensity from hyperemia to necrosis, depending on the irritant and exposure. It is particularly marked in the palpebral aperture when due to vapors or dusts, and in the lower fornix area when due to liquids. It is accompanied by watery discharge in mild cases and by mucopurulent discharge in severe cases.

Atopic conjunctivitis is mediated by histamine release, typical of allergic reactions. At its most severe state, it can result in swelling of the eyelids, superficial necrosis of the eyelids, involvement of regional lymph nodes, and formation of a pseudomembrane of the conjunctiva.

Prevention

Irritant conjunctivitis can be prevented by applying standard prevention strategies, including industrial hygiene evaluation and control methods, medical surveillance, and medical monitoring. Prevention of atopic conjunctivitis is more difficult because reactions vary significantly among individuals and may be provoked by very low concentrations of airborne allergens. The precise allergen should be identified, preferably without challenging atopic individuals. Either it should be removed from the workplace or atopic workers may have to be transferred to areas where it is not used. Use of protective eyewear that forms an airtight seal around the eyes or face (if the reaction is not limited to conjunctivitis) may be useful, but care must be taken to avoid contamination.

Other Issues

Even though conjunctivitis is usually not considered a serious medical problem, it can have a significant adverse effect on workers. It can be very annoying to the individual worker and can reduce workers' attentiveness, which can, in turn, result in accidents and impaired performance.

People with a history of atopy, allergic rhinitis, or asthma are at increased risk. However, the magnitude of the risk is unknown. Therefore, in the absence of knowledge about reaction to specific allergens, excluding such people from certain jobs or workplaces a priori, based only on a history of atopy, is not warranted.

SV

Further Reading

Duane DD, Jaeger EA, eds. *Clinical Ophthalmology*. Vol 4: *External Disease*. Philadelphia, Pa: Harper & Row; 1987.

Friedlander MH, Okumoto M, Kelly J. Diagnosis of allergic conjunctivitis. *Arch Ophthalmol.* 1982;100:1275.

Kanski JJ. *Clinical Ophthalmology: A Systemic Approach.* 2nd ed. London, England: Butterworths; 1989.

Rosen JP. Baker's rhinoconjunctivitis: a case report. *N Engl Reg Allergy Proc.* 1987;8(1):37-38.

Contact Dermatitis, Allergic

ICD-9 692
SHE/O

Identification

Allergic contact dermatitis is defined as the development of a T-cell-mediated, antigen-specific response of a patient to a hapten or antigen applied to the skin. The development of dermatitis requires an antecedent period of exposure, during which the patient develops an immune response. For most antigens, only a small percentage of the population is capable of developing the immune response.

The clinical spectrum is broad. Pruritus (itching) is the primary symptom. Typically, there are grouped or linear vesicles or bullae. Erythema (redness), edema (swelling), exudation, crusting, and scaling are typically present. As with irritant contact dermatitis, lesions occur in exposed or contact areas.

Occurrence

Most contact allergens produce sensitization in only a small percentage of exposed individuals. Exceptions to this rule are poison ivy, poison oak, and poison sumac, from which more than 70% of those who are exposed develop allergic contact dermatitis. (*See* Contact Dermatitis, Irritant, for data on incidence rates for occupational skin diseases.)

Causes

Contact allergens commonly found in the workplace are nickel salts, epoxy resins, chromium salts, paraphenylenediamine, and formaldehyde. Sometimes airborne substances cause allergic contact dermatitis.

Pathophysiology

Allergic contact dermatitis is a form of delayed-type hypersensitivity. It results from the development of T-cell-mediated immunity in individuals to contact allergens. The incubation period after initial exposure to an antigen ranges from 5 to 21 days. The reaction time after subsequent reexposure is 1 to 3 days. The typical reaction of a sensitized person after reexposure to a contact sensitizer is the appearance of an eczematous dermatitis in 1 to 3 days and its disappearance within 2 to 3 weeks. With heavy exposure or exposure to potent sensitizers, lesions appear more quickly (within 6 to 12 hours) and heal more slowly.

Prevention

The most effective preventive measure is removing the contact allergen from the workplace, especially if a high percentage of workers has become sensitized to it. Other measures include preventing exposure by using gloves and protective clothing, closed systems, and special ventilation.

Other Issues

Patch testing is an in vivo bioassay used to determine the presence of contact allergy. Patch tests are performed by placing nonirritating concentrations of contact sensitizers into a suitable vehicle in contact with the skin for 48 hours. Evidence for a delayed-type hypersensitivity reaction (erythema, edema, vesiculation, bullae formation, or frank necrosis) is sought 30 minutes after the patches are removed. Additional (delayed) readings (from 1 to 5 days) after patch removal improves test sensitivity. Properly performed, patch testing is an invaluable tool in evaluating workers with contact dermatitis. Considerable experience with patch testing is required to perform and interpret the test adequately.

MB, KA

Further Reading

Adams RM. *Occupational Skin Disease.* Philadelphia, Pa: WB Saunders; 1990.

Breathnach SM. Immunologic aspects of contact dermatitis. *Clin Dermatol.* 1986;2:5-17.

Fisher AA. *Contact Dermatitis.* 3rd ed. Philadelphia, Pa: Lea & Febiger; 1986.

Contact Dermatitis, Irritant

ICD-9 692
SHE/O

Identification

Irritant contact dermatitis is defined as cutaneous inflammation that develops as a result of the direct effect of chemicals on the skin. Prior sensitization is not required, and antigen-specific immune responses cannot be detected. Irritant contact dermatitis will develop in all workers exposed to adequate concentrations for an adequate length of time. The clinical spectrum of irritant contact dermatitis is very broad, ranging from slight erythema to large bullae, necrosis, and ulceration. Mild irritation is characterized by localized erythema, edema, vesicle and papule formation, and crusting and scaling. Thickening of the skin may occur with chronic scratching. Exposure to strong irritants causes acute injury to the epidermis and sometimes to the dermis and may cause the formation of bullae and ulceration. Irritant contact dermatitis usually affects exposed surfaces, mainly the hands and forearms. If an irritant soaks into clothing, the skin underlying the soaked clothing will become affected.

Occurrence

Occupational skin diseases, most of which are contact dermatitis, are the second most common occupational disease reported to BLS in the U.S. Available data sources do not contain information on the occurrence of specific dermatologic conditions. The reported incidence rate among private-sector workers in the U.S. in 1987 for establishments employing 11 or more employees was 7.4 cases per 10 000 worker-years. The highest incidence rates were consistently found in agriculture, manufacturing, and construction, with 30.0, 16.4, and 7.3 cases per 10 000 worker-years, respectively. Incidence rates based on workers' compensation claims differ, sometimes more than twofold, from those measured from nationally reported sources, but the rank order of major industrial groupings remains the same. The annual cost of occupational skin diseases, including lost production, medical costs, and lost wages, is estimated to range from $222 million to $1 billion.

The overall incidence rate has declined by half over the past decade. In 1987, over half of all cases were reported in the manufacturing sector and another fifth in service industries. The large proportion of cases in service industries can be attributed to the large number of workers who are employed in services, even though the sectorwide incidence rate is below the national average.

Within agriculture, all three major subsectors (agricultural production, agricultural services, and forestry) have high rates of occupational skin disease. In one survey conducted in California, cases were most often attributed to plants (52%), chemicals (20%), and food products (13%).

Within manufacturing, the highest incidence rate occurs in the leather and leather products industries (31.8 cases per 100 000 worker-years), transportation products (26.5), food and kindred products (22.3), and chemicals and allied products (20.8). At the three-digit Standard Industrial Classification (SIC) level, the highest incidence rates were in the leather tanning and finishing industry (127.4), followed by meat products (57.2), ophthalmic goods (51.4), engine and turbine manufacturing (47.6), and screw machine products (41.6). These elevated rates are likely associated with chromate exposure in leather tanning, fungal infection of chronically wet hands in the meat products industry, an unknown agent in ophthalmic goods, and exposure to metalworking fluids or organic solvents in the others.

Causes

Among the substances causing irritant contact dermatitis are strong alkalis and acids, soaps and detergents, and many organic compounds. Aggravating factors include reduced humidity at work, friction, occlusion, and excess environmental heat. Sometimes airborne elements, such as dusts, fibers, acids, alkalis, gases, or vapors, cause irritant contact dermatitis.

Pathophysiology

Contact irritants cause direct injury or necrosis of epidermal cells and cells within the dermis. Injury leads to the release of inflammatory mediators, including histamines, prostaglandins, leukotrienes, and cytokines. Inflammatory mediators attract a mixed cellular infiltrate into the skin, which may cause further damage and is responsible for the typical histological picture seen in acute irritant dermatitis.

Prevention

Prevention is of primary importance in irritant contact dermatitis. Skin contact with the irritant should be eliminated or significantly reduced. This goal can be accomplished by substituting a less irritating substance, implementing engineering measures such as enclosing the process in which the irritant is used, changing work practices to reduce the likelihood of worker exposure to the irritant, and using gloves and protective clothing. Cleanliness of the worker and good housekeeping of the workplace are important, as are good ventilation, optimal temperature (17°C to 22°C) and humidity (about 50%), education of workers, and labeling of irritant substances. Availability of showers and wash facilities and time to use them at the workplace are important.

Dermatitis from metalworking fluids, especially from water-soluble oils, is common. Exposure can be controlled by changing fabrication methods, keeping fluids as clean as possible, and judiciously selecting and using germicides. Personal protective devices, such as gloves and aprons, must be selected in accordance with specific job requirements as well as with the properties of the irritants from which protection is needed. Barrier creams are often ineffective.

In agriculture, including forestry, substitution and other engineering controls are difficult to implement for vegetable toxins, because these toxins are inherent in the product (eg, melons, tomatoes) or the environment (eg, poison ivy, poison oak). Exposure can be reduced by substituting mechanical for manual handling of these products or by using personal protective devices.

Other Issues

Fibrous glass causes a pruritic rash on exposed areas, usually the hands and forearms. The rash is often nonspecific (erythema, lichenification, and small papules). Diagnosis requires a high index of suspicion. A history of occupational exposure to fibers must be sought, such as in industries where fibrous glass or asbestos is found. The small fibers are not normally visible to the naked eye but may be visualized by a microscopic examination of tape stripping samples from the affected area or by an examination of a biopsy specimen with polarized light.

Atopy is a constellation of clinical findings, including asthma, hay fever, and atopic eczematous dermatitis, occurring either in an individual or in the individual's family. There are no pathognomonic features of atopic disease; atopy is the major genetic condition predisposing to irritant contact dermatitis. Because workers with a history of atopic dermatitis are 13 times more likely to develop irritant contact dermatitis in the workplace, preplacement screening should always include eliciting a history of atopy. Exposure of atopic workers to irritants should be prevented or severely limited, but workers with atopic dermatitis should be excluded from a job only if avoidance of irritants is impossible.

MB, KA

Further Reading

Adams RM. *Occupational Skin Disease.* Philadelphia, Pa: WB Saunders; 1990.

Fisher AA. *Contact Dermatitis.* 3rd ed. Philadelphia, Pa: Lea & Febiger; 1986.

Maibach HI, Gellin GA. *Occupational and Industrial Dermatology.* 2nd ed. Chicago, Ill: Year Book Medical Publishers; 1988.

Mathias CGT, Morrison JH. Occupational skin diseases, United States: results from the Bureau of Labor Statistics annual survey of occupational injuries and illnesses, 1973 through 1984. *Arch Dermatol.* 1988;124:1519-1524.

O'Malley MA, Mathias CG. Distribution of lost-work-time claims for skin disease in California agriculture: 1978-1983. *Am J Ind Med.* 1988;14:715-720.

O'Malley M, Thun M, Morrison J, Mathias CGT, Halperin WE. Surveillance of occupational skin disease using the supplementary data system. *Am J Ind Med.* 1988;13:291-299.

Taylor JS, ed. Occupational dermatoses. *Dermatol Clin.* 1988;6:1-156.

US Department of Labor, Bureau of Labor Statistics. *Occupational Injuries and Illnesses in the United States by Industry, 1987.* Washington, DC: U.S. Government Printing Office; 1989.

Coronary Artery Disease—*See* Ischemic Heart Disease

Cumulative Trauma Disorders—*See* Carpal Tunnel Syndrome, Hard-Arm Vibration Syndrome, Peripheral Nerve Entrapment Syndromes, and Tendinitis (Tenosynovitis)

Emphysema

ICD-9 492

Identification

Pulmonary emphysema is a disorder of lung anatomy defined as a permanent, abnormal increase in the size of airspaces distal to the terminal bronchiole accompanied by destruction of the lung tissue. Because the tissue destruction is nonuniform, the orderly appearance of the airspaces is disrupted and may be lost altogether.

The clinical presentation of emphysema can vary from mild shortness of breath to severe breathlessness and respiratory failure. The level of dyspnea depends on the demands placed on the respiratory system. When the disease has progressed to the point where the forced expiratory volume in 1 second (FEV_1) is about 40% of what is predicted, carbon dioxide retention and cor pulmonale (right heart failure) may complicate the picture. Because emphysema is a process destructive to lung tissue, the diffusing capacity ($D_{L_{CO}}$) is typically reduced in moderate to severe disease. Both emphysema and bronchitis can result from the same exposures; therefore, many people with one condition will also have signs or symptoms of the other.

X-ray changes are absent in early emphysema. Later, the characteristic changes are of two general types: (a) an increase in the volume of the thorax occupied by the lung, and (b) a decrease of the overall pulmonary vascular pattern.

Occurrence

Emphysema, with or without associated chronic bronchitis, is associated with about 3% of hospital deaths in North America. It is the most costly disease that the U.S. Veterans Administration currently treats.

The precise number of workers with emphysema as a result of occupational exposure (in whole or in part) is not known. Groups known to be at increased risk include workers exposed to cadmium oxide or coal mine dust. Other exposures contributing to emphysema may be difficult to isolate because of concurrent tobacco smoking.

Causes

The most common agent indicated in causing emphysema is tobacco smoking. However, there are several important occupational causes that, alone or in combination with cigarette smoking, will produce emphysema. Since the 1950s, reports have described emphysema due to cadmium fumes, particularly in workers exposed to cadmium oxide for long periods of time. In addition, emphysema occurs in coal workers. If these workers also smoke, there is more emphysema and it is more advanced when compared with matched controls. Emphysema also occurs occasionally in relation to other pneumoconioses, such as asbestosis and silicosis. Nitrogen oxides can cause emphysema in laboratory animals, but these findings have not been confirmed in humans.

Pathophysiology

Emphysema is defined and classified in anatomical terms, depending on how the acinus (the portion of the lung distal to the terminal bronchioles) is involved. This classification is available in textbooks of pathology or medicine.

However, the severity rather than the type of emphysema is the important variable in relation to clinical dysfunction. Decreases in expiratory flow rates and increases in residual volume are not well correlated with the severity of emphysema. An increase in the total lung capacity with a decrease in the diffusing capacity shows the highest correlation with anatomic changes.

Prevention

Aggressive control of exposures associated with increased risk of emphysema is the only reasonable preventive strategy. No current screening tool is sufficiently sensitive to identify the development of emphysema at an early stage. Surveillance of exposed workers for associated symptoms and lung function changes may, in some instances, be appropriate. In some special situations, personal protective devices such as respirators may also be appropriate.

Other Issues

There is a need to determine the prevalence of emphysema in various occupational groups. This determination should include study of groups having a high prevalence of obstructive airways disease and studies in which the type of agent, severity of exposure, or evidence from experimental animals suggests that obstructive airways disease may occur. Interactions between cigarette smoking and various workplace agents need to be examined more closely, and interactions between nontobacco particulate exposures in the genesis of emphysema should be studied.

Homozygous persons with alpha-1-antitrypsin deficiency are at significant risk for development of emphysema and should not work in areas where there is exposure to dust, respiratory irritants, or causes of emphysema. Heterozygotes do not appear to be at abnormally increased risk, however.

DCC

Further Reading

Becklake MR. Chronic airflow limitation: its relation to work in dusty occupations. *Chest.* 1985;88:608-617.

Cockcroft A, Seal RME, Wagner JC, Lyons JP, Ryder R, Andersson N. Post-mortem studies of emphysema in coalworkers and non-coalworkers. *Lancet.* 1982;ii:600-603.

Hensley MJ, Saunders NA. *Clinical Epidemiology of a Chronic Destructive Pulmonary Disease.* New York, NY: Marcel Dekker; 1989.

Merchant JA, ed. *Occupational Respiratory Diseases.* Washington, DC: U.S. Government Printing Office; 1986. DHHS (NIOSH) Publication 86-102.

Parkes WR. *Occupational Lung Disorders.* 2nd ed. London, England: Butterworths; 1982.

Snider GL, Kleinerman J, Thurlbeck WM, Bengali ZN. The definition of emphysema. *Am Rev Respir Dis.* 1985;132:182-185.

Encephalopathy, Toxic

ICD-9 349.82, 323.7
SHE/O

Identification

Toxic encephalopathy is a syndrome occurring after exposure to toxic chemicals and is characterized by fatigue and the central nervous system (CNS) symptoms of affect lability, irritability, depression, and memory disturbances. A WHO working group has proposed a scheme for classifying the syndrome (*see* Table 1 at the end of this entry). People affected by *acute intoxication* present with dizziness, light-headedness, incoordination, and balance problems; they may also show signs of CNS depression, ataxia, and psychomotor impairment from a few minutes to a number of hours following exposure to organic solvents or inhalant anesthetics. A more severe and potentially fatal *acute toxic encephalopathy*, which may follow from hours to days after exposure to heavy metals, particularly lead, can occur with seizures, coma, abnormal reflexes, and electroencephalographic (EEG) slowing. Chronic encephalopathy is insidious in onset, occurring with continued exposure to the toxin as the neurological damage progresses from mild to severe. In *organic affective syndrome*, the earliest form of chronic toxic encephalopathy, neurobehavioral disturbances predominate. It is manifested by depression and other mood changes, sleep disturbances, apathy, impotence, fatigability, psychomotor impairments, and concentration difficulties. *Mild chronic toxic encephalopathy*, which presents with similar neurobehavioral disturbances but with greater frequency and severity, begins with mood disorders and progresses to impairment of cognitive functions, with deficits in neuropsychological functions (visuopartial ability, abstract concept formation, and short-term memory learning ability) and mood abnormalities that impair social and work activities. *Severe chronic toxic encephalopathy* (dementia) is a permanent impairment of CNS function, with severe cognitive impairment, short-term memory difficulties, diminished attention span, inability to adapt to situations requiring the learning of new information, and, often, personality changes, impairment of reasoning judgment and abstract thinking, and depression. Diagnosis depends on careful medical and occupational/environmental histories and on diagnostic procedures that identify causes of organic mental syndromes. Mild and severe chronic toxic encephalopathy are generally irreversible; symptoms of acute toxic encephalopathy and organic affective syndrome may subside following termination of exposure.

Occurrence

Knowledge about occurrence of encephalopathy or of any other occupational neurological disease is limited to case series and partial listings of compensation awards. In 1986, a total of 8723 workers' compensation awards for occupational conditions of the nervous system among both public and private sector employees were reported to BLS from 23 states. The vast majority of these (98%) were for "diseases of the nerves and peripheral ganglia"; the remainder either were unclassified or affected the CNS. Most (63%) occurred in the manufacturing sector. Of related interest, the same data source reported 8300 compensation awards for mental disorders, occurring predominantly among public sector employees (43%), workers in the services industry (16%), and workers in manufacturing (13%). The portions caused by either exposure to workplace toxins or other causes are unknown.

Causes

Metals (eg, lead), organic solvents (eg, carbon disulfide), gases, and pesticides have been implicated in acute and chronic encephalopathic syndromes (*see* Table 2 at the end of this entry). People with more intense exposure and more prominent disruption of function appear to be at increased risk for persistent functional loss. For example, viscose rayon workers exposed to carbon disulfide have excessive irritability, fatigue, memory loss, and problems with intellectual processes requiring visualization; these symptoms are often followed by peripheral neuropathy. Workers exposed to mixed solvents have demonstrated impairments in reasoning, visuoconstructive abilities, short-term memory, motor coordination, speed, and attention, as well as EEG abnormalities. Nonoccupational causes of toxic encephalopathy include drug abuse, alcoholism, and the ingestion of lead-based paint, household solvents, and pesticides. Other clinical syndromes characterized by cognitive impairment include nontoxic causes such as Alzheimer's disease, hydrocephalus, multiple cerebrovascular infarctions, depression, and trauma.

Pathophysiology

Delirium and dementia are global cerebral disturbances, but the pathophysiology of organic mental disorders is unclear. Researchers have observed edema involving astrocytes, axons, and myelin sheaths in organotin intoxication. Cerebral swelling in the absence of myelin degeneration was observed in lead-induced encephalopathy in animal studies. Encephalopathy is a common symptom in chronic carbon disulfide poisoning based on diffuse vascular damage with or without focal lesions. Brain atrophy has been diagnosed in patients with suspected chronic organic solvent poisoning.

Prevention

Exposure to neurotoxins should be minimized. Effective industrial hygiene monitoring and control of respiratory, gastrointestinal, and dermal routes of chemical entry are discussed in part 1. Biological monitoring of workers for specific chemical exposures may be used to check on individual exposure levels, but careful attention must be paid to the interpretation of individual results. For more information, the American Conference of Governmental Industrial Hygienists (ACGIH) has issued biological exposure indices (BEIs) for a number of chemicals.

Other Issues

Drug abuse and alcohol abuse are two variables that confound the diagnosis of encephalopathy. Also, workers who have previously experienced somatic symptoms following a toxic exposure are at special risk for developing symptoms of encephalopathy as a posttraumatic stress disorder. Severe chronic toxic encephalopathy should be considered an entity distinct from other progressive dementias, such as Alzheimer's disease, and also should be evaluated with caution in older persons because of the deterioration in intellectual functioning that occurs with age.

RF, SP

Further Reading

Baker EL, White RF, Murawski BJ. Clinical evaluation of neurophysiological effects of occupational exposure to organic solvents and lead. *Int J Ment Health.* 1985;14(3):135-158.

Johnson BL, ed. *Prevention of Neurotoxic Illness in Working Populations.* New York, NY: John Wiley & Sons; 1987.

Norton S. Toxic responses of the central nervous system. In: Klaasen CD, Amdur MD, Doull J, eds. *Casarett and Doull's Toxicology.* 3rd ed. New York, NY: Macmillan; 1986:359-386.

Schottenfeld RS, Cullen MR. Occupational-induced post-traumatic stress disorders. *Am J Psychiatry.* 1985;142:198-202.

Spencer PS, Schaumburg HH, eds. *Experimental and Clinical Neurotoxicology.* Baltimore, Md: Williams & Wilkins; 1980.

White RF, Feldman RG, Travers PH. Neurobehavioral effects of toxicity due to metals, solvents, and insecticides. *Clin Neuropharmacol.* 13(5):392-412.

Table 1.—Toxic Central Nervous System Disorders

Acute Organic Mental Disorders

- Acute Intoxication
  - Pharmacological effect: duration of minutes to hours
  - No sequelae
  - Clinical: acute CNS depression, psychomotor impairment
  - Agents: solvents

Table 1.—Toxic Central Nervous System Disorders (cont.)

- Acute Toxic Encephalopathy
  - Rare
  - Pathophysiology: cerebral edema, CNS capillary damage
  - May cause permanent deficits
  - Clinical: coma, seizures
  - Agents: lead

Chronic Organic Mental Disorders

- Organic Affective Syndrome
  - Mood disturbance: depression, irritability, loss of interest in daily activities
  - Pathophysiology: unclear
  - Course: days to weeks
  - No sequelae
  - Agents: solvents, metals, pesticides (eg, lead, carbon disulfide)

- Mild Chronic Toxic Encephalopathy
  - Clinical symptoms: fatigue, mood disturbances, memory complaints, attention complaints
  - Course: insidious onset; duration of weeks to months
  - Pathophysiology: unclear; reversibility uncertain
  - Reduced CNS function
    - Psychomotor function (speed, attention, dexterity)
    - Short-term memory
    - Other abnormalities common
  - Agents: solvents, metals, pesticides (eg, styrene, carbon disulfide)

Table 1.—Toxic Central Nervous System Disorders (cont.)

Severe Chronic Toxic Encephalopathy
- Clinical manifestations
  - Loss of intellectual abilities of sufficient severity to interfere with social or occupational functioning
  - Memory impairment
  - Other
    - Impairment of abstract thinking
    - Impairment of judgment
    - Other disturbances of cortical function
    - Personality change
- Course: insidious onset, irreversible
- Pathophysiology: unclear
- Reduced CNS function
  - Types of abnormalities similar to mild chronic toxic encephalopathy but more pronounced
  - Some neurophysiological and neuroradiological tests abnormal
- Agents: controversial—solvents, lead, carbon disulfide

Table 2.—Exposures Associated with Encephalopathy

Neurotoxin	Major Uses or Sources of Exposure
Metals	
Arsenic	Pesticides
	Pigments
	Antifouling paint
	Electroplating industry
	Semiconductors
	Seafood
	Smelters

Table 2.—Exposures Associated with Encephalopathy (cont.)

Neurotoxin	Major Uses or Sources of Exposure
Lead	Solder Illicit whiskey Storage battery manufacturing plants Foundries, smelters Lead-based paint Lead shot Insecticides Auto body shops Lead-stained glass Lead pipes
Manganese	Iron, steel industry Metal-finishing operations of high-manganese steel Manufacturers using oxidation catalysts Manufacturers of fireworks, matches Manufacturers of dry cell batteries Fertilizers Welding operations
Mercury	Scientific instruments Amalgams Photography Taxidermy Pigments Electrical equipment Electroplating industry Felt making Textiles
Solvents	
Carbon disulfide	Manufacturing of viscose rayon Preservatives Electroplating industry Paints, varnishes Rubber cement

Table 2.—Exposures Associated with Encephalopathy (cont.)

Neurotoxin	Major Uses or Sources of Exposure
Perchloroethylene	Paint removers Degreasers Extraction agents Dry-cleaning and textile industries
Toluene	Rubber solvents, glues Paints, lacquers Manufacturers of benzene Cleaning agents Automobile, aviation fuels Paint thinners
Trichloroethylene	Degreasers Painting industry Paints, lacquers Varnishes Process of extracting caffeine from coffee Adhesive in shoe and boot industry Rubber solvents Dry-cleaning industry
Gases	
Carbon monoxide	Exhaust fumes of internal combustion engines, incomplete combustion Acetylene welding
Insecticides	
Organophosphates	Agricultural industry

Esophageal Cancer

ICD-9 150

Identification

The first symptom of esophageal cancer is progressive difficulty in swallowing, first for solids and then for liquids. Substernal or back pain is a later sign, generally implying an unfavorable prognosis. Anorexia and weight loss are not common early in the course of disease. Barium swallow x-rays are the primary diagnostic modality. Endoscopic brushing or biopsy are used to make the histological diagnosis.

Occurrence

Approximately 9300 new cases of esophageal cancer occurred in 1986, with ratios of 3:1 for men to women and of 3:1 for blacks to whites of the same gender. Age-adjusted mortality rates are comparable to incidence rates. From 1981 to 1985 in the United States, mortality rates (per 100 000) were 4.7 for white males, 1.4 for white females, 15.0 for black males, and 4.3 for black females.

Causes

Chronic irritation is thought to be the main predisposing factor. Implicated occupational exposures include asbestos and ionizing radiation. Woodworkers, shoemakers, rubber workers, and workers exposed to solvents may have an elevated risk. Some studies have shown workers in the construction trades, including painters, laborers working with wood, and roofers, to have excess risk. No estimates are available for the proportion of all esophageal cancers that may be due to occupational exposures. Alcohol and cigarette use are the primary nonoccupational risk factors. Hot fluids such as tea may be a risk factor worldwide. Acid reflux (Barrett's esophagus) is thought to be another underlying or predisposing cause.

Prevention

Primary prevention includes eliminating or reducing exposure to agents with known or suspected carcinogenicity. Secondary prevention involves treating reflux, whether through smoking cessation, dietary modification, or other means, but these methods have not been shown to lead to lower rates of esophageal cancer. No effective treatment strategies are available. Both surgical and radiation treatments lead to 5-year survival rates of less than 8%.

HF

Further Reading

Neugut AI, Wylie P. Occupational cancers of the gastrointestinal tract. In: Brandt-Rauf PW, ed. *Occupational Cancer and Carcinogenesis. State of the Art Reviews in Occup Med.* 1987;2:109-135.

Sleisenger MH, Fordtran JS. *Gastrointestinal Disease.* 2nd ed. Philadelphia, Pa: WB Saunders; 1983.

Extrinsic Allergic Alveolitis—*See* Hypersensitivity Pneumonitis

Eye Injury

ICD-9 870, 871, 918, 921, 930

Identification

Occupational eye injuries include any work-related trauma to the eye and supportive tissue, mechanical or otherwise, that results in physical damage or in temporary or permanent vision impairment.

Mechanical trauma includes the following: (a) Superficial injuries: these are usually painful and accompanied by tearing and photophobia (sensitivity to light). (b) Penetrating injuries: these are usually caused by sharp, small objects that may enter and damage internal structures in the eye and threaten vision. Sometimes these injuries are not obvious, especially if the object is small or has penetrated the eye completely. (c) Blunt object trauma: this can affect all structures surrounding the eye and the eye itself. External damage to the eye is usually obvious, but internal lesions can easily be overlooked.

Burns include the following: (a) Thermal burns: the blink reflex of the lids is usually quick enough to protect the eye from internal thermal burns, so eyelids are usually affected. Damage may range from minor skin edema to necrosis, depending on the heat source, duration of exposure, and area of contact. Intraocular damage occurs as heat cataract or retinal damage. (b) Chemical burns: these are caused by, in order of seriousness, alkalis, acids, and organic solvents. Because injuries can result in permanent damage in seconds, they are true emergencies requiring immediate action. (c) Radiation burns: the most common of these are due to ultraviolet (UV) radiation. They occur after "welder's flash" from arc welding. Chronic cumulative exposure to solar UV radiation can be a significant risk factor for developing cataracts.

Occurrence

About one fourth of all eye injuries are work related, based on an eye injury registry; eye injuries account for about one eighth of all compensated occupational injuries. Compared with non-occupational eye injuries, work-related eye injuries have the lowest rate of enucleation and loss of light perception. Almost all work-related eye injuries occur in men; nearly 60% occur among people in their 20s.

Occupational eye injuries occur most often in manufacturing and construction (about 80%), with fewer in agriculture, mining, and transportation. Metalwork, excavating and foundation work, and logging are high-risk occupations.

Chemical burns occur most often in manufacturing, followed by services, retail trade, and construction. About five sixths of chemical burns affect men, and two thirds of those affected are in their 20s. In 1986, a total of 2170 workers' compensation awards for welder's flash burns (50% in manufacturing, 20% in construction) and 3306 awards for other eye injuries (30% in manufacturing, 21% in the services industries, and 14% in retail trade) were reported to the BLS from 23 states.

Causes

About three fourths of occupational eye injuries are caused by flying objects associated with machines or hand tools; about 80% of all eye injuries are due to superficial mechanical trauma. About 10% are caused by chemicals. Other causes are UV burns of the cornea, other eye burns, blunt object trauma, penetrating wounds, and posttraumatic infection. About half of chemical burns to the head and neck also affect the eyes. Foreign bodies in the eye, contusions, and open wounds of the eye and orbit are very common. About 40% of these injuries result in permanent visual impairment.

Pathophysiology

Mechanical injuries result from physical trauma to the eye and surrounding structures. Injury caused by chemical burns varies with the type of chemical. Many alkalis can penetrate through and between the cells of the cornea to enter the anterior chamber and other intraocular tissues before being neutralized. Substances with a pH above 11 are very dangerous to the eye. As a general rule, acid burns of the eye are less severe than alkali burns. Organic solvents cause considerable eye irritation but rarely cause permanent damage. Injury caused by UV radiation can be acute or chronic. Acute injury, as in welder's flash, is a first-degree burn to the conjunctiva.

Prevention

Primary prevention of mechanical trauma from flying physical objects can be achieved by engineering controls designed to prevent objects from becoming airborne. This can be accomplished by identifying the source and changing or shielding the process. For chemical burns, chemicals should be handled in such a way as to minimize splashing and spills. Radiation burns can be prevented by changing a welding process (eg, using submerged arc welding) and providing welding booths to protect nearby people or providing shade from the sun for outdoor workers. Workplaces with potential for eye injury should have a written and well-communicated eye safety policy for employees. All at-risk workers should wear protective eyewear, available at no cost to workers. All ophthalmic equipment should meet ANSI standard Z87.1-1989, Standard Practice for Occupational and Educational Eye and Face Protection; eyewear that meets this standard is easily identifiable by a "watermark" etched onto one of the lenses. A supervisor, health and safety committee member, union representative, or other person should spot-check to ensure that at-risk workers are wearing protective eyewear and that engineering controls on injury sources, such as flying objects, chemical spills or splashes, and UV radiation, are in place and properly functioning.

Vision screening should be conducted on-site periodically. Screening helps detect uncorrected visual acuity and other eye problems, and it provides an opportunity for education. Preparation for emergencies should include a plan for action and installation of eye/face and drench showers at places where chemical burns may occur.

Other Issues

Although eyewear is essential, it is not "fail-safe," and it only works if it is worn. It is more likely to be worn if it is required. In a survey of workers who had sustained occupational eye injuries, 41% were wearing some form of eye protection at the time of injury. Of those not wearing eye protection, 60% were not required to do so by their employers.

Some professional athletes (eg, boxers, basketball players) are at increased risk for eye injury, especially from blunt object trauma. Some of these injuries can be prevented by the use of thumbless boxing gloves and mandatory protective eyewear. Recreational athletes, such as racquetball and basketball players, also are at higher risk.

Some health professionals, including dentists, dental hygienists, and laser users, are also at increased risk. However, protective eyewear is often not used among these groups. Efforts are needed to improve compliance, including education and peer pressure.

Individuals who have had lenses surgically replaced are at increased risk for both near UV and blue light injury, because the lens that blocks these sections of the spectrum is no longer present.

SV

Further Reading

Taylor HR, West SK, Rosenthal FS, et al. Effects of ultraviolet radiation on cataract formation. *N Engl J Med.* 1988;319:1429-1433.

US Department of Labor, Bureau of Labor Statistics. Accidents involving eye injuries. Washington, DC: U.S. Government Printing Office; 1984. Report 597.

White MF Jr, Morris R, Feist RM, et al. Eye injury: prevalence and prognosis by setting. Eye Injury Registry of Alabama. *South Med J.* 1989;82(2):151-158.

Eye Irritation—*See* Conjunctivitis

Eye Strain

ICD-9 368.13

(Asthenopia)

Identification

Eye strain presents with headaches and periorbital pain after ordinary work under dim light, or after or during excessive amounts of close or persistently vigilant work.

Occurrence

Eye strain commonly occurs in workers who need to perform prolonged visual monitoring. It appears to be more common among older workers. It occurs more frequently among women than men, primarily because of differences in job assignments. Although most prevalence estimates are based on subjective symptoms, these symptoms are well correlated with objective measurements of impairment.

Causes

Eye strain is caused by tasks requiring prolonged attention to visual detail. It has been found among video display terminal (VDT) operators, microscopists, and workers who monitor radar screens. Prolonged use of VDTs may also cause a temporary increase in myopia in conjunction with eye strain. However, when illumination level, luminance, and contrast are adequate, these problems are not observed.

Pathophysiology

The mechanism for eye strain is prolonged contraction of ciliary or recti muscles. Occupational stimuli for eye strain include prolonged reading, gazing at distant objects (including use of a microscope), and prolonged work at a VDT. Poor posture of the head and neck can be a contributing factor. Whether acute symptoms develop into permanent impairment is unknown.

Prevention

Work requiring prolonged visual vigilance should be redesigned to allow for increased variety in visual requirements. This could be accomplished in part with the use of frequent work breaks (at least once per hour). Workstations for VDT operators or radarscope monitors should be designed to reduce glare and reflection; to provide for adequate illumination, luminance, and contrast; and to allow for proper posture. Keyboards should be movable to a comfortable position, and the visual distance to screens should be adjustable. Regular on-site eye exams are helpful, not only to detect eye strain but also to ensure proper visual acuity.

Other Issues

Individuals predisposed to eye strain include those with under-corrected refractive error, especially hyperopia and astigmatism; intolerance of new correction with astigmatism, anisometropia, and aniseikonia; muscle imbalance; improperly set bifocals; and improperly adjusted frames. Middle-aged workers are more likely to have problems adjusting to varying focal distances and may require special glasses while working on VDTs.

SV

Further Reading

DeGroot JP, Kamphius A. Eyestrain in VDT users: physical correlates and long-term effects. *Hum Factors.* 1983;25(4):409-413.

Goussard Y, Martin B, Stark L. A new quantitative indicator of visual fatigue. *IEEE Trans Biomed Eng.* 1987;34(1):22-29.

Howarth PA, Istance HO. The validity of subjective reports of visual discomfort. *Hum Factors.* 1986;28(3):347-351.

Iwaski T, Kurimoto A. Objective evaluation of eye strain using measurements of accommodative oscillation. *Ergonomics.* 1987;30(3):581-587.

Rose L. Workplace video display terminals and visual fatigue. *J Occup Med.* 1987;29(4):321-324.

World Health Organization. Visual display terminals and workers' health. *WHO Offset Publ.* 1987;99:1-206.

Fatty Liver Disease and Cirrhosis

ICD-9 571.5, 571.8

Identification

Fatty liver disease may occur without symptoms or merely with simple right upper-quadrant pain, fullness, and discomfort, along with some nausea—often only after heavy meals. On the other hand, it may present acutely with fever and chills as fatty liver hepatitis. The progression to cirrhosis may be imperceptible. Complications, including bleeding and ascites, may be the presenting complaints. Diagnosis may be made on the basis of (a) elevations of alanine aminotransferase (ALT) or aspartate aminotransferase (AST), reflecting liver injury; (b) abnormal results of liver function tests, such as postchallenge (or postprandial) serum bile acids, or of tests of the ability of the liver to metabolize foreign substances such as caffeine, antipyrine, or indocyanine green; or (c) abdominal CT scanning or sonography, revealing fatty infiltration of the liver. Unequivocal diagnosis is made by liver biopsy. Physical examination of the liver is difficult, but a palpable liver with a rounded edge may be found. The development of cirrhosis is associated with a firm liver with a rounded or a nodular rim, spider nevi, and ascites. Associated physical findings from hormonal changes in men include a change in body hair and testicular atrophy.

Occurrence

No incidence data are available. An overall death rate is estimated at 11.1 cases per 1 million per year, or 1.4% of all deaths in 1983, with men twice as likely as women to develop disease. Liver cirrhosis is the 10th leading cause of death and the 9th leading cause of years of potential life lost before the age of 65. The proportion of these cases in which hepatotoxin exposure plays a contributing role is unknown but is generally assumed to be low. The prevalence of fatty liver disease in autopsies series in the U.S. and Europe ranges from 21% to 50% in unselected populations and up to 80% in populations with a substantial probability of heavy alcohol consumption. It is estimated that alcohol causes or contributes to 70% of cirrhosis deaths in the U.S.

Causes

A large number of organic and some inorganic compounds have been implicated in human disease. Many more are implicated by animal studies, although these compounds have been studied only inadequately in humans. Halogenated hydrocarbons are the best known, and their hepatotoxicity is well described: The shorter the base chain, the less saturated the chain, the more halogen atoms per molecule, the higher the mass number of the halogen, and the greater the electronegative bond between substituent and base chain, the greater the associated hepatotoxicity. Carbon tetrachloride (CCl_4) is a greater hazard than chloroform ($CHCl_3$); trichloroethylene is greater than trichloroethane; ethylene dibromide is greater than ethylene dichloride. Other agents in this series include dichloromethane, carbon chloride, dichloroethylene, trichloroethylene, and trichloroethane. 1,1,1-Trichloroethane is the last agent of the series so far detected for which human toxicity is likely. Other agents include phosphorous, organic and inorganic arsenic, hexachlorobenzene, nitrobenzene, naphthalene, pentachlorophenol, proprionitrile, acetylene, butadiene, acrylate monomers, vinyl halothane, polychlorinated biphenyls, chlordecone, methylene diamine, chlordane, ethylene glycol, and toluene. Probable causes include styrene, fluorocarbons, and glycol ethers. In addition, acute lead intoxication has been associated with fatty liver disease.

Risk factors for the development of cirrhosis also include diabetes mellitus, obesity (more than 20% overweight), alcohol consumption, and interaction with hydrocarbons that have no intrinsic hepatotoxicity (eg, simultaneous exposure to acetone).

Increasing age is a risk factor for liver toxicity from some agents. Obesity alone is thought to progress to cirrhosis only rarely. Alpha-1-antitrypsin deficiency haplotypes, apart from the homozygous Z state, are thought to increase the probability of developing cirrhosis. Reasons for progression of fatty liver disease to cirrhosis are unknown.

Pathophysiology

Mechanisms are generally classified as direct or indirect. Direct toxins do not require metabolism for activation. Indirect toxins require activation through the mixed-function oxidase system; their phase II products then interfere with the specific pathways that lead to structural injury. A primary factor in both is probably glutathione depletion. Two separate mechanisms are implicated in the toxicity of common agents such as carbon tetrachloride: decoupling of lipid transport out of cells and direct cellular toxicity (lipid peroxygenation). Cytotoxic and cholestatic effects may occur. There is still controversy over whether the presence of fat intracellularly is toxic in itself.

Prevention

Prevention of exposure prevents disease. Because of the low vapor pressure of some agents, the likelihood that they will induce disease without skin contact is thought to be minimal. For some agents, skin absorption is probably the primary mode of toxicity. Appropriate contact prevention must then be considered. General preventive measures include maintenance of ideal body weight, control of diabetes, and abstention or moderation in alcohol consumption. Inadequate information is available to determine whether some individuals are at greater risk than others because of different metabolic pathways. Antioxidants have not been shown to reduce the incidence of disease in humans. There is some suggestion that colchicine and propranolol may reduce the progression of disease and the incidence of bleeding, respectively. The value of sclerotherapy is controversial.

MH

Further Reading

Horton AA, Fairhurst S. Lipid peroxidation and mechanisms of injury. *Crit Rev Toxicol.* 1987;18:27-79.

Schiff L, Schiff ER. *Diseases of the Liver.* Philadelphia, Pa: JB Lippincott; 1987.

Zimmerman HJ. *Hepatotoxicity.* New York, NY: Appleton-Century-Crofts; 1979.

Frostbite—*See* Cold-Related Disorders

Gastric Cancer—*See* Stomach Cancer

Glomerulonephritis

ICD-9 580, 582

Identification

Glomerulonephritis presents as proteinuria with or without nephrotic syndrome (edema, albuminuria >3.5 g/24 h, hypoalbuminemia, and hyperlipidemia). Urinary abnormalities include the presence of casts and red blood cells. Clinical progression varies greatly according to the pathological picture. Some types of glomerulonephritis are self-limited, some respond to therapy, and others progress relentlessly to renal failure. Glomerular disorders underlie most cases of end-stage renal failure.

Occurrence

Some types of glomerulonephritis, such as postinfectious and focal glomerulosclerosis, are most common in children. Females predominate in childhood postinfectious glomerulonephritis, but the sexes are affected equally as adults. The proportion of glomerulonephritis caused by occupational exposures is unknown, but it is probably small.

Causes

Glomerulonephritis is caused by one of several immune processes triggered by factors that are usually unidentified but that may include infections, neoplasms, and toxic exposures. Occupational exposures are the suspected precipitants of both immune complex disease and anti-glomerular basement membrane (GBM) antibody, although, except in cases of mercury-induced nephrotic syndrome, the evidence is circumstantial. Occupational associations include the following:

Hydrocarbon solvents. Several investigators report frequent exposure to hydrocarbon solvents among people with glomerulonephritis. Concern initially involved Goodpasture's syndrome (a rare immunologic lung and kidney disease) but has been extended to more common glomerular disorders. At least nine case-control studies have examined the relationship between organic solvents and glomerulonephritis; eight of these studies have found an association. It remains unclear what types of solvents, exposure levels, or host susceptibility factors may underlie the findings.

Silica. Case reports have described glomerulonephritis in workers who are heavily exposed to silica. Silicosis patients develop immunologic changes resembling lupus erythematosus, and the accompanying glomerular disease resembles that of lupus nephritis.

Inorganic mercury. Mild albuminuria appears in some workers exposed to mercury vapor at levels associated with tremor. Frank nephrotic syndrome occurs in approximately 3% of workers exposed to $\geq 500\ \mu g/m^3$ TWA (occupational legal limit is $50\ \mu g/m^3$ TWA). Membranous glomerulonephritis has also been described in people who take mercurial diuretics or use mercury skin creams.

Nephrotic syndrome. Nephrotic syndrome occurs with other heavy metals, in approximately 10% of rheumatoid patients taking gold salts, and as an idiosyncratic response to lithium and bismuth.

Occupations at risk for glomerulonephritis include chlor-alkali workers; manufacturers of thermometers, barometers, and other instruments; fur preservers; and gold extractors, all of whom may be exposed to mercury, as well as miners, foundry workers, grinders, and sandblasters, all of whom may be exposed to silica.

Other, mostly nonoccupational causes of glomerulonephritis include (a) immunologic disorders (lupus erythematosus, periarteritis nodosa, Henoch-Schonlein purpura); (b) neoplasms (multiple myeloma, Hodgkin's disease); (c) infections (malaria, schistosomiasis); (d) drug toxicity (penicillamine, bismuth, mercurials, gold salts); (e) snake bites and bee stings; (f) renal vein thrombosis; (g) amyloidosis; and (h) unknown familial factors.

Pathophysiology

Goodpasture's anti-GBM antibody syndrome is believed to result from deposition of circulating IgG on the basement membrane in a linear pattern, initiating immunologically mediated damage of the basement membrane. Immune complex disorders are believed to involve deposition of antigen-antibody complexes in the glomeruli, initiating damage.

Prevention

Some people may be susceptible to initiation of the immune response that causes renal disease. However, tests are not currently available to identify such people. Prevention must thus be based on controlling exposures to minimize disease in all individuals. Some evidence exists that people with glomerulonephritis associated with mercury and solvents improve when they are removed from exposure. It is unclear whether removal from exposure improves the prognosis for people with disease caused by other agents, such as silica.

MT

Further Reading

Churchill DN, Fine A, Gault MH. Association between hydrocarbon exposure and glomerulonephritis: an appraisal of the evidence. *Nephron.* 1983;33:169-172.

Daniell WE, Couser WG, Rosenstock L. Occupational solvent exposure and glomerulonephritis. *JAMA.* 1988;259:2280-2283.

Lauwerys R, Bernard A, Viau C, Buchet JP. Kidney disorders and hepatotoxicity from organic solvent exposure. *Scand J Work Environ Health.* 1985;11(suppl 1):83-90.

Osorio AM, Thun MJ, Novak RF, Van Cura J, Avner ED. Silica and glomerulonephritis: case report and review of the literature. *Am J Kidney Dis.* 1987;9:224-229.

Hand-Arm Vibration Syndrome

ICD-9 443.0
SHE/O

Identification

Hand-arm vibration syndrome (HAVS) is also known as vibration-induced white finger, traumatic vasospastic disease, or secondary Raynaud's phenomenon of occupational origin. It is a disorder of the blood vessels and nerves in the fingers that is caused by vibration transmitted directly to the hands ("segmental vibration") by tools, parts, or work surfaces. Reduced handgrip strength has also been noted in some groups of exposed workers.

The condition is primarily characterized by numbness, tingling, and blanching (loss of normal color) of the fingers. Initially, there is intermittent numbness and tingling; blanching is a later sign, first in the fingertip and eventually over the entire finger. Symptoms usually appear suddenly and are often precipitated by exposure to cold. Attacks usually last 15 to 60 minutes, but in advanced cases, they may last 1 or 2 hours. Recovery begins with a red flush ("reactive hyperemia"), usually starting at the wrist and palm and moving down to the fingers.

The grading system for severity was first proposed by Taylor and Pelmear[1] and was recently revised[2] to take account of the fact that the injuries to nerves and blood vessels appear to develop independently. The revised grading system defines up to four stages, based on (a) the frequency of tingling, numbness, and blanching, and (b) the extent of the loss of function in each hand. With continuing exposure to vibration, signs and symptoms become more severe and are eventually irreversible. In later stages, there is reduced sensitivity to heat and cold, with accompanying pain. Tactile sensitivity and neuromuscular control are impaired and finger joints become increasingly stiff, so precise manual tasks, such as picking up small objects and fastening buttons and zippers, become difficult. In the most advanced and severe cases, the fingers have a dusky, cyanotic (bluish) appearance, and there is complete obliteration of the arteries.

There are no reliable, objective diagnostic tests. The available tests may differentiate affected from unaffected workers on a group basis, but they have poor validity on an individual basis. They include tests of peripheral vascular function (plethysmography, arteriography, and skin thermography) and neurological function (two-point discrimination, depth sense, pinprick, touch, and temperature tests), as well as radiographs of the fingers and hands. Thermal and vibratory perception thresholds (*see* Carpal Tunnel Syndrome) have been shown to be higher in vibration-exposed subjects with advanced neurological symptoms but not in those with vascular symptoms.[3]

The differential diagnosis includes ruling out peripheral nerve compression at the wrist, forearm, or shoulder (*see* Carpal Tunnel Syndrome and Peripheral Nerve Entrapment Syndromes); nonoccupational vascular or connective tissue diseases (*see* Causes below); and lacerations and fractures of the hands and fingers.

Occurrence

NIOSH has estimated that 1.45 million workers in the U.S. are potentially exposed to vibrating hand tools or other sources of segmental vibration and are therefore at risk of developing HAVS.[4] The industries with the largest numbers of workers probably exposed are construction, farming, and truck and automobile manufacturing. Any worker using a power hand tool (eg, chain saw, pneumatic drill, chipping hammer or jack hammer, grinder, buffer, or polisher) should be considered at risk. Prevalence estimates range from 20% to 100% among workers exposed for at least 1 year, but the intermittent nature of symptoms in the earlier stages leads to substantial underreporting.[5] Both the intensity (acceleration level) and the duration of exposure determine how short the latency is until the first episode and how often the attacks occur. Three longitudinal studies have shown that, following the introduction of antivibration saws, prevalence rates as high as 90% in logging and forestry workers have decreased, on average, by four fifths.

Causes

The cause of this syndrome is the direct physical transmission of vibration from a mechanical object to the hand and arm. This occurs through the use of vibrating hand tools or through other hand or arm exposure to segmental vibration, such as that transmitted through a truck or bus steering wheel or a part held to a grinding wheel. Exposure for as little as 1 year is sufficient to initiate onset. Latency from first exposure to onset of blanching has been estimated to range from 1 to 17 years, depending on the duration, frequency, and intensity of vibration and on work practices. Therefore, the shorter the latent period, the more severe the expected syndrome if exposure continues. Chronic exposure to cold temperatures, especially during vibration exposure, exacerbates the effects of vibration.

Occupational exposure to vinyl chloride monomer has been reported to produce a secondary Raynaud's phenomenon. No data are available regarding the effects of combined exposure to segmental vibration and vinyl chloride or other peripheral neurotoxins such as lead (*see* Peripheral Nerve Entrapment Syndromes).

Primary Raynaud's syndrome occurs spontaneously in 5% to 10% of the general population; the female-to-male ratio has been estimated at 5:1.[6] Nonoccupational secondary Raynaud's phenomenon may be associated with acute injury (frostbite, fracture, or laceration) or with medical conditions such as connective tissue diseases (eg, scleroderma, rheumatoid arthritis), vascular disorders (eg, Buerger's disease, arteriosclerosis), neurogenic causes (eg, poliomyelitis), or a long history of high blood pressure. Nicotine is considered to aggravate or precipitate the condition, because it is a vasoconstrictor that reduces the blood supply to the hands and fingers.

Pathophysiology

Both circulatory and neurological effects result from exposure of the hand to vibration, although the exact physiological cause is not known. Vibration appears to cause direct injury to peripheral nerves, resulting in the numbness of fingers. Paresthesia (decreased sensation) of the hands may be secondary to constriction of the blood vessels, causing ischemia (loss of blood supply) to the peripheral nerves. Other physiological and chemical changes in the blood vessels have also been documented, although their causal role is not clear at this time.

An additional mechanism, the tonic vibration reflex (TVR), appears to contribute to soft tissue damage through the effect of vibration on tendon function.[7] Vibration interferes with the sensitivity of the nerve endings that sense the force exerted by a tendon. When the hand holds a vibrating object and this sensory feedback to the central nervous system is disrupted, the muscles are signaled to exert higher force than is necessary to grip the object. This increases the strain on the tendons (*see* Tendinitis); the tighter grip also increases the amount of vibration transmitted to the hand, resulting in greater nerve and blood vessel damage as a result. The nerve endings in the tendons then lose even more sensitivity and the control reflex is even further disrupted, thus continuing this vicious circle.

Prevention

Primary prevention consists of measures to reduce exposure to sources of vibration. Redesign of production processes and work methods might make it possible to minimize the use of vibrating hand tools or equipment; for example, improved quality of castings could reduce the need for later grinding and polishing. Where vibration cannot be eliminated from the workplace, engineering controls, work practices, and administrative controls should be considered to reduce the intensity and duration of exposure.

At the source of exposure, engineering controls consist of the redesign of power hand tools to minimize vibration generated or transmitted during operation.[8] Mechanical isolation and damping should be used to reduce the acceleration (intensity) of the vibration transmitted to the hand and arm. It is often recommended that tools should be selected that vibrate well above the natural resonant frequency of the hand and arm (30 to 300 Hz), which is where the hand and arm are most vulnerable. Recent data suggest, however, that these high frequencies may be more hazardous than was previously understood.

In the path of exposure, installation of a tool on an articulating arm or an overhead suspended balancer will help to reduce vibration transmission.[4] This principle also applies to the exposure that occurs when workers hold parts to be ground or polished against a grinding wheel. Vibration transmission to the hand will also be reduced if workers can use less force to grip objects and operate tools; this can be accomplished by improving friction between the tool or part and hand (or glove), installing a slip clutch or other means of reducing torque, and reducing the weight of the tool or part. Padded gloves or pads on tool handles may achieve some vibration isolation, although their effectiveness should be demonstrated for the particular tool and task; commercially available isolation materials vary tremendously, even in a laboratory setting, in their effectiveness across the range of typical tool vibration frequencies.[9] In any case, gloves should fit well to avoid increasing the grip force required to hold and operate the tool. Work practices and administrative controls may include running power tools at lower speeds, ensuring regular tool maintenance, engaging in frequent rest breaks or alternate work without vibration exposure, and keeping the hands warm and dry and the body core temperature stable.

Three "consensus standards" exist in the U.S. for control of HAVS; these are published by ACGIH,[10] ANSI,[11] and ISO.[12] Each considers the frequency spectrum of the vibration exposure to be a key determinant of the limits on intensity and duration of exposure. NIOSH chose not to incorporate such "frequency weighting" into its criteria document;[4] instead, it recommended prospective medical and exposure monitoring, with the explicit goal of developing a better epidemiological basis for any future quantitative exposure limits.

Secondary preventive measures include encouraging workers to report symptoms to their physicians or to the workplace medical service. All workers who use vibrating hand tools should be examined for signs and symptoms of HAVS; work histories should specifically include questions about previous exposure to segmental vibration. Workers with preexisting signs or symptoms should not be assigned to work with vibrating tools. Exposed workers and their supervisors should be informed of the symptoms of HAVS. If tingling, numbness, or blanching occur, workers should seek medical attention promptly and should be reassigned to work with little or no exposure to vibration. The jobs of affected workers should be evaluated for implementation of engineering controls. Health care providers should be trained in interview and clinical examination procedures necessary to identify occupational HAVS.[4,6]

Other Issues

Workers with exposure to vibration, through hand tools or other sources, should be encouraged to reduce or stop smoking, because nicotine has a separate effect on peripheral circulation that makes blood vessels more vulnerable to the effects of vibration.

LP, KR

References

1. Taylor W, Pelmear PL, eds. *Vibration White Finger in Industry.* London, England: Academic Press; 1975.

2. Gemne G, Pyykko I, Taylor W, Pelmear P. The Stockholm Workshop Scale for the classification of cold-induced Raynaud's phenomenon in the hand-arm vibration syndrome (revision of the Taylor-Pelmear scale). *Scand J Work Environ Health.* 1987;13:290.

3. Ekenvall L, Gemne G, Tegner R. Correspondence between neurological symptoms and outcome of quantitative sensory testing in the hand-arm vibration syndrome. *Br J Ind Med.* 1989;46:570-574.

4. National Institute for Occupational Safety and Health. *Criteria for a Recommended Standard: Occupational Exposure to Hand-Arm Vibration.* Washington, DC: U.S. Government Printing Office; 1989. DHHS (NIOSH) publication 89-106.

5. National Institute for Occupational Safety and Health. *Vibration Syndrome.* Washington, DC: U.S. Government Printing Office; 1983. Current Intelligence Bulletin 38. DHHS (NIOSH) publication 83-110.

6. Brammer AJ, Taylor W. *Vibration Effects on the Hand and Arm in Industry.* New York, NY: Wiley-Interscience; 1982.

7. Armstrong TJ, Fine LJ, Radwin R, Silverstein BA. Ergonomics and the effects of vibration in hand-intensive work. *Scand J Work Environ Health.* 1987;13:286-289.

8. Andersson ER. Design and testing of a vibration attenuating handle. *Int J Ind Ergonomics.* 1990;6:119-125.

9. Hampel GA, Hanson WJ. Hand vibration isolation: a study of various materials. *Appl Occup Environ Hyg.* 1990;5:859-869.

10. American Conference of Governmental Industrial Hygienists. 1990 notice of intended changes: hand-arm (segmental) vibration (HAVS). *Appl Occup Environ Hyg.* 1990;5:464-470.

11. American National Standards Institute. *Guide for the Measurement and Evaluation of Human Exposure to Vibration Transmitted to the Hand.* New York, NY: ANSI; 1986. No. S3.34.

12. International Standards Organization. *Guidelines for the Measurement and the Assessment of Human Exposure to Hand-Transmitted Vibration.* Geneva, Switzerland: International Standards Organization; 1986. ISO 5349.

Further Reading

Wasserman DE, Taylor W. Occupational vibration. In: Rom WN, ed. *Environmental and Occupational Medicine.* Boston, Mass: Little, Brown; 1983: chap 68.

Hard Metal Disease—*See* Cobalt-Induced Interstitial Lung Disease

Headache

ICD-9 784.0

Identification

Head pain can be either a symptom of an acute neurological emergency or a chronically recurrent annoyance. Usually headache symptomology is not seen in isolation in exposed workers but is consistently related to a number of other subjective symptoms. Each complaint of headache must be carefully analyzed according to its quality, location, pattern of onset, and frequency in relation to the job history in order to reach the correct diagnosis and to target preventive measures.

Hypoxia or hypercapnia may bring on an acute headache by causing vasodilation; this is usually the mechanism underlying headaches caused by acute exposures to toxic chemicals. Cyclic and more chronic forms of headaches, such as migraines and cluster headaches and those resulting from transient acute cerebrovascular insufficiency, have also been observed in exposed workers. Cyclic migraine headaches are unilateral in a single attack and begin with a prodromal of visual or other neurological symptoms. Cluster headaches resemble migraines in their unilaterality and are characterized by a nonthrobbing, aching pressure. Other than obtaining a comprehensive history and description of the headache symptomology from the worker and removing the worker from potential exposure and seeing whether the headache disappears, there are no specific diagnostic tests to determine the work-relatedness of the symptom to the exposure. Diagnostic tests such as computerized axial tomography and magnetic resonance imaging can show an intracranial lesion, such as a brain tumor, abscess, subdural hematoma, or obstructive hydrocephalus, and can thus aid in ruling out nonoccupational causes of headaches.

Occurrence

No reliable occurrence data exist for occupationally induced headache (*see* Occurrence section of Peripheral Neuropathy).

Causes

Acute and chronic occupational exposure to metals, such as arsenic, lead, nickel, tin, and tellurium; to organic solvents; to gases; and to certain insecticides commonly produce headaches (*see* Table 1 at end of entry). Transient headache concomitant with malaise, depression, nausea, dry throat, and fever following chronic exposure to zinc oxide fumes is symptomatic of metal fume fever as can occur with exposure to a number of metals (*see* Metal Fume Fever). Inhalation of organic solvents (carbon disulfide, methyl-n-butyl ketone, n-hexane, perchloroethylene, toluene, and trichloroethylene) may produce headaches as a part of a prenarcotic syndrome experienced in conjunction with cognitive impairments, depression, and/or anxiety, which, with prolonged exposure, often develops into toxic encephalopathy. Headaches brought on by carbon monoxide exposure progress in severity as exposure concentration (measured by the percentage of carboxyhemoglobin present in the blood) increases; what begins as a slight headache with tightness across the forehead turns into a throbbing frontal headache and then into a severe frontal and occipital headache (*see* Asphyxiation). Acute, high-dose exposure to trichloroethylene can produce head and facial pain due to effects on the trigeminal nerve, although in chronic exposures it more usually produces a subtle loss of sensation in the trigeminal nerve distribution. In both instances, there may be an accompanying degree of headache.

Nonoccupational vasodilation headaches can be due to alcoholic hangover; caffeine withdrawal; and exposure to histamines, nitrates, monosodium glutamate, and tyramine. Headache may be a symptom of essential hypertension. There are numerous other causes of acute headache, including trauma; sinusitis; glaucoma; temporal arteritis; trigeminal neuralgia; and extracranial inflammation such as cellulitis of the scalp, periostitis, or osteomyelitis of the skull. Headache is present with a brain tumor when the tumor has reached a critical size and causes intracranial pressure to increase. Oral contraceptives containing progestin and estrogen may produce a constant dull headache in susceptible women. A muscle tension headache may be due to stress, anxiety, and depression. Muscle contraction or muscle tension headache is found in 40% of patients complaining of head pain; 75% of these patients are women, and 40% report a family history of headache. Common migraine may be precipitated by the ingestion of certain foods containing tyramine (cheeses) and nitrate (sausage, bologna); it is seen most often in premenopausal women.

Pathophysiology

The mechanisms for headache following metal exposure are brain swelling (tin, lead), transient hypoxia, and vasodilation (zinc, tellurium, manganese, nickel). Carbon monoxide also produces headache by vasodilation. The afferent fibers of the trigeminal nerve mediate most of the pain sensations in the head, but fibers of the glossopharyngeal, vagal, and upper cervical nerves can also mediate head pains. Because of the wide distribution and anastomoses of nerve fibers, referred pain is common in headache syndromes.

Prevention

Exposure to neurotoxins and other causes of headache should be minimized. Effective industrial hygiene monitoring and control of respiratory, gastrointestinal, and dermal routes of chemical entry are discussed in part 1. The symptoms of headache may be ameliorated with analgesics, muscle relaxants, sedatives, antidepressants, or antianxiety drugs (*see* Asphyxiation, Stress, Collective Stress Disorder, and Building-Related Illness).

Other Issues

People with compromised cardiac function, hypertension, or other vascular problems may be at increased risk for headache following exposure to chemicals such as carbon monoxide, zinc, and tellurium because of the vasodilation effects of these chemicals. Also, people on diuretic or vasodilation medications may have increased frequency of headaches.

Headache is very common in the working environment, but, as described above, it has many different causes. Regarding its association with specific workplace chemical hazards, one must recognize the potential effects of the chemicals in use and be aware of workers' complaints, both individually and collectively.

RF,SP

Further Reading

Feldman RG. Headache. In: Wilkins RW, Levinsky NG, eds. *Medicine: Essentials of a Clinical Practice.* Boston, Mass: Little, Brown; 1983:837-841.

Taub A. Headache. In: Feldman RG, ed. *Neurology: The Physician's Guide.* New York, NY: Thieme-Stratton; 1984:129-146.

Table 1.—Toxic Exposures Associated with Headache

Neurotoxin	Major Uses or Sources of Exposure
Metals	
Arsenic	Pesticides Antifouling paint Seafood Semiconductors Pigments Electroplating industry Smelters

Neurotoxin	Major Uses or Sources of Exposure
Lead	Solder Illicit whiskey Storage battery manufacturing plants Auto body shops Lead-based paint Foundries Lead shot Insecticides Smelters Lead pipes Lead-stained glass
Nickel	Electroplating industry Paints Alloys Surgical and dental instruments Nickel-cadmium batteries Inks Coinage
Tin	Canning industry Solder Polyvinyl plastics Coated wire Silverware Electronic components Fungicides
Tellurium	Coloring agent in glazes, glass Rubber vulcanization Electronics industry Foundries Semiconductors Thermoelectric devices
Solvents	
Carbon disulfide	Manufacturers of viscose rayon Preservatives Rubber cement Electroplating industry Paints Textiles Varnishes

Neurotoxin	Major Uses or Sources of Exposure
Methyl-n-butyl ketone	Paints Varnishes Quick-drying inks Lacquers Metal-cleaning compounds Paint removers
n-Hexane	Lacquers Printing inks Pharmaceutical industry Rubber cement Stains Glues
Perchloroethylene	Paint removers Degreasers Extraction agent for vegetables and mineral oils Dry-cleaning industry Textile industry
Toluene	Glues Paints Automobile, aviation fuels Paints, thinners Cleaning agents Manufacturers of benzene Lacquers Gasoline
Trichloroethylene	Degreasers Painting industry Paints Varnishes Process of extracting caffeine from coffee Lacquers Dry-cleaning industry Adhesive in shoe and boot industry Rubber solvents

Neurotoxin	Major Uses or Sources of Exposure
Gases	
Carbon monoxide	Exhaust fumes of internal combustion engines, incomplete combustion Acetylene welding
Methane	Natural gas heating fuel Enclosed areas; mines, tunnels
Anesthetic gases (waste)	Operating rooms Dental offices
Insecticides	
Chlordecone	Agricultural industry

Hearing Loss, Noise-Induced

ICD-9 388.1
SHE/O

Identification

Noise-induced hearing loss is a chronic disorder with insidious onset. There is no pain and seldom any discomfort. It is best diagnosed by pure tone audiometric testing, which usually shows a characteristic pattern (*see* Fig. 1). Shifts in hearing threshold may be temporary at first and often accompanied by tinnitus (ringing in the ear). There is a gradual decrease in the ability to hear: It seems as though other people are talking more softly, and one needs to turn up the volume of radio or television. With repeated exposure to noise, the loss becomes permanent and progresses to the point where conversation becomes difficult, especially with groups of people or when background noise is present. The hearing loss is usually bilateral and is not amenable to medical intervention.

There is some disagreement about safe levels of exposure to noise. The permissible exposure limit (PEL) used by OSHA and MSHA in the U.S. Department of Labor is 90 dB(A)* for 8 hours, with a 5-dB(A) exchange rate and a maximum PEL of 115 dB(A) for continuous noise. (The exchange rate is the relationship between allowable exposure level with each halving or doubling of duration. A 5-dB exchange rate permits the exposure level to be increased by 5 dB(A) for each halving of duration.) Both NIOSH and ACGIH recommend a PEL of 85 dB(A) for 8 hours, with a 5-dB exchange rate and a maximum PEL of 115 dB(A) for continuous noise. Most European countries, as well as the U.S. EPA, use the more conservative 3-dB exchange rate. EPA recommends a maximum 8-hour PEL of 75 to 80 dB(A).

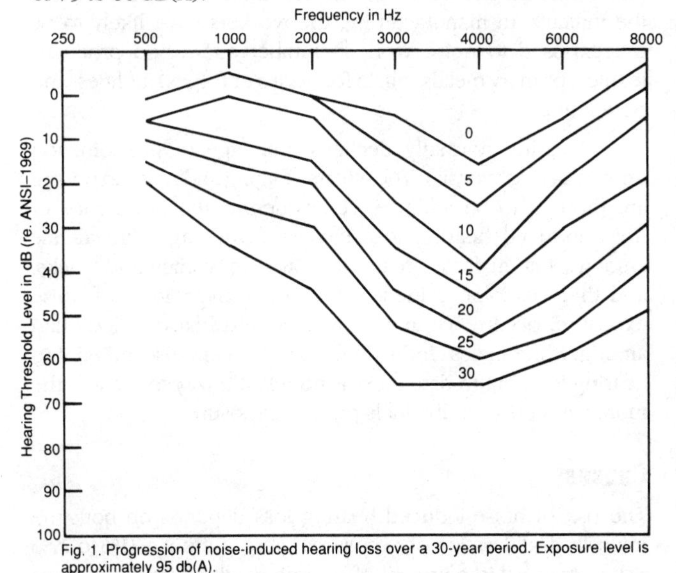

Fig. 1. Progression of noise-induced hearing loss over a 30-year period. Exposure level is approximately 95 db(A).

---

* dB(A) means decibels measured on the A scale of the sound-level meter. The A scale correlates well with the ear's response to sound in terms of both annoyance and vulnerability to hearing damage.

There has also been some controversy about the point at which a noise-induced hearing loss becomes a significant or "material" impairment. Currently, there is general consensus that material impairment is present when the average hearing threshold is greater than 25 dB at 1000, 2000, and 3000 hertz (Hz). (Hz or hertz is the contemporary term for cycles per second.)

Occurrence

Occupational noise-induced hearing loss is estimated to occur in 1 million people in the U.S. About 9 million workers are currently exposed to noise above 85 dB(A). About half of these are in manufacturing and utilities; a fifth are in transportation; and the remainder are in agriculture, mining, construction, and the military. In manufacturing, the workers most likely to be overexposed to noise work in lumber and wood products, textiles, primary metals (eg, in foundries or forges), utilities, and petroleum.

Hearing normally declines with age (a phenomenon known as presbycusis). The effects of aging and noise exposure are thought to be additive; consequently, the prevalence of noise-induced hearing loss increases with age. Prevalence among white males is greater than among women and blacks, and there is some evidence that, for a given level of noise exposure, occurrence among women and blacks is less than among white males. Individuals vary in their susceptibility to hearing loss, but to date there is no reliable way to identify the more susceptible individuals prior to exposure.

Causes

The risk of noise-induced hearing loss depends on both the intensity and duration of exposure. There is about a 21% to 29% risk of material impairment of hearing resulting from exposure to 90 dB(A) over a working lifetime. The risk at 85 dB(A) is 5% to 15%, and at 80 dB(A) the risk is negligible. Occasionally, a single traumatic exposure to noise in the range of 150 dB(A) is sufficient to cause hearing loss. More often, however, hearing loss is caused by repeated exposure to noise above 85 dB(A) over long periods.

There are many nonoccupational causes of hearing loss, including middle-ear infections, impacted cerumen, and sensorineural hearing losses associated with diseases such as meningitis and Ménière's disease. Nonoccupational noise-induced hearing loss may also result from frequent exposure to loud music and noisy hobbies (particularly sport shooting). However, most hazardous exposures are occupational.

Pathophysiology

The usual mechanism of noise-induced hearing loss is gradual destruction of cochlear hair cells through metabolic or vascular means or through a combination of mechanical, metabolic, and vascular processes. When acute traumatic exposures produce hearing loss in a single sudden blast, sound pressure mechanically dislodges sensory cells located in the cochlea (the sensory organ located in the inner ear).

The audiometric manifestation of excess noise exposure is a permanent sensorineural hearing loss beginning in the high frequencies (4000 Hz and higher) and eventually spreading to the middle and low frequencies. Figure 1 shows a hypothetical example of noise-induced hearing loss for an individual subject to an average daily exposure of approximately 95 dB(A) over 30 years. Sound frequency in Hz is displayed on the horizontal axis and hearing threshold level in dB along the vertical axis, with years of exposure as the parameter.

Prevention

The most effective way to prevent noise-induced hearing loss is to use engineering measures to reduce the generation of noise at its source. Source controls entail redesigning the noisy equipment or process. Another effective method is to control the transmission of sound by applying mufflers; enclosing the source; or using absorptive materials, baffles, or barriers along the path. The third method is to control workers' exposure through protective devices such as earplugs, earmuffs, or canal caps (sometimes called "semi-aurals"). These devices can provide up to about 30 dB of attenuation, with better protection at the higher frequencies. Field testing indicates, however, that attenuation values in actual use are approximately one third of the laboratory values and that variability is considerably greater. Correct sizing, fitting, and wearing practices are crucial to the effectiveness of hearing protectors. But workers will often choose not to wear these devices because of discomfort, difficulties in hearing speech and warning sounds, and other problems.

In addition to the requirements for engineering controls, OSHA requires hearing conservation programs consisting of noise measurement, audiometric testing, hearing protective devices, employee training and education, and record keeping (29 CFR 1910.95). Workers in construction, mining, and agriculture are not covered by the hearing conservation amendment, but the first two categories of workers are covered by other, less comprehensive regulations. There is currently no noise standard for agricultural workers.

Workers' compensation claims are a minor incentive for employers to control noise. The total cost of workers' compensation claims between 1977 and 1987 for noise-induced hearing loss was estimated at $800 million. The average award per worker, however, amounted to only about $2500. Consequently, enforcement of regulations concerning noise exposure and hearing conservation programs is an important means of preventing noise-induced hearing loss.

Other Issues

Noise exposure, when combined with vibration, organic solvents, or carbon monoxide, may act synergistically to affect hearing.

There are many extra-auditory effects of exposure to noise, including impaired communication on the job, occupational stress, and increased risk of injury. Data on the relationship between noise exposure and the various extra-auditory effects are often inconclusive.

AS

Further Reading

Berger EG, Ward WD, Morrill JC, Royster LH, eds. *Noise and Hearing Conservation Manual.* 4th ed. Akron, Ohio: American Industrial Hygiene Association; 1986.

Suter AH. The development of federal noise standards and damage risk criteria. In: Lipscomb DM, ed. *Hearing Conservation in Industry, Schools, and the Military.* Boston, Mass: College Hill Press; 1988;45-66.

U.S. Department of Labor, Occupational Safety and Health Administration. Occupational noise exposure: hearing conservation amendment, final rule. *Federal Register.* 1983;46:9738-9785.

Heat-Related Disorders

ICD-9 992, E900

Identification

The heat-related illnesses can be grouped into six major types, which differ in symptoms, prognosis, and treatment: heat stroke, heat exhaustion, heat cramps, heat rash, heat syncope, and heat fatigue.

Heat Stroke. Heat stroke, or hyperthermia, is a life-threatening disorder associated with working under very hot environmental conditions. Heat stroke usually results from a failure of the body temperature-regulating system, which may result in a body temperature of 40° to 42°C (104° to 108°F). The high body temperature is usually accompanied by hot, dry skin; mental confusion; convulsions; and unconsciousness. Death frequently results; the fatality rate for heat stroke may be as high as 50%. A body core temperature above 42°C (108°F) for more than a few hours is usually fatal. Thus, heat stroke constitutes a medical emergency. Procedures to reduce the body temperature must be initiated as early as possible. Early recognition and treatment of heat stroke will decrease the risk of death or damage to the brain, liver, and kidneys. An approved first-aid method for lowering the body temperature is to remove the outer clothing, wet the skin with water, and fan vigorously. This procedure will maximize body cooling by evaporation and will prevent further body temperature increase while the patient is being transported to a hospital. Body temperature should be monitored to ensure that it is reduced but does not fall below normal.

Heat Exhaustion. Heat exhaustion results from the reduction of body water content or blood volume. The condition occurs when the amount of water lost by sweating exceeds the volume of water drunk during the heat exposure. The major signs and symptoms of heat exhaustion include fatigue; extreme weakness; nausea; headache; faintness; and a cool, pale, clammy skin. In some cases, the individual may faint. The body temperature, however, is usually normal or only slightly elevated. Treatment consists of removal to a cool area, recumbent rest, and cool fluids by mouth. Recovery usually occurs within a few hours or overnight. Generally there are no permanent after-effects. Heat exhaustion can be prevented by adequate heat acclimatization and fluid intake sufficient to replace the water lost in sweat.

Heat Cramps. Heat cramps are painful cramps of the leg, arm, or abdominal muscles. They occur when individuals take in less salt than the amount of salt lost in their sweat during hard physical work and high heat loads. The cramps may occur during work or in the evening after work. If they do not disappear spontaneously, they will disappear after salted fluid is ingested by mouth. Individuals not acclimatized to heat may require an increased salt intake, but usually a normally salted diet is adequate to prevent heat cramps.

Heat Rash. Heat rash, commonly known as "prickly heat," is often experienced in hot, humid conditions that keep the skin and clothing wet from unevaporated sweat most of the day and even at night. The extent of the condition may range from slight, involving only small areas of the skin, to extensive, involving the entire torso. When large areas of the skin are involved, sweat production is compromised, the ability to regulate body heat by sweat evaporation is reduced, and the capacity to work in the heat decreases. Even after the affected area of skin is healed, it takes 4 to 6 weeks before normal sweat production returns. If the skin can be kept clean and dry for at least 12 hours each day, severe heat rash can usually be prevented. Treatment may thus require reducing the length of time the individual is exposed to hot, humid conditions each day. Infections of the involved skin must be prevented.

Heat Syncope. Heat syncope is alarming to the patient but is the least serious of heat-induced disorders. Heat syncope is characterized by dizziness and/or fainting while standing immobile in the heat for an extended period. The condition occurs primarily in individuals who are not acclimatized to the heat, and it results from the pooling of blood in the dilated vessels of the skin and lower parts of the body with a resulting decrease in blood flow to the brain. Treatment consists of removal of the individual to a cooler area, if possible, and recumbent rest. Recovery is usually prompt and complete.

Heat Fatigue. Heat fatigue is a set of behavioral responses to acute or chronic heat exposure. The behavioral responses to acute heat exposure include impairment in (a) the performance of skilled sensorimotor tasks, (b) cognitive performance, and (c) alertness. The behavioral responses to chronic heat exposure include reductions in performance capacity, standards of performance, and concentration. These signs and symptoms of acute or chronic heat exposure reflect the discomfort, physiological strain, psychosocial stress, and perhaps hormonal changes associated with working and living in hot climates. These aspects of heat stress are not well understood or documented.

Occurrence

Heat stress is a potential safety and health hazard in many industries. An estimated 5 to 10 million workers in the U.S. are exposed for at least part of the year to hot work conditions that can seriously threaten their health. Some hot industries include mining; ferrous and nonferrous metal founding; smelting; baking; laundering; roofing and road repair; construction; hazardous waste cleanup; fire fighting; agriculture; and glass, rubber, and chemical manufacturing. These industries all have sources of heat and humidity that can compromise the worker's ability to maintain a normal body temperature. When the body temperature-regulating mechanism fails, the worker's health is seriously jeopardized.

The incidence of occupational heat-related disorders in the U.S. is not known. However, a survey of workers' compensation cases in 23 states for 1979 showed an incidence rate of 9.2 cases per 100 000 workers in the agriculture industry, 6.4 cases in construction, and 5.0 cases in mining. Other industrial divisions showed fewer than 3 incident cases per 100 000 workers. In 1986 in these states, there were a total of 3 reported occupational fatalities due to excessive heat. In California, in 1973 there were 422 cases of lost-time worker illness because of "heat and humidity," the most frequently reported physical agent; 47 workers were hospitalized and 3 died.

Causes

The causes of heat stress are metabolic and environmental. Metabolic heat load is a result of physical activity. Environmental heat load is a function of air temperature, humidity, radiant heat, and air velocity, all of which affect heat exchange and, eventually, core body temperature. Heat exchange by convection depends on air temperature and air velocity. Humidity affects heat loss by speeding or slowing the evaporation of sweat. Radiant heat exchange is dependent on the temperature of a heat source and the distance from the source.

Pathophysiology

The core body temperature for humans must be maintained at a rather constant level, even when individuals are required to perform heavy physical labor in hot environmental conditions. Some of the critical body temperatures are listed in Table 1.

Table 1.—Critical Body Temperatures

Type of Temperature	°C	°F
Normal core temperature	37.5 ± 1	99.5 ± 1.8
Skin pain threshold	45	113
Survival-limit core temperature	> 42	> 107.6
Fever	> 41	> 108.8
Heavy exercise	39	102.2

A relatively small increase in body temperature of less than 3.5°C (6.5°F) above normal can mean the difference between good health and death from hyperthermia. Maintaining an acceptable body temperature while working in hot conditions can be a major challenge.

An acceptable body temperature is achieved by balancing the amount of heat produced by the body and the amount of heat lost to or gained from the environment. The rate at which the body gains or loses heat is determined by the temperature, humidity, and movement of air and by radiation. The amount of body heat that needs to be lost is influenced by the level of physical work done. The harder one works, the more body heat one must lose to the environment.

Body heat is lost to or gained from the environment by three heat-exchange methods: convection, radiation, and evaporation of sweat. When the air temperature is lower than the skin temperature, the body will lose heat by convection. However, when the air temperature is higher than the skin temperature, the body will gain heat by convection. The rate of heat loss or gain by convection depends on the air temperature, and the rate of exchange increases with air velocities up to about 5 miles per hour.

The amount of heat the body gains or loses by radiation depends on the temperature of the objects surrounding the worker (eg, furnaces, hot or molten metals, and the walls of mines or of rooms). In some deep mines with high wall temperatures, the radiant heat load can present serious health problems.

Heat is always lost from the body when sweat evaporates from the skin. The rate of evaporative heat loss from the skin is faster when the air humidity is low and slower when it is high. Sweat production of 1 quart per hour is not unusual for strenuous work in very hot environments. This is usually sufficient to prevent overheating of the body. Increased air velocities up to about 5 mph increase the rate of evaporation and thus the rate of heat loss from the body.

The amount and type of clothing worn can drastically alter the heat exchange between the air and the body. Clothing systems that are semipermeable or impermeable to air and water vapor (eg, fire fighters' ensembles, protective clothing for hazardous waste cleanup, and reflective clothing) will interfere with the usual heat exchange mechanisms and can increase the heat load. Moreover, because chemical protective clothing and ensembles interfere with normal avenues of heat exchange, wet bulb globe temperature (WBGT) and other heat-stress indices cannot be used with such clothing to assess the heat load. Instead, the adjusted air temperature should be used to calculate the level of heat stress for workers wearing impermeable clothing systems. The water lost in the sweat can be replaced. Under extremely hot conditions, it may be necessary to use cooled garments to prevent overheating.

Acclimatization (to heat) is a gradual adjustment of the body to a prescribed stepwise increase of workload and/or exposure to workplace heat stress over several days. Without this acclimatization, a worker who is placed abruptly on full work load in a hot environment will experience such adverse health effects as fatigue, light-headedness, dizziness, nausea, distress, fainting, and even death. The first few days of working in a hot environment are most critical The major physiological benefits derived from acclimatization include a lower heart rate during work, a lower body temperature, an increase in sweat production (which increases evaporative cooling capacity), and a decrease in salt lost in the sweat (which decreases the need for extra salt intake). With these adjustments, work can be performed in the heat with no greater strain than would result from working in a cooler environment. A satisfactory daily acclimatization routine would be to limit occupational heat exposure to one third of the workday during days 1 and 2, to one half of the workday during days 3 and 4, and to two thirds of the workday during days 5 and 6. An acclimatization procedure must be repeated after days off due to illness or a return from a vacation of 1 week or more, because the acclimatized status can be lost fairly rapidly while the individual is away from the heat exposure.

Heat acclimatization does not come without effort: A person must work in the heat at the activity level required by the job. Performing sedentary work in the heat will not acclimatize an individual to hard physical labor in a hot environment. Similarly, if the level of heat stress is increased, the worker will not be acclimatized to the higher heat load. The better the worker's physical fitness, the faster the heat acclimatization process will be.

A cardinal rule is that the fluid intake during heat exposure must at least equal the amount of water lost in the sweat, which may amount to 4 to 8 quarts per day.

Age appears to reduce the capacity to work in the heat. Both acute and chronic illnesses also reduce tolerance to heat, as do many therapeutic or psychotropic drugs.

Excessive heat has an adverse affect on male fertility if the temperature of the testes is raised. Heat stress is also a risk factor for accidental injuries. Simultaneous exposure to heat and toxic chemicals may have a synergistic effect.

Table 2.—Heat Stress Control Strategies for Various Sources of Heat

Dominant Source of Heat	Control Strategy*
Body heat production during work	Decrease physical workload, increase number and length of rest periods, and use power tools and equipment.
Radiative heat exchange	Interpose line-of-sight barriers between workers and radiative heat source, insulate heat source, use heat-reflective clothing, and cover exposed body parts. Increase the distance from the source or decrease the time of exposure.
Convective heat exchange	If air temperature exceeds 35°C (95°F), reduce air temperature if possible; otherwise, reduce air movement across skin and wear clothing. If air temperature is below 35°C (95°F), increase air movement over skin and reduce amount of clothing.
Evaporative heat exchange	Decrease humidity, increase airflow over the skin, and reduce amount of clothing.

* In extremely hot environments, use auxiliary body-cooling garments and heat-reflective clothing.

Prevention

Heat-induced illnesses can be prevented through the application of heat-stress control strategies. The control strategy best suited for a particular situation depends on the dominant source of heat present: body heat production during work, or radiative, convective, or evaporative heat exchange. The major heat-stress control strategies for the various heat sources are listed in Table 2. The main underlying principle is to identify and control sources of heat.

Assessment of heat stress requires measurement of the metabolic heat load, air temperature (dry bulb temperature), humidity (natural wet bulb temperature), and radiant heat (black globe temperature). Metabolic heat is often estimated.

At present, there is no occupational health standard in the U.S. limiting exposure to heat. However, NIOSH and ACGIH have proposed recommended limits, based on the WBGT index. The NIOSH standard is a sliding scale that differs for acclimatized and unacclimatized workers. It includes an alert limit for unacclimatized workers and a ceiling limit that is not to be exceeded without the use of special heat-protective or body-cooling clothing or equipment. ACGIH has proposed a similar TLV.

Several factors are important in preventing and controlling heat-induced illnesses: heat acclimatization, water balance, obesity, age, gender, drugs, physical work capacity, electrolyte (salt) balance, health status (acute or chronic disorders), level of inherent heat tolerance, and training to cope with heat stress. Heat acclimatization was discussed under Pathophysiology above. The more critical of the remaining factors that affect prevention and control of heat stress are discussed briefly below. Dietary factors are not included because they are not known to affect heat tolerance. Except for fluid intake, which must be increased in the heat, a diet adequate for temperate conditions is also adequate for a hot environment.

The replacement of body water lost as sweat is critical to the prevention of heat-induced illnesses. To replace the 4 to 8 quarts of sweat that may be produced in hot environments, a worker will require ½ to 1 cup of water intake each 20 minutes of the workday. Because the normal thirst mechanism is not strong enough to drive a worker to drink that much water, a worker must make a conscious effort to follow a daily water intake schedule. Water that is about 13°C (55°F) is more acceptable to drink than ice water or warm water. Individual disposable cups are preferable to drinking fountains. A body-water deficit of only 2 quarts will trigger early signs and symptoms of heat-induced disorders. It is neither necessary nor advisable to increase salt intake except for treatment of heat cramps.

Other Issues

Obesity can be an important factor in inducing heat-related disorders. The additional weight of the fat will increase the energy required to perform a task and thus increase the heat produced in the body. The skin fat layer acts as insulation and reduces the rate of body heat loss. Obesity generally results in a lower physical work capacity and cardiovascular reserve, both of which are important factors in heat tolerance.

Elderly people and infants have been shown to be less heat tolerant than people in other age groups during periods of hot weather, possibly because of the incompletely developed heat regulatory mechanism in infants and the presence of debilitating disorders in elderly people. Within the usual age range for the working population, moderate levels of physical work and heat stress do not constitute a special hazard for older workers. However, physical work capacity and cardiovascular reserve do generally decrease with age, and the incidence of heat stroke in very hot work environments is higher among older workers. Consequently, older workers are at a higher risk of developing heat-induced illness in work situations involving a high level of physical work and of heat.

Many therapeutic and psychotropic drugs have adverse effects on heat tolerance. Drugs affect heat tolerance if they have anticholinergic or monoamine oxidative reactions that influence central nervous system activity or cardiovascular reserve, or if they cause body dehydration or electrolyte depletion. Excessive alcohol intake (which is often associated with heat stroke) interferes with nerve and cardiovascular functions and dehydrates the body. The ingestion of alcohol should be prohibited during work in hot environments.

AH

Further Reading

ACGIH Threshold Limit Values and Biological Exposure Indices for 1989-90. Cincinnati, Ohio: American Conference of Governmental Industrial Hygienists; 1989.

Dukes-Dobos FN. 1981 mortality and morbidity of occupational heat stress. Proceedings of the American Industrial Hygiene Association Conference; May 25-29, 1981; Portland, Ore.

Jensen RC. 1983 Worker's compensation claims attributed to heat and cold exposure. *Professional Safety.* September 1983;19-24.

Fatalities from occupational heat exposure. *MMWR.* July 20, 1984;33:410-412.

National Institute for Occupational Safety and Health. *Criteria for a Recommended Standard: Occupational Exposure to Hot Environments.* Rev. 1986. Cincinnati, Ohio: National Institute for Occupational Safety and Health; 1986. DHHS (NIOSH) publication 86-113.

National Institute for Occupational Safety and Health. *The Industrial Environment: Its Evaluation and Control.* Cincinnati, Ohio: National Institute for Occupational Safety and Health; 1973. DHEW (NIOSH) publication 74-117.

National Institute for Occupational Safety and Health. *Working in Hot Environments*. Rev. 1986. Cincinnati, Ohio: National Institute for Occupational Safety and Health; 1986. DHHS (NIOSH) publication 86-112.

Rachootin P, Olsen J. The risk of infertility and delayed conception associated with exposure in the Danish work place. *J Occup Med.* 1983;25:394-402.

Hepatic Angiosarcoma

ICD-9 155.0
SHE/O

Identification

Primary symptoms include abdominal pain, weakness, weight loss, anorexia, and abdominal swelling. Physical examination may reveal jaundice, hepatomegaly, and ascites. Elevations of serum alanine aminotransferase, bilirubin, and alkaline phosphatase are common. The diagnostic procedure of choice is angiography.

Occurrence

Hepatic angiosarcoma is a rare tumor, with an annual U.S. incidence of fewer than 1 per 100 000 population. In the past, 10% or more of all cases were due to occupational exposure to vinyl chloride monomer (VCM). Although angiosarcoma of the liver has occurred primarily among polymerization workers, cases have also been reported among polyvinyl chloride fabrication workers exposed to low levels of VCM and among residents of neighborhoods near VCM fabrication or polymerization plants. Latency has ranged from 11 to 37 years (mean 19) after vinyl chloride exposure. Hepatic angiosarcoma also has occurred in agricultural workers exposed to arsenical pesticides.

Causes

Vinyl chloride monomer, thorotrast (thorium dioxide, used in angiography and liver-spleen scans from 1930 to 1955), and organic arsenic exposure (in pesticides or medications) have been implicated.

Pathophysiology

Vinyl chloride is a mutagen in short-term assays and a known animal carcinogen. In VCM-exposed workers, histological lesions seen in the liver include (a) focal hepatocytic hyperplasia, (b) sinusoidal dilation with hyperplasia, and (c) focal mixed hyperplasia. There appears to be a progression from focal hepatocytic hyperplasia, to sinusoidal cell hyperplasia or focal mixed hyperplasia, to developing fibrosis (first parenchymal, subsequently portal). The final stage of transformation is sinusoidal cell dysplasia with malignant transformation. This tumor has developed in individuals with preexisting focal mixed dysplasia but not with focal hepatocytic hyperplasia.

Prevention

Primary prevention involves exposure control through engineering. Concerning secondary prevention, indocyanine green clearance is thought to be the best indicator of underlying liver disease that may predispose to the subsequent development of this tumor. Liver biopsies are appropriate for further diagnostic testing. Development of cellular abnormalities mandates removal from further exposure. Regarding tertiary prevention, liver transplantation has not been undertaken for this tumor.

MH

Further Reading

Neugut AI, Wylie P. Occupational cancers of the gastrointestinal tract. In: Brandt-Rauf PW, ed. *Occupational Cancer and Carcinogenesis. State of the Art Reviews in Occupational Medicine.* 1987;2:109-35.

Symposium on vinyl chloride. *Ann NY Acad Sci.* 1975;245:1.

Hepatic Porphyrias

ICD-9 277.1

Identification

Porphyria is a disturbance of porphyrin metabolism characterized by a significant increase in the formation and excretion of porphyrins or their precursors. Initial neurological symptoms include weakness, paresthesia, cogwheeling, and myotonia. Hyperpigmentation, scarring, hirsutism, and skin fragility also are present. Laboratory findings include elevated levels of uroporphyrin, penta- and hepta-carboxyporphyrin, and delta-aminolevulinic acid in the urine. Blood levels of hexachlorobenzene can be measured. Unfortunately, the agreement reached among different laboratories measuring blood levels of polychlorinated biphenyls (PCBs) is poor.

Occurrence

Epidemics of hepatic porphyria have occurred in Turkey and Louisiana from exposure to hexachlorobenzene; in Japan, Taiwan, and Michigan from exposure to PCBs; and in New Jersey from exposure to 2,4-dichlorophenol (2,4-D) and 2,4,5-trichlorophenol (2,4,5-T). Sporadic cases have also been reported.

Causes

Implicated agents include hexachlorobenzene, PCBs, and tetrachlorodibenzodioxin (TCDD). Organochlorine compounds in general should be suspected.

Pathophysiology

Organochlorine compounds interfere with porphyrin metabolism by inhibiting hepatic and erythrocyte uroporphyrinogen decarboxylase. Excretion of penta- and hepta-carboxyporphyrin is increased, as is cellular concentration of toxins. Depending on the specific cause, blood levels of hexachlorobenzene can be measured.

Prevention

Primary prevention can be achieved by preventing exposure to the causes. This can be accomplished by implementing the guidelines described in part 1. Chelation with ethylene diamine tetra acetic acid (EDTA) has been tried with variable success on people already affected.

Other Issues

Alcohol and other hepatotoxins and enzyme inducers, genetic factors, iron, and lead may worsen disease. Important for laboratory diagnosis is the fact that agreement among laboratories measuring blood levels of PCBs is poor.

MH

Further Reading

Silbergeld EK, Fowler BA. Mechanisms of chemical-induced porphyrias. *Ann NY Acad Sci.* 1987;514:1-351.

Strik JJ, Koeman JH. *Chemical Porphyria in Man.* Amsterdam, The Netherlands: Elsevier Press; 1979.

Hepatitis A Virus (HAV) Infection

ICD-9 070.0, 070.1
SHE/O

Identification

Hepatitis A virus (HAV) infection may present with or without symptoms. Symptoms commonly include jaundice, fever, loss of appetite, nausea, and fatigue. Adults are much more likely to develop severe clinical illness than children. The transition from good to ill health generally takes less than 36 hours. The time period from viral exposure to clinical illness ranges from 15 to 49 days. Measures on liver injury tests (such as alanine aminotransferase [ALT] and aspartate aminotransferase [AST]) are commonly elevated. Specific diagnosis may be made by demonstrating IgM antibody to the hepatitis A virus (anti-HAV), which persists for several months after the acute infection. The presence of IgG anti-HAV, without anti-IgM, implies past infection and immunity.

Occurrence

Cross-sectional studies have revealed that over 40% of adults in the U.S. have evidence of prior infection with HAV. Approximately 28 500 cases of hepatitis A were reported in the U.S. in 1988, representing half of all reported hepatitis cases. The actual number of cases is thought to be several times the reported number. Most cases occur among young adults. Disease tends to occur in waves, with epidemic spread. In general, acute HAV infection is not thought to lead to chronic infection, although isolated cases have been reported after epidemics among the military.

Causes

Transmission of HAV occurs via the fecal-oral route. Epidemics commonly occur in day-care centers and food preparation establishments. Shedding of the virus precedes development of clinical illness by approximately 14 days, a prolonged asymptomatic state that contributes substantially to the spread of infection.

Pathophysiology

Infection with HAV, an RNA virus of 25 to 29 nm diameter, leads to the development of liver injury. Development of specific immunity leads to clearing of infection.

Prevention

Hand washing generally reduces risk of HAV transmission, as does adequate staffing and separate space for diapering and hand washing in child-care facilities. Risk of fecal-oral transmission can be reduced by wearing gloves when working with infected patients. Immunoglobulin (Ig) administered before exposure or during the incubation period protects against clinical illness; its prophylactic effectiveness is greatest when it is administered early in the incubation period.

Preexposure prophylaxis with Ig is recommended for all susceptible travelers to developing countries; people living in such countries for long periods should receive Ig regularly. Postexposure prophylaxis with Ig is recommended (a) for household and sexual contacts of people with hepatitis A; (b) for staff and attendees of day-care centers with children in diapers if one or more children or employees are diagnosed as having hepatitis A or if cases are recognized in two or more households of center attendees; (c) for people with close contact with patients when a school- or classroom-centered outbreak occurs; (d) for residents and staff in prisons and facilities for the developmentally disabled, if any have close contact with patients with hepatitis A when outbreaks occur; and (e) for people exposed to feces of infected patients in cases of hospital outbreaks. Routine Ig administration is not indicated under the usual office or factory conditions for people exposed to a fellow worker with hepatitis A. Casual contact at work does not result in virus transmission.[1]

MH, BSL

Reference

1. Centers for Disease Control. Protection against viral hepatitis: recommendations of the immunization practices advisory committee. *MMWR.* Feb 9, 1990;39(No. RR-2):3-5.

Further Reading

Maynard J. Hepatitis A. In: Last JM, ed. *Maxcy-Rosenau: Public Health and Preventive Medicine.* Norwalk, Conn: Appleton-Century-Crofts; 1986.

Schiff L, Schiff ER. *Diseases of the Liver.* Philadelphia, Pa: JB Lippincott; 1987.

Hepatitis B Virus (HBV) Infection

ICD-9 070.2, 070.3
SHE/O

Identification

Initial presentation of hepatitis B virus (HBV) infection ranges from a completely asymptomatic state to acute liver failure. Symptoms commonly include jaundice, fever, loss of appetite, nausea, and fatigue. Acute yellow dystrophy may rapidly develop into encephalopathy with uncontrolled bleeding. The incubation period ranges from 28 to 160 days. Diagnosis is based on the demonstration of signs of liver injury (ALT, AST) and on markers of infection with the HBV: hepatitis B surface antigen

(HB_sAg) and/or IgM antibody to hepatitis B core antigen (anti-HB_c). Antibody to HB_sAg (anti-HBs) develops in people who resolve infection or who have received hepatitis B vaccine. Levels of this antibody tend to decrease over time. The hepatitis B e antigen (HB_eAg) is currently thought to be the best marker of infectivity and can be found during both acute and chronic infection.

Occurrence

Although 23 200 cases were reported in the U.S. in 1988, it is estimated that almost 300 000 new hepatitis B infections occur in the U.S. each year, despite the availability of hepatitis B vaccine since 1982. Between 5% and 10% of people who are acutely infected become chronic carriers. Up to 0.3% of blood donors in the U.S. may be chronic carriers, whereas the rate is thought to be 1% among hospital workers. The chronic-carrier state is defined by the presence of positive HB_sAg on two occasions at least 6 months apart or by the presence of HB_sAg and anti-HB_c (IgM class negative). Those at occupational risk for HBV infection include medical and dental workers, related laboratory and support personnel, and public service employees who have contact with blood, as well as staff in institutions and classrooms for people who are mentally retarded.[1] The prevalence of HB_sAg among U.S. health care workers with frequent blood contact is 1% to 2%; among health care workers with infrequent or no blood contact, the prevalence is 0.3%; and among staff of institutions for people with developmental disabilities, the prevalence is 1%. Among the general U.S. population, prevalence is 0.9% among blacks and 0.2% among whites. The cumulative risk for HBV infection among susceptible high-risk hospital workers is 15% to 30%.

Since 1985, the number of hepatitis B cases among some groups of high-risk health care workers has declined significantly, probably as a result of hepatitis B immunization.

Causes

Infection occurs when a susceptible person comes into contact with body fluids of an individual infected with HBV. All health care workers with potential exposure to body fluids are at risk, including not only dialysis nurses and phlebotomists but also other groups such as emergency medical technicians and hospital cleaning personnel. The best predictor of risk is the frequency of exposure to infected blood, taking into account frequency of needlesticks, prevalence of HBV infection in the specific patient population, and invasiveness of procedures (such as hemodialysis). Overt exposures, such as by needlesticks, account for approximately 20% of transmissions, so other factors, such as contact with nonintact skin (preexisting cuts, scratches, or dermatitis), contact with mucous membranes (oral, genital, or ocular mucosa), or contact with environmentally contaminated fomites, must be important. Potential exposure to infected body fluids—not job title, classification, or industry—is the risk factor for disease.

Hepatitis B virus is also often transmitted through sexual contact, through intravenous drug use (sharing needles), and from mother to infant at the time of birth. Household contacts of carriers are also at increased risk of becoming infected. People with HBV infection may develop coinfection or superinfection with the delta virus, an incomplete virus. The mortality and morbidity of such combined infection is greater than those of HBV alone. Risk factors for the development of the carrier state include age and immune status.

Hepatitis B virus is a major cause of chronic hepatitis, cirrhosis, and primary liver cancer worldwide. In Africa and Southeast Asia, where childhood HBV infection is common, the frequent development of cirrhosis is associated with an equally frequent development of liver cancer. Among Taiwanese civil servants with hepatitis B, the probability of developing liver cancer was shown to be two to three times higher than it was among individuals without HBV infection or cirrhosis.

Pathophysiology

Hepatitis B virus belongs to a newly defined group of DNA viruses, the Hepadna viruses, that lead to liver injury. The other members of the group are all obligate animal pathogens. The development of specific immunity, as evidenced by the development of antibodies, leads to clearing of the infection.

Prevention

Three strategies for prevention are available: preexposure prophylaxis, postexposure prophylaxis, and exposure prevention.

Two types of products are available for prophylaxis against hepatitis B. Hepatitis B vaccines provide active immunization against HBV infection, and their use is recommended for both preexposure and postexposure prophylaxis. Hepatitis B immune globulin (HBIG) provides temporary, passive protection and is indicated only in certain postexposure settings.

The CDC "Protection Against Viral Hepatitis" guidelines state the following with regard to preexposure vaccination of people with occupational risk:

HBV infection is a major infectious occupational hazard for health-care and public-safety workers. The risk of acquiring HBV infection from occupational exposures is dependent on the frequency of percutaneous and permucosal exposures to blood or blood products. . . . If . . . tasks involve contact with blood or blood-contaminated body fluids, . . . workers should be vaccinated. Vaccination should be considered for other workers depending on the nature of the task. Risks among health-care professionals vary during the training and working career of each individual but are often highest during the professional training period. For this reason, when possible, vaccination should be completed during training in health professions schools before workers have their first contact with blood. In institutions for the developmentally disabled, staff who work closely with clients should be vaccinated. Risk is associated not only with blood exposure but also may be consequent to bites and contact with skin lesions and other infective secretions. Staff in nonresidential day-care programs attended by known HBV carriers should also be considered for vaccination.[2]

Two recombinant hepatitis B vaccines are licensed in the U.S. These vaccines are safe and effective and, in those who develop adequate anti-HB_S, are virtually 100% protective against the disease. Risk factors for failure to develop an adequate antibody titer to the antigen include obesity, increasing age, and delivery of the antigen into adipose tissue rather than into muscle. Antibody status should be checked several months after vaccination; if antibody titers are insufficient, a second set of injections should be administered. Even after second attempts, not everyone develops adequate antibody levels for prevention of disease. There is no justified concern for the simultaneous transmission of other blood-borne diseases, such as human immunodeficiency virus (HIV) infections.

Postexposure prophylactic treatment to prevent hepatitis B infection should be considered in several situations, including occupational percutaneous or permucosal exposure to HBsAg-positive blood. In any exposure, a regimen combining HBIG with hepatitis B vaccine is recommended.

CDC's "Universal Body Fluid Precautions" discusses methods of prevention for all blood and body fluid-borne disorders, including HBV infection.[3] OSHA has proposed a standard that incorporates most of these guidelines. The use of gloves alone by health care workers may not prevent transmission of the disease; dentists have been shown to have persistently elevated infection rates despite the appropriate use of gloves. This may be true, at least in part, because glove trauma, with loss of barrier integrity, may be more frequent among dentists, who subject gloves to a higher degree of physical stress than do other health professionals.

Postexposure prophylaxis with HBIG and hepatitis B vaccine is generally recommended as a secondary prevention measure.

BSL, MA

References

1. U.S. Centers for Disease Control. Guidelines for prevention of transmission of human immunodeficiency virus and hepatitis B virus to health-care and public-safety workers. *MMWR.* June 1989;38:S-6.

2. U.S. Centers for Disease Control. Protection against viral hepatitis. Recommendations of the Immunization Practices Advisory Committee (ACIP). *MMWR.* 1990;39:RR-2.

3. U.S. Centers for Disease Control. Guidelines for the prevention of body fluid-borne diseases. *MMWR.* 1987;36:47-71.

Further Reading

Alter MJ, Hadler SC, Margolis HS, et al. The changing epidemiology of hepatitis B in the United States. *JAMA.* 1990;263:1218-1222.

Hadler SC. Hepatitis B virus infection and health care workers. *Vaccine.* 1990;8(suppl):S24-S28.

Maynard J. Hepatitis B. In: Last JM, ed. *Maxcy-Rosenau: Public Health and Preventive Medicine.* Norwalk, Conn: Appleton-Century-Crofts; 1986.

Schiff L, Schiff ER. *Diseases of the Liver.* Philadelphia, Pa: JB Lippincott; 1987.

Hepatitis, Non-A, Non-B

ICD-9 070.4
SHE/O

Identification

Two distinct forms of non-A, non-B hepatitis have been identified: (a) a parenterally transmitted (PT) form that is common in the U.S. and often progresses to chronic liver disease, and (b) an enterically transmitted (ET) form that is not endemic in the U.S. and has not been shown to progress to chronic liver disease. Acute symptoms of both forms are indistinguishable from hepatitis A or hepatitis B, and physical examination also cannot distinguish between the different forms of infectious hepatitis. Diagnosis of non-A, non-B hepatitis is made by serologically excluding the diagnoses of hepatitis A, hepatitis B, delta hepatitis, cytomegalovirus, Epstein-Barr virus, and other causes of liver inflammation. Each form of non-A, non-B hepatitis has characteristic pathological features on biopsy.

Occurrence

The PT form of non-A, non-B hepatitis is thought to account for 20% to 40% of all viral hepatitis in the U.S. Its epidemiological features are thought to be similar to those of hepatitis B. At high risk are transfusion recipients, parenteral drug users, and dialysis patients. Health care work associated with frequent blood contact has also ben identified as a risk factor. The role of sexual or other person-to-person contact in the transmission of this disease has not been well defined.

Causes

Recent studies indicate that most PT non-A, non-B hepatitis is caused by the hepatitis C virus (HCV). A serological test that detects antibody to this virus (anti-HCV) is commercially available. When using this assay for diagnosis, there may be a prolonged interval between onset of disease and detection of anti-HCV. In addition, some individuals with HCV infection cannot be detected by the currently available assay. When results are positive, this antibody test does not distinguish between acute and chronic infection or between ongoing infection and recovery. The ET form of non-A, non-B hepatitis appears to be caused by the hepatitis E virus. Although this virus has recently been cloned, a serological assay is not yet available.

Pathophysiology

Of most interest among those with the PT form of this disease is that chronic active hepatitis has been shown in 60% of patients biopsied, and cirrhosis has been shown in 10% to 20% of those biopsied. Among nonfulminant cases of the ET form, liver biopsies have shown cholestasis atypical of acute viral hepatitis.

Prevention

The PT form can be prevented in the workplace by taking "universal precautions" (primarily to eliminate blood contact). There are no screening tests. The role of immunoglobulins in prevention is not clear. There are no vaccines.

BSL, MA

Further Reading

Alter MJ, Hadler SC, Judson FN, et al. Risk factors for acute non-A, non-B hepatitis in the United States and association with hepatitis C infection. *JAMA*. 1990;264:2231-2235.

Alter MJ, Purcell RH, Shih JW, et al. Detection of antibody to hepatitis C virus in prospectively followed transfusion recipients with acute and chronic non-A, non-B hepatitis. *N Engl J Med*. 1989;321:1494-1500.

Reyes G, Purdy MA, Kim JP, et al. Isolation of a cDNA from the virus responsible for enterically transmitted non-A, non-B hepatitis. *Science*. 1990;247:1335-1339.

Hepatitis, Toxic

ICD-9 570, 573.3
SHE/O

Identification

Toxic hepatitis may present in a continuum of forms—arbitrarily defined as acute, subacute, or chronic disease—and in icteric (with jaundice) or anicteric (without jaundice) forms. Symptoms may be nonspecific, such as fatigue and irritability, or referrable to the gastrointestinal tract (anorexia, nausea, right upper-quadrant pain, and jaundice). Acute disease, generally after short-term, severe exposure to hepatotoxins, develops within 24 to 48 hours and reaches maximum severity within 36 to 60 hours. Subacute disease results from repeated exposure to smaller doses of toxic agents and is associated with mild to minimal symptoms that are often only intermittent. Chronic injury may first present with organ failure, with either bleeding from esophageal varices or encephalopathy. Results of liver function tests, such as aspartate aminotransferase (AST) and alanine aminotransferase (ALT), are frequently abnormal. The more acute the presentation, the greater the likelihood of abnormal results on liver function tests. Dye clearance tests, such as indocyanine green clearance, may yield abnormal results in up to 95% of individuals with any of the three forms. Where exposure to a known hepatotoxin is identified, fine-needle liver biopsy is the appropriate diagnostic procedure. Where no known hepatotoxin can be identified but disease appears associated with an exposure, a controlled challenge with the suspected offending agent is sometimes performed, although there are attendant risks of massive necrosis. Except in rare instances, this is not a recommended procedure. The fully informed patient must weigh the benefits of specific diagnosis against the risks of this diagnostic procedure.

Occurrence

No incidence figures are available for either occupational or nonoccupational agents.

Causes

Several forms of hepatotoxicity have been described. In almost all individuals, when exposure is severe enough, agents with "predictable dose- and time-dependent" toxicity lead to disease—generally zonal hepatocellular alterations without inflammation. These agents are classified as Type I agents. Type II agents lead to hepatic injury in only a small percentage of the exposed population without dose- or time-dependent toxicity.

Pathophysiology

Causative mechanisms are difficult to distinguish from effects of injury. Changes noted include lipid accumulation as well as decreased protein synthesis and lipid peroxidation. Morphological patterns may be cholestatic or necrotic with inflammation. A variety of agents and situations may potentiate hepatotoxicity; these include alcohol, agents without intrinsic hepatotoxicity such as acetone, and glutathione deficiency. There is some suggestion that alpha-1-antitrypsin deficiency, even with haplotypes other than Z, leads to increased susceptibility for hepatotoxicity after exposure to toxic agents.

Prevention

Engineering controls may reduce exposure to hepatotoxins to an acceptable concentration. Dermal exposure may be prevented through the appropriate use of barriers, such as gloves. Type II hepatotoxicity may be preventable only through reduction in use of potential hepatotoxins. Exposure to known hepatotoxins should lead to the introduction of screening programs for the detection of subclinical injury. This is best achieved by first using a very sensitive test and then using a very specific test. For example, the use of a test with high sensitivity and low specificity will characterize the largest percentage of individuals as abnormal; the subsequent use of a specific test will identify a large number of false-positive results. This strategy will reduce both false-positive and false-negative results.

Liver injury and function tests have varying degrees of sensitivity and specificity, depending on the acuteness or chronicity and the severity of the injury as well as on the nature of the toxin. Many physicians, when they are seriously concerned about the presence of hepatotoxins, use at least one liver injury test (such as ALT or AST) and one liver function test (such as serum bile acids or indocyanine green clearance).

No treatment for hepatotoxicity has been demonstrated, although experimental evidence does suggest the usefulness of administering acetylcysteine within hours after ingestion of acetaminophen. No other agents are considered useful. Liver transplantation, though still considered by some to be an experimental procedure, is associated with an 80% 3-year survival rate in individuals surviving the initial transplantation.

MH

Further Reading

Farber E, Fisher MM. *Toxic Injury of the Liver* (parts A and B). New York, NY: Marcel Dekker; 1979, 1980.

Tamburro CH. Chemical hepatitis. *Med Clin North Am.* 1979;63:545-566.

Zimmerman HJ. *Hepatotoxicity.* New York, NY: Appleton-Century-Crofts; 1979.

High Blood Pressure—*See* Hypertension

Histoplasmosis

ICD-9 115
SHE/O

Identification

Histoplasmosis is primarily a pulmonary disease that may at times disseminate widely throughout the body. It is caused by the dimorphic fungus *Histoplasma capsulatum*, which grows as mycelia (mold) in soil and as a yeast in living tissue.

Occurrence

Histoplasmosis may occur in many areas of the world. The fungus lives in soil that has been enriched by bird or bat droppings, especially in temperate climates with high humidity. In the U.S., histoplasmosis is endemic in the Ohio and Mississippi River Valley areas, but cases and epidemics have occurred throughout the eastern half of the country. Infection occurs when the soil containing *Histoplasma* becomes dry and is disturbed, thus raising dust containing fungal elements. It is estimated that 500 000 infections occur yearly in the U.S. alone.

Workers with increased risk of exposure to *Histoplasma* include agricultural workers, bulldozer operators, construction workers, demolition crews, field equipment users and maintenance crews, forest workers, laboratory technicians, linemen, and railway or road clearing/maintenance crews. Equipment or materials containing contaminated soil may introduce the fungus into an unsuspected environment.

Histoplasmosis is widely distributed in temperate zones but is most heavily concentrated in the U.S. Focal infections are common over wide areas of the Americas, Europe, Africa, eastern Asia, and Australia; clinical disease is far less frequent, and severe progressive disease is rare. In endemic areas, primary infection develops early and equally in both sexes, increasing from childhood to 30 years of age; differences by sex are not usually observed with the exception of the chronic pulmonary form, which is more common among middle-aged men, and in occupationally exposed cases where differences by sex are related to patterns of employment.

Prevalence of histoplasmosis ranges from 15% to 20% in the U.S. overall and increases to over 90% in endemic areas such as Arkansas, Kentucky, Missouri, Tennessee, Illinois, Indiana, Ohio, Oklahoma, Alabama, Kansas, Louisiana, Maryland, West Virginia, Mississippi, and Texas. Half a million new infections occur each year in the U.S., with peak incidence in areas of the Ohio, Missouri, and Mississippi River valleys. The percentage of a population with positive skin tests for histoplasmosis is greatest among farm dwellers, followed by rural dwellers, and, lastly, urban dwellers. Over 95% of individuals who have been infected with *H. capsulatum* can recall no clinically distinctive illness and remain free of complications of the disease. Approximately 15 000 people per year are hospitalized with histoplasmosis in the U.S.

For 1976, the overall case-fatality rate was 2.9%. Mortality in the U.S. is approximately 150 deaths per year; 90% of individuals with primary pulmonary histoplasmosis recover. Occasionally histoplasmosis develops into adult respiratory disease syndrome, in which death may follow quickly unless an early diagnosis is made and treatment with amphotericin B is rendered. Untreated, the chronic cavitary form of histoplasmosis results in progressive pulmonary disability and death in 50% of affected individuals within 5 years.

Outbreaks have occurred in families or in groups of workers with common exposure to bird or bat droppings or to recently disturbed contaminated soil. Workers at risk for developing histoplasmosis are those who come into close contact with soil, particularly soil enriched with avian and bat feces. Such workers include farmers cleaning chicken coops, bat-infested lofts, and pigeon roosts; construction workers and other workers involved in earth-moving operations; workers involved in road construction, bridge scraping, tree cleaning, landscaping, or grave digging; and workers involved in cleaning or dismantling contaminated buildings.

Cases of histoplasmosis have been reported in cave explorers and in excavators as well. Histoplasmosis also occurs in dogs, cats, rats, opossums, skunks, foxes, and other animals. There is no evidence of animal-to-human transmission nor of human-to-human transmission.

Causes

Infection occurs by inhalation of dust containing spores or hyphal fragments of the fungus. Thus, people who work in dusty environments that have been inhabited by birds or bats are at increased risk of infection. Histoplasma has been isolated in residential and commercial buildings, storm cellars, belfries, chicken houses, barns, and other structures in which birds or bats have been long-term occupants. Cases result from work in these areas that creates airborne dust. Epidemics have occurred in people working in, or passing through, bat caves and bird roost sites. Living or working in proximity to a bird roost site containing *Histoplasma* increases the risk of infection.

Pathophysiology

Infection occurs by inhalation of dust containing the fungus. Fortunately, in most people the resultant infection is asymptomatic, resulting only in conversion of the skin test and, usually, in focal parenchymal and hilar calcifications.

Disease may take several forms. People who inhale large numbers of fungal elements within a relatively short time may develop "epidemic histoplasmosis," an acute, usually self-limiting, febrile illness with cough and chest pain. The chest x-ray has a characteristic pattern of diffuse lesions resembling a snowstorm. Acute pulmonary disease may have symptoms similar to those of the common cold or a "flu-like" illness that may linger for weeks to months. Chest films may show infiltrative-type lesions. This form is also usually self-limiting. The chronic pulmonary disease is clinically and radiologically indistinguishable from that of pulmonary tuberculosis. It often follows a slowly progressive pattern with productive cough, malaise, and weight loss. Radiologically, the disease is apical in location and often bilateral. Cavitation is common. Treatment is necessary; a 10% to 13% relapse rate is not unusual.

In the very young, the aged, or immunocompromised individuals, the fungus may spread to virtually any organ of the body. Without therapy, this disseminated form is rapidly fatal. Signs and symptoms may be quite varied and reflect the various organ systems involved. The classic pattern is fever, hepatosplenomegaly, anemia, and wasting.

Amphotericin B has been the therapy of choice. Ketoconizole and related compounds have shown promise in treatment.

Prevention

Before beginning any activity that might create dust, it is important to identify potentially hazardous areas that may harbor the fungus. Soil and/or accumulated droppings should be submitted for culture. If found to contain *H. capsulatum*, the area should be decontaminated with 5% formalin. In large open areas, an alternative is to apply 2 to 3 feet of fill dirt over the contaminated area. The fungus appears unable to reach the surface, at least within several years, unless bird or bat droppings continue to accumulate. In some instances, however, these methods are not feasible.

Saturating the area with water to avoid dust is helpful. Masks and protective clothing that can be decontaminated after use should be required. Equipment or materials that have been used in contaminated areas may become covered with dust or dirt containing the fungus. If such equipment or materials are used or repaired in other areas, workers at the new location are at risk of infection. When there is work in an environment that is known to be or suspected of being contaminated, it should be kept in mind that individuals already exhibiting a positive skin test reaction to histoplasmin are presumed to have a greater resistance to infection than individuals with a negative response.

EC

Further Reading

Chick EW. Epidemiologic aspects of the pulmonary mycoses. *Semin Respir Med.* 1987;9:123-129.

Chick EW, Compton SB, Pass T, et al. Hitchcock's Birds, or the increased rate of exposure to histoplasmosis from blackbird roost sites. *Chest.* 1981;80:434-438.

George RB, Burford JG. Histoplasmosis. In: Einstein HE, ed. *Handbook on Fungal Diseases.* Park Ridge, Ill: American College of Chest Physicians; 1981:34-42.

Human Immunodeficiency Virus (HIV) Infection

ICD-9 042
SHE/O

Acquired Immunodeficiency Syndrome [AIDS]

Identification

Acquired immunodeficiency syndrome (AIDS) is characterized by a profound immunosuppression due predominantly to a depletion of the T4 helper/inducer subset of T lymphocytes. This syndrome represents the late clinical stage of infection with the human immunodeficiency virus (HIV), which most often results in progressive damage to the immune and other organ systems.[1] Within a few days after being infected with HIV, some people experience a mild, viral-like illness, with fever and lethargy. Infected people begin to produce anti-HIV antibodies within 3 months after onset of infection. Following this transient illness, an infected person can remain free of symptoms for an extended period. During this time, the virus continues to replicate and slowly destroys T4 lymphocytes, eventually causing a severe immunosuppression that results in the development of opportunistic infections and certain cancers. Onset may be insidious, with lymphadenopathy, anorexia, chronic diarrhea, weight loss, fever, chronic weakness, or disturbed cognitive or motor function. Individuals often develop fungal, bacterial, or viral infections of the skin and mucous membranes.

AIDS per se is characterized by severe, life-threatening, opportunistic disease. A diagnosis of AIDS is based on consideration of anti-HIV antibody status, specific clinical signs, opportunistic infections, and/or malignancies, which differ for different age groups. Common conditions include *Pneumocystis carinii* pneumonia, toxoplasmosis, multiple gastrointestinal infections, Kaposi's sarcoma, and certain types of lymphomas. For surveillance purposes, CDC published a revised case definition for AIDS in 1989.[2] AIDS-related complex (ARC) refers to conditions caused by HIV that otherwise do not fit the formal case definition. As the number of women infected with HIV grows, the list of AIDS-defining opportunistic diseases may be extended to include certain infections and cancers of the female reproductive system.

Occurrence

HIV infection occurs worldwide. Affected populations and incidence rates vary within and among countries. In the U.S., the number of people infected with HIV is estimated to be between 800 000 and 1.2 million. From 1981, when AIDS was first reported to CDC, through November 1990, more than 157 000 persons have been reported with AIDS. Among adults, 59% of the cases have occurred in homosexual or bisexual men, 22% in intravenous (IV) drug users, 7% in homosexual men who are IV drug users, 5% in heterosexuals, 2% in people receiving transfusions/blood components, 1% in individuals with hemophilia/coagulation disorder, and 3% in people with undetermined exposures. To date, 62% of all known AIDS patients in the U.S. have died. The case-fatality rate, over several years, may well approach 100%.

The prevalence of HIV infection is higher among blacks and Hispanics than among whites and, in the U.S., higher among white men than among white women. HIV infection is significantly more common among black and Hispanic children than among white children. The number of HIV-infected women and infants is increasing due largely to IV drug use and sexual activity with infected IV drug users. In some countries (primarily in sub-Saharan Africa), the prevalence of HIV infection is approximately equal between men and women.

Although AIDS patients have been reported in all states, approximately two thirds of them live in five states (22% in New York, 19% in California, 9% in Florida, 7% in Texas, and 7% in New Jersey), with cases predominating in large urban centers. The HIV epidemic, which for many years had been concentrated in these urban areas, is currently increasing in other urban centers as well as in rural areas of the South and Midwest.

Causes

The cause of this infection is the human immunodeficiency virus, a member of the lentivirus family of retroviruses. Transmission of HIV can occur through any of three routes: (a) heterosexual or homosexual contact with an HIV-infected person, (b) parenteral exposure to infected blood or blood products (such as by IV injection using a needle contaminated with HIV), or (c) perinatally, from an HIV-infected mother to her fetus or infant. Although HIV has been isolated from a number of body fluids, only blood, semen, vaginal secretions, and breast milk have been implicated in its transmission. There is no evidence of its transmission through food, insects or other vectors, or casual contact. Studies of more than 700 household contacts of more than 200 infected individuals have revealed that HIV did not spread to any household contact who was not the sexual partner of or an infant born to the HIV-infected person. It is presumed that HIV-infected individuals are able to transmit the virus at any time after infection has begun.

Occupational risk of infection results primarily from exposure to (a) infected blood or blood products, (b) fluids contaminated with infected blood or other body fluids, or (c) concentrated preparations of HIV in laboratory or commercial settings. Reports of HIV infection following percutaneous exposures to blood or body fluids containing blood from HIV-infected patients have clearly demonstrated that there is a risk for occupational transmission of HIV in health care and laboratory settings. Fourteen prospective studies have been published; these include 1962 health care workers who sustained percutaneous exposure to HIV-infected blood. Six seroconversions were documented in these studies for a 0.3% rate of transmission following a single percutaneous exposure. In these studies, no seroconversions were found in health care workers who had mucous membrane exposures to HIV-infected blood or exposure to fluids other than blood. In addition to health care workers enrolled in longitudinal studies, there have been further case reports of HIV infection following needlestick or mucous membrane exposure. Two laboratory workers in facilities that commercially produce high volumes of HIV have reportedly been infected with HIV. At least four of the known occupationally related HIV infections occurred during emergencies or in outpatient settings. Depending on the geographical area, health care workers and other public safety or emergency service workers, such as fire and police department personnel, may be exposed to a significant number of HIV-infected people at work. One study of HIV infection in emergency department patients in an inner-city hospital revealed that 119 of 2302 consecutive adult patients (5.2%) were seropositive for HIV. Of these HIV-infected individuals, 92 (77%) had unrecognized HIV infection.

It is likely that the number of exposures will increase as the number of HIV-infected people continues to rise. These data underscore the need for universal precautions to prevent HIV infection in health care settings (*see* Prevention below).

Pathophysiology

Regardless of the route of HIV transmission, the virus ultimately infects certain cells of the immune system and the central nervous system, predominantly T4 lymphocytes and cells of the monocyte/macrophage lineage that express the CD4 molecule on their surface. HIV binds to the CD4 receptor and is brought into the cell. Once inside the cell, the viral genomic RNA is transcribed into DNA by the viral reverse transcriptase and is integrated as a provirus into the host cell genome. Once integrated, the provirus may remain dormant for extended periods. Upon activation of the cell, HIV is stimulated to replicate and viral particles are released by the infected cell. Although there is a dramatic decline in the numbers of T4 lymphocytes in people with AIDS, the exact mechanisms of HIV-induced cell killing in vivo are not known.

The duration from start of HIV infection to onset of AIDS symptoms ranges from a minimum of 4 months (for transfusion-associated infection) to a maximum of 12 years. The median incubation period is estimated to be 10 to 11 years.

Neurological abnormalities occur to varying degrees in approximately 60% of AIDS patients. The virus is thought to be carried into the brain by infected monocyte/macrophages that may release products that are toxic to neurons or that lead to infiltration of the brain with inflammatory cells.

Prevention

Preventive measures are designed to minimize or eliminate contact with body fluids that are contaminated with the virus. Preventive measures for nonoccupational exposure have been published elsewhere.[1] In occupations that involve contact with blood or body fluids (such as health care workers, public safety workers, fire and police department personnel, other emergency service workers, and some sanitation workers), precautions must be taken to minimize contact with such fluids. NIOSH has published guidelines for the education of workers and the provision of safety equipment.[3]

Because HIV infection cannot always be precisely identified and because other blood-borne infections, such as hepatitis B, also represent a risk to health care workers, CDC recommends that blood and body fluid precautions be used consistently for *all* patients. These universal precautions, which are designed to decrease the occupational risk of all blood-borne infections in the health care setting, include the following recommendations:

1. Wash hands or other skin surfaces immediately before and after patient contact, following contamination with blood or other body fluids, and after removing gloves.

2. Use appropriate barrier mechanisms (gloves, gowns, masks, and protective eyewear) to prevent skin and mucous membrane exposures to blood; to body fluids containing blood; to semen; to vaginal secretions; to tissues; and to cerebrospinal, synovial, pleural, peritoneal, and amniotic fluids. Although universal precautions do not apply to breast milk, gloves may be worn by health care workers who are frequently exposed to breast milk, such as in breast-milk banking. Similarly, although saliva has not been implicated in HIV transmission, special precautions are recommended for dentistry, because both contamination of saliva with blood and trauma to health care workers' hands are common during dental procedures.

3. Handle needles and syringes with caution. Needles should not be recapped, bent, or broken, but should be placed in a puncture-resistant container designed for disposal.

4. To minimize the need for emergency mouth-to-mouth resuscitation, mouthpieces or other ventilation

devices, such as Ambu bags, should be available for use where the need for resuscitation is predictable.

5. Health care workers with exudative lesions or weeping dermatitis should refrain from all direct patient care and from handling patient-care equipment until the condition resolves.

Because antibodies to HIV may not develop for several months following infection, mandatory HIV testing of patients will not identify all HIV-infected patients and may create a false sense of security in health care workers. Universal precautions remain the preventive measure of choice to avoid blood-borne pathogens in the health care setting.

More specific precautions have been published by CDC for morticians, dialysis and laboratory personnel, individuals performing invasive procedures, dental care workers, fire and emergency service workers, and law enforcement and correctional workers. Facilities involved in producing large volumes of HIV or highly concentrated virus are subject to much more stringent biosafety regulations. Individuals working in such facilities should be acutely aware of the attendant guidelines for such activities, which are available from CDC[4] or from state health departments. OSHA has also proposed a regulation covering workers under its jurisdiction to prevent infection by HIV, hepatitis B virus, and other blood-borne infectious pathogens.[5]

All health care workers, laboratory workers who handle HIV, and emergency and sanitation workers with potential contact with contaminated body fluids should be educated about the possibility and risks of infection and about appropriate precautions. Such education should follow the guidelines outlined above.

There currently exists a wide range of opinion regarding the use of zidovudine (AZT) as a postexposure prophylaxis following an occupational exposure to HIV. The U.S. Public Health Service has issued a statement concerning this use, which recognizes that some physicians and institutions have offered AZT to workers following an occupational exposure to HIV. However, limitations of current knowledge preclude a recommendation at this time.[6]

Counseling people who have been exposed to HIV, occupationally or otherwise, is essential. The purposes of counseling are to inform individuals about HIV transmission, to initiate and sustain changes in behavior to reduce the risk of infecting others, and to refer people to treatment facilities.

To enroll an exposed worker in the CDC prospective surveillance system, telephone (404) 639-1644. To enroll a massively exposed worker in the National Institute for Allergy and Infectious Disease study of postexposure prophylaxis with zidovudine, telephone (800) 537-9978.

ZFR, DJH

References

1. Benenson AS, ed. *Control of Communicable Diseases in Man.* 15th ed. Washington, DC: American Public Health Association; 1990.
2. U.S. Public Health Service, Centers for Disease Control. Guidelines for prevention of transmission of human immunodeficiency virus and hepatitis B virus to health-care and public-safety workers. *MMWR.* June 23, 1989;38(S-6).
3. National Institute for Occupational Safety and Health. Guidelines for prevention of transmission of human immunodeficiency virus and hepatitis B virus to health-care and public-safety workers. Washington, DC: U.S. Government Printing Office; 1989. DHHS (NIOSH) publication no. 89-107.
4. U.S. Public Health Service, Centers for Disease Control. 1988 Agent summary statement for human immunodeficiency virus and report on laboratory-acquired infection with human immunodeficiency virus. *MMWR.* April 1, 1988;37(S-4).

5. U.S. Department of Labor, Occupational Safety and Health Administration. Proposed rulemaking. Occupational exposure to bloodborne pathogens. 54 FR 23042-139, May 30, 1989.

6. U.S. Public Health Service, Centers for Disease Control. Public Health Service statement on management of occupational exposure to human immunodeficiency virus, including considerations regarding zidovudine postexposure use. *MMWR.* January 26, 1990;39(RR-1).

Further Reading

Curran JW, Jaffe HW, Hardy AM, Morgan WM, Selik RM, Dondero TJ. Epidemiology of HIV infection and AIDS in the United States. *Science.* 1988;239:610-616.

Fauci AS. The human immunodeficiency virus: infectivity and methods of pathogenesis. *Science.* 1988;239:617-622.

U.S. Public Health Service, Centers for Disease Control. Revision of the CDC surveillance case definition for acquired immunodeficiency syndrome. *MMWR.* August 14, 1987;36(S-1).

U.S. Public Health Service, Centers for Disease Control. Human immunodeficiency virus infection in the United States: a review of current knowledge. *MMWR.* December 18, 1987;36(S-6).

Homicide and Assault

ICD-9 E960-E969

Identification

Work-related homicide results from an assault to which the victim was exposed because of his or her work. Injury is usually inflicted by firearms or other weapons. Identifying the homicide as being work related and performing the pathological examination (autopsy) are mostly the responsibility of the coroner's (or medical examiner's) office.

Occurrence

Average annual incidence of work-related homicide is approximately 2 per 100 000 working people. The rate in males is about four times higher than in females. There appear to be no appreciable differences in incidence by age between the ages of 16 and 64. For males, work-related homicides are most frequent in the retail trades, personal services, and public administration industries. Rates are also high in the business and repair industries. For females, rates are highest in the personal service industries. Work-related homicide rates for males are highest in service occupations—specifically, police; taxi drivers; security guards; and supervisors, proprietors, and sales personnel in food and beverage industries. Among females, waitresses are at highest risk.

Causes

Firearms account for more than three fourths of all work-related homicides, proportions being slightly higher for males than for females. Over 90% of supervisors or proprietors of food and dairy stores, convenience stores, liquor stores, or eating and drinking places who were murdered on the job died from gunshot wounds. Risk factors include working in neighborhoods or areas with high crime rates, working alone or being otherwise isolated, and working in the presence of cash or other valuables. The peak hour of fatal assault is 11:00 pm to midnight, and two thirds of all homicides with known injury times occurred between 3:00 pm and 3:00 am. About one fourth of homicides with known injury time were among occupations having frequent public contact: supervisors or proprietors; waiters, bartenders, and waitresses; sales clerks; taxi drivers; and police.

Pathophysiology

The damage in fatal work-related assaultive injuries is highly variable but lethal, including the crushing and destruction of superficial tissues and internal organs. All body parts can be involved, but the most lethal among such injuries are those involving damage to the brain and thoracic organs. Death may be immediate or delayed, depending on the degree of damage and the victim's constitutional factors. Information on nonfatal assaultive work-related injuries is incomplete, so a comparison of survival rates by type of body damage cannot be made.

Prevention

The epidemiology of fatal work-related assaultive injuries suggests that a number of preventive strategies may be available. Inasmuch as OSHA does not recognize homicide in the workplace as a work-related injury, efforts must be made to influence OSHA to develop protective standards. With the available data, however, it appears that several countermeasures can be instituted, including efforts to control crime. Controlled access to firearms of all types is one such countermeasure. Among taxi drivers, attempts to isolate the driver physically from the passenger or to adopt policies against picking up passengers in suspected high-crime areas or during late-night hours may be effective. Cashiers or others handling money in 24-hour stores, gas stations, liquor stores, and the like should be provided with bulletproof barriers. Reducing the amount of money exchanged or available and placing the cashier's counter in a highly visible location also may reduce risk.

Other Issues

Special studies are needed to determine the nature of the exposures and other possibilities for countermeasures. Some employers have been prosecuted for homicide when their gross negligence caused the death of an employee.

JFK

Further Reading

Davis H. Workplace homicides in Texas males. *Am J Public Health*. 1987;77:1290-1293.

Hales T, Seligman P, Newman S, Timbrook C. Occupational injuries due to violence. *J Occup Med*. 1988;30:483-487.

Kraus J. Homicide while at work: persons, industries, and occupations at high risk. *Am J Public Health*. 1987;77:1285-1289.

Hyperbaric Injury

ICD-9 E902.9, 993

Identification

Hyperbaric injuries and diseases result from exposure to increased atmospheric pressure. Such pressure occurs during underwater diving, in pressurized underground or underwater caissons, or therapeutically in clinical hyperbaric chambers. Increased atmospheric pressure is used in caissons, sewers, and other tunnel work to purge ground water from the worksite. Diving exposures may include scientific research and oceanography, commercial oil field construction, repair work at sea, and underwater salvage.

Disorders include barotrauma, inert gas narcosis (commonly referred to as "nitrogen narcosis," "the rapture of the deep," or, in the United Kingdom, "the narks") or high-pressure nerve syndrome (only occurs during very deep diving), and decompression sickness ("the bends" or caisson worker's disease). Noise-induced hearing loss and hypothermia are also associated with work under increased pressure or underwater.

Barotrauma is traumatic tissue damage resulting either from bodily compression or from decompression. Tissue itself, like water, is relatively incompressible and resistant to injury. Air, however, is highly compressible. Thus, barotrauma occurs along the boundaries of air-containing structures. The external ear canal, middle ear space, and sinuses, as well as potential spaces, such as dental fillings, lungs, or intestines, may all be adversely affected by changes in pressure, resulting in loss of hearing, pain, and other forms of traumatic injury.

The most dramatic form of barotrauma is air embolism, in which air, nitrogen, or another gas is introduced into the vascular tree, usually via the pulmonary circulation. Bubbles may migrate to the brain and precipitate acute, severe neurological injury, resulting in death. As little as 0.4 cc of blood air foam delivered to the brain stem may be fatal if respiratory control centers are damaged.

Inert gas narcosis results from the toxic effects of inert gases on neural cells. Because pure oxygen under pressure is toxic to lungs (where it may cause pneumonitis progressing to fibrosis) and to the central nervous system, either compressed air or oxygen diluted with other inert gases (eg, helium, neon) is used during diving or other work in a hyperbaric environment. In tunnels and caisson work, where pressure rarely exceeds 3 atmospheres, compressed air (composed of 20% oxygen, 79% nitrogen, and 1% other gases) is almost always used. At higher pressures or during dives deeper than 30 meters and for longer than 3 hours, nitrogen narcosis may occur. This syndrome is similar to acute ethyl alcohol intoxication; symptoms include numbness, loss of coordination, and mental confusion. Effects are short-term and apparently reversible. Mental confusion while under water, however, could obviously be disastrous. At 100 meters or more, divers may lose consciousness from nitrogen narcosis.

For dives deeper than about 60 meters, inert gases other than nitrogen are used—primarily helium, but there have been experiments with other mixtures. For even deeper dives, however, a high-pressure nerve syndrome results from the effects of high pressure, aggravated by too rapid a descent. This syndrome can include nausea, vomiting, uncontrollable shaking, and loss of coordination.

Decompression sickness includes both acute effects, which can result in long-term disability, and chronic effects, primarily aseptic osteonecrosis. Decompression sickness, commonly referred to as "the bends," results from too rapid decompression from a hyperbaric environment to sea level. Decompression sickness can also result from ascent to heights, as in mountainous regions or during air travel in unpressurized aircraft. Symptoms of decompression sickness include skin itching, vague to extreme joint pain, vertigo, breathing difficulty, blindness, paralysis, and convulsions leading to death. Onset can be insidious or rapid and can occur during decompression or up to 24 hours later.

Aseptic osteonecrosis (sometimes referred to as dysbarism-related osteonecrosis) may result from a single episode or from repeated episodes of inadequate decompression. The shoulders, the femoral heads, the femur or humerus, and, rarely, the knees may be affected. Occurrence is usually bilateral and at several anatomic sites. Nitrogen bubbles may result in an inadequate blood supply, which can cause gradual destruction of bone and subsequent loss of cartilage and joint function, leading to lifelong disability.

Occurrence

Occurrence of hyperbaric disorders is highly variable by geographic location. In general, it is more common among tunnel workers than among divers. In 1986, there were 507 compensated injuries in 23 states in the U.S due to changes in atmospheric pressure. Most (89%) occurred in the transportation and public utilities industries, including sewer services.[1]

In a group of Milwaukee sewer workers studied in 1974, 35% had osteonecrosis. In a group of Gulf Coast commercial divers, the rate of positive x-ray findings for osteonecrosis was 27% in 1972. More recently, the rate of osteonecrosis was reported as 17% among compressed air workers and 4.2% among commercial divers. Prevalence increased with age, work experience, increased pressure (or depth), and with the number of acute episodes of decompression sickness.[2]

Causes

Hyperbaric disorders are caused by increased pressure, excessively rapid decompression, and toxic effects of inert gases. Decompression-related disorders can result from excessive pressure, rapid surfacing following an underwater dive, exposure to decreased atmospheric pressure (eg, in an airplane, in a mountainous region, or otherwise at increased elevation), or too rapid decompression from a hyperbaric worksite (tunnel or caisson) or chamber.

Pathophysiology

At sea level, atmospheric pressure is 14.7 pounds per square inch (psi) or 1 kilogram per square centimeter (kg/cm^2), also referred to as one atmosphere (ATA). When a diver descends underwater, the surrounding pressure on the diver is directly proportional to depth and increases by one ATA every 10 meters. Thus, the total pressure on a diver at 10 meters is 2 ATA, at 20 meters it is 3 ATA, and so on. Depth gauges used by divers in the U.S. and U.K. usually read "fsw," or feet of sea water, which is easily converted into pressure measurements.

As pressure increases, the volume of a fixed quantity of gas decreases, according to Boyle's law. At 1 ATA, 1000 cc of gas is compressed to 500 cc at 2 ATA, and 250 cc at 4 ATA. Thus, changes in volume are not linear with depth. The largest changes in gas volume occur during the initial stages of a dive or pressurization—during the first 10 meters. During descent or pressurization, real or potential air spaces within the body must be able to reach equilibrium. Injury occurs when the body's air spaces fail to reach equilibrium during pressurization, resulting in relative vacuums within these air spaces. During decompression, the inability of the air to escape may result in damage from overdistension.

Any potential or real air space may be subject to changes in pressure. Thus, conjunctival hemorrhage may occur from failure to ventilate a sealed face mask, skin injuries may result from sealed dry suits, and intestinal distension or perforation or a hernia may result from swallowing air. Air in the bowels may be sufficient to limit lung expansion.

Rarely, dental pain results from improperly filled teeth where the amalgam incompletely fills the cavity socket. During pressurization, the nerve root and pulp expand into this space, resulting in severe pain and, occasionally, loss of an amalgam on descent, sometimes with explosive force.

Changes in pressure may cause ear damage (ear squeezes). External ear canal injury is usually the result of air being trapped in the canal by a tight-fitting hood, foreign body, or, rarely, tortuosity and blockage by osteomas. During pressurization, the relative vacuum in the external ear canal may precipitate pain or injure the tympanic membrane.

The most common form of ear squeeze occurs in the middle ear, which is normally ventilated by the Eustachian tube. During pressurization, a diver usually tries to exhale against a closed glottis (Valsalva's maneuver) to force air into the middle ear. Failure to do so may result in implosion of the tympanic membrane (with consequent hearing loss) or hemorrhage into the middle ear with severe pain and, rarely, disruption of the middle ear bones.

Round window blowout is a more profound form of pressure-related ear injury and may occur when a pressurized worker unsuccessfully attempts the Valsalva maneuver. Pressure may be transmitted to the round window of the middle ear, causing rupture with sudden roaring (tinnitus), deafness, and occasionally severe dizziness. This can be a permanent injury.

Alternobaric vertigo, or dizziness secondary to pressure change, is almost always associated with decreases in pressure and is usually transient. It results from unequal clearing of the middle ears, which the vestibular apparatus interprets as spin, resulting in severe dizziness, disorientation, or panic.

The sinuses likewise must be ventilated during changes in pressure. Inability to ventilate can result in hemorrhage, severe localized pain, and subsequent chronic sinus dysfunction. Chronic sinusitis, sinus infection, or allergies may increase the risk of sinus injury.

The pathophysiology of inert gas narcosis is poorly understood but probably results from toxic effects on nerve cells. These toxic effects are apparently a function of the affinity of inert gases for fat. Thus, the relative potential for inert gases to cause narcosis can be ranked according to their fat affinity: helium, neon, hydrogen, nitrogen, argon, krypton, and xenon.

Decompression sickness results from inert gas that has been dissolved in tissue under pressure forming bubbles when it comes out of solution. Air emboli thus formed may appear anywhere in the body, blocking blood flow or damaging capillaries or other blood vessels with consequent tissue damage that can lead to death.

Prevention

Primary prevention can be achieved by eliminating the need for work under pressure, if possible, or by strict adherence to empirically derived protocols for compression and decompression, commonly referred to as decompression tables. None of the decompression tables presently in use by any navy, commercial oil company, diving agency, construction company, or regulatory agency can eliminate completely the risk of decompression sickness.

The Autodec III-02 Table, employing oxygen for decompression, is recommended[3] as superior to tables for caisson workers adopted by OSHA. Individual day-to-day variation in work conditions and a diver's physical condition result in variable on- and off-gassing of nitrogen during a pressure exposure. Therefore, workers and worksites must be evaluated continuously and individually.

Secondary prevention requires regular medical monitoring of people who work under pressure. Medical examinations should focus on x-ray examination of joints, the chest cavity, sinuses, and teeth and assessment of the ears, cardiac and central nervous system functions, and symptoms of osteonecrosis. Emergency treatment of decompression sickness requires onsite decompression chambers, trained personnel prepared for prompt response in case of an emergency, and transportation to a previously identified center for more extensive support if needed.

Other Issues

People with chronic sinus infection, sinusitis, allergies, asthma, obesity, or improperly filled teeth are at increased risk of injury.

People with right-to-left cardiac shunt are at increased risk of air emboli. Gas bubbles form more easily in venous than in arterial circulation, because it is at lower pressure. In normal cardiac function, these bubbles are filtered in the pulmonary circulation. But if there is a right-to-left cardiac shunt, the pulmonary circulation is bypassed and gas bubbles may pass into the arterial tree and form gas emboli.

Work under pressure is frequently associated with higher pay, especially for divers. Therefore, if a worker becomes symptomatic, or is labeled as "bendable," there can be a significant economic penalty, which is a significant disincentive to disclose symptoms. Acceptable alternative employment should be available to symptomatic workers to prevent progression of hyperbaric disorders.

If oxygen is present in a hyperbaric environment, there is a greater risk of fire than at atmospheric pressure because of its higher partial pressure. Therefore, attention must be paid to controlling fuels and sources of ignition.

Noise levels at construction sites—whether on the surface, under water, or in a caisson—are frequently elevated. Coupled with the sound of in-rushing air and within the confined environment of a caisson, diving bell, or clinical hyperbaric chamber, noise exposure may be hazardous. For example, when a sound-reducing muffler is not used, the noise level in a clinical chamber may exceed 130 dBA. Even brief exposure to this level of noise may result in transient hearing loss, and long-term exposure may result in permanent loss. Divers rarely experience these noise levels except when they are in a compressed air environment such as a diving bell. However, sound transmission in water is excellent, and concussion injuries may result from underwater blasts and explosions. Noise-induced hearing loss can be prevented by following the guidelines in the entry on Hearing Loss, Noise Induced.

The effects of toxic gases and CO_2 under pressure are significantly greater than at atmospheric pressure. For example, a relatively safe concentration of 30 ppm of carbon monoxide (CO) at the surface, resulting from exhaust gases of a compressor venting into an intake, may supply potentially lethal CO levels (equivalent to 150 ppm at 132 feet of seawater [5 ATA]), to a diver breathing this gas. Other gases, such as methane (CH_4) from decaying organic matter, carbon dioxide (CO_2), and oxides of nitrogen (NO_x) from diesel-powered generators as well as contaminants such as oil vapors may result in problems.

Because of the increased toxicity of these common air contaminants, close attention to the cleanliness and safety of the in-flowing gas supply is essential. Injuries and illnesses in divers and other workers have resulted from failure to pay attention to such simple things as a change in wind direction or the movement of a diving barge during a tide change. Frequent air sampling—usually for CO, CO_2, CH_4, NO_x, and oil vapors—of the air supplied to a diver or a clinical chamber followed by prompt corrective action is essential.

RG

References

1. U.S. Bureau of Labor Statistics, Supplementary Data System, 1986.
2. Davidson JK. Dysbaric disorders: aseptic bone necrosis in tunnel workers and divers. *Baillieres Clin Rheumatol.* 1989;3(1):1-23.

Further Reading

Downs GJ, Kindwall EP. Aseptic necrosis in caisson workers: a new set of decompression tables. *Aviat Space Environ Med.* 1986;57(6):569-574.

Hypersensitivity Pneumonitis

(Extrinsic Allergic Alveolitis)

ICD-9 495
SHE/O

Identification

Hypersensitivity pneumonitis is a descriptive term characterizing a continuously expanding group of allergic interstitial lung disorders elicited in response to inciting organic dust antigens. Recently, the term has also been expanded to include certain simple inorganic chemical exposures as well. The disease has certain distinguishing characteristics and occurs in acute and chronic forms:

Acute. The acute form is clinically characterized by chills, fever, cough, shortness of breath without wheezing, and malaise occurring 4 to 10 hours after antigen exposure. It usually subsides within 18 to 24 hours. Chest x-ray findings during the acute period may reveal a fine reticulonodular pattern bilaterally. Pulmonary function tests indicate a mainly restrictive pattern with decrease in forced vital capacity, FEV_1, and total lung capacity. There may be a decrease in oxygen saturation and in diffusion capacity as a result of ventilation profusion mismatches. Often, small airways obstruction also is present. Serological studies reveal precipitating antibodies present in over 90% of cases. Bronchoalveolar lavage studies reveal large numbers of lymphocytes.

Chronic. Prolonged exposure to an organic dust causing hypersensitivity pneumonitis can lead to permanent impairment with irreversible lung function changes and pulmonary fibrosis. Some patients insidiously develop symptoms, including cough and sputum production, shortness of breath, fatigue, and weight loss. Both restrictive and obstructive defects may be seen. Other patients demonstrate a progressive restrictive impairment with hypoxemia and decrease in lung volume, diffusion capacity, and compliance. Even in the absence of further exposure, fibrosis may progress in the chronic form of the disease. The chronic form may be associated with weight loss, anorexia, and chest x-ray findings of diffuse interstitial fibrosis (with or without honeycombing). Acute episodes usually no longer appear when this form occurs.

Table 1 at the end of this entry shows the disease manifestations in known or suspected offending agents of hypersensitivity pneumonitis.

Occurrence

The incidence and prevalence of hypersensitivity pneumonitis are not known, but the disease is less common than occupational asthma. Patients with chronic disease may be misclassified as having idiopathic pulmonary fibrosis, and those with acute disease may be misclassified as having occupational or environmental asthma or toxic inhalation syndromes. Little information is available regarding the complex relationship between the antigen, exposure intensity, and disease. In a population similarly exposed to a potential sensitizing agent, the proportion of individuals with detectable disease ranges from 3% to 15%. Seroepidemiological studies have shown that up to 50% of asymptomatic people with similar exposures have detectable circulating antibodies in response to antigen exposure with no evidence of disease present. Hence, there are many important variables determining disease development, which probably include genetics; immunologic abnormalities of the host; the role of chronic inflammation in the lungs; and potentiating effects of other toxins, such as air pollutants, tobacco smoke, and other workplace irritants.

Given the above caveats, prevalence studies of farmers in the United Kingdom have revealed a 4% to 9% prevalence of farmer's lung in several counties. In the U.S., a prevalence survey in Wyoming revealed a 4% prevalence, and one in Wisconsin revealed a 3.2% prevalence in farming populations. Bagassosis occurs wherever sugar cane is processed, including Puerto Rico, Cuba, India, Italy, Peru, the United Kingdom, the U.S. (about 5000 exposed workers), and China, but prevalence estimates are generally unavailable. Pigeon breeder's disease has a 6% to 15% prevalence rate among the 75 000 pigeon breeders in the U.S. and the 250 000 breeders in Belgium.

Hypersensitivity pneumonitis has been described in office workers and has been reported to be secondary to contaminated air-conditioning systems, with one outbreak having a 15% incidence rate and another, less than 1%.

Causes

Substances that account for many cases of this disease are shown in Table1.

Pathophysiology

The clinical response to inhaled antigen challenge depends on the presence of airway hypersensitivity, the nature of the organic material inhaled, the particle size (which determines deposition site), and the dose of exposure. The morphology seen on biopsy during the acute phase reveals noncaseating granulomas with foreign-body giant cells, large numbers of lymphocytes, and foamy macrophages in at least 50% of the cases. The pathological picture is of an interstitial pneumonitis that has a predominantly polymorphonuclear reaction in the acute phase with some eosinophils. In the subacute form, the cell populations change and interstitial inflammation now consists more of lymphocytes and plasma cells. The bronchioles may be involved in the acute and subacute phase with bronchiolar wall damage. In the chronic phase, after repeated acute insults, there is persistent inflammation and diffuse interstitial fibrosis, and honeycombing or bronchiolectasis may develop. The predominant cell type in chronic interstitial pneumonitis is a lymphocyte with a variety of mononuclear and histiocytic forms.

Several characteristics of offending organisms promote their role as agents in hypersensitivity pneumonitis. The relative nondigestibility of thermoactinomycete antigens by lysosomal enzymes presumably facilitates a foreign-body response and prolonged persistence of antigen in bronchopulmonary macrophages. Direct activation of complement via alternate pathways as well as complement consumption via the classic pathway has been shown with *Micropolyspora faeni.* The same spores may serve as adjuvant for induction of specific cell-mediated immunity.

Prevention

Avoidance of repeated antigen exposure is the only preventive measure that can be effectively taken to control disease. Such avoidance usually results in complete remission of symptoms and return to normal lung function in the early stages of disease. Clearly, complete avoidance of exposure is not likely in many jobs that involve handling substances that cause this disorder. Also, product substitution is often not viable. Engineering controls and work practices that may decrease the growth of biological contaminants (such as molds) can be effective in controlling the disease. In addition, personal protective devices, such as respirators or airstream helmets, may be used to prevent exposure.

There is limited information indicating that certain well-designed negative-pressure respirators may protect farmers from developing farmer's lung disease, but the most effective prevention remains reducing exposure through changes in work practices or engineering. For example, both maple bark disease and bagassosis have been well controlled or virtually eliminated in certain areas of the world after simple alterations of material handling were instituted, resulting in diminished growth of the contaminating, disease-causing organisms.

Other Issues

There is no proven method to identify individuals who are particularly susceptible to these disorders. However, an individual who has had one episode of a specific disorder is, having been sensitized, more likely to have another episode of that same disorder. Serology results are an index of exposure but not of disease.

DCC

Further Reading

Fink JN. Hypersensitivity pneumonitis. In: Merchant JA, ed. *Occupational Respiratory Diseases*. Washington, DC: U.S. Government Printing Office; 1986:481-497. DHHS (NIOSH) publication 86-102.

Merchant JA. Agricultural exposure to organic dusts. In: Rosenstock L, ed. *Occupational Pulmonary Disease. State of the Art Reviews in Occupational Medicine*. 1987;2(2):409-425.

Parkes WR. Disorders caused by organic agents (excluding occupational asthma). In: Parkes WR, ed. *Occupational Lung Disorders*. Boston, Mass: Butterworths; 1982:359-414.

Stankus RP, Salvaggio JE. Infiltrative lung disease: hypersensitivity pneumonitis. In: Santer M, Talmage P, Frank M, et al., eds. *Immunological Diseases.* 4th ed. Boston, Mass: Little, Brown; 1988.

Table 1.—Disease Manifestations of Hypersensitivity Pneumonitis

Offending Agent	Antigen Source	Disease
Thermophilic Actinomycetes		
Micropolyspora faeni	Moldy hay	Farmer's lung
	Mushroom composts	Mushroom worker's lung
Thermoactinomyces candidis	Humidifier and air-conditioning ducts	Humidifier lung
T. sacchari	Bagasse	Bagassosis
T. vulgaris	Moldy hay	Farmer's lung
Actinomycetes		
Streptomyces olivaceus	Thatched-roof dust	New Guinea lung
S. albus	Soil	—
Fungi		
Alternaria sp.	Moldy wood pulp	Wood pulp worker's disease
Mucor stolonifer (*Alternaria* sp.)	Moldy paprika pods	Paprika slicer's lung
Aspergillus clavatus	Moldy barley and malt	Malt worker's lung
A. fumigatus	Moldy barley and malt	Malt worker's lung
	Moldy hay	Farmer's lung (Finland)
A. glaucus (*A. umbrosis*)	Moldy hay	Farmer's lung (Finland)
Pullularia sp.	Water and steam	Sauna-taker's lung
	Moldy wood dust	Sequoiosis
Cryptostroma corticale	Moldy maple bark	Maple bark disease
Basidiomycetes		

Table 1.—Disease Manifestations of Hypersensitivity Pneumonitis (cont.)

Offending Agent	Antigen Source	Disease
Merulius lacrymans	Dry rot	—
Cryptococcus neoformans	House dust in Japan	—
Graphium sp.	Moldy wood dust, especially redwood	Sequoiosis
Cephalosporium sp.	Humidifier water	Humidifier lung
Rhizopus rhizopodiformis	Moldy wood planks	Wood trimmer's disease (Sweden)
Penicillium casei	Cheese mold	Cheese washer's lung
	Humidifier water	Humidifier lung
P. frequentans	Moldy cork dust	Subcrosis
Animal Proteins		
Bovine and porcine proteins	Pituitary snuff	Pituitary snuff-taker's lung
Rat serum proteins	Rat urine	—
Gerbils	Unknown	—
Avian Proteins		
Pigeon serum proteins	Pigeon droppings	Pigeon breeder's disease
Parrot serum proteins	Parrot droppings	Budgerigar fancier's disease
Chicken proteins	Chicken products	Feather plucker's lung
Duck proteins	Feathers	Duck fever
Thermotolerant Bacteria		
Bacillus subtilis	Wood dust	—
B. subtilis enzymes	Detergent	
B. sereus	Humidifier water	Humidifier lung

Table 1.—Disease Manifestations of Hypersensitivity Pneumonitis (cont.)

Offending Agent	Antigen Source	Disease
Amoebae		
Naegleria gruberi	Humidifier water	Humidifier lung
Acanthamoeba castellani	Humidifier water	Humidifier lung
A. polyphagi	Humidifier water	Humidifier lung
Insect Products		
Sitophilus granarius (wheat weevil)	Contaminated grain	Miller's lung (wheat weevil disease)
Vegetable Derivatives		
Ramin (*Gonystylus bancanus)*	Ramin dust	
	Sawdust, redwood, maple, red cedar	Sequoiosis
	Dried grass and leaves	Thatched-roof disease
	Coffee dust	Coffee worker's lung
Reactive Simple Chemicals		
Altered proteins (neo-antigens) or hapten protein conjugates	Toluene diisocyanate (TDI)	TDI hypersensitivity pneumonitis
	Trimellitic anhydride (TMA)	TMA hypersensitivity pneumonitis

Table from Stankus RP, Salvaggio JE. Infiltrative lung disease: hypersensitivity pneumonitis. In: Santer M, Talmage P, Frank M, et al., eds. *Immunological Diseases.* 4th ed. Boston, Mass: Little, Brown; 1988.

Hypertension

ICD-9 405

Identification

The upper limit of normal blood pressure is usually considered to be 140/90 mm Hg. The diagnosis of hypertension is made by determination of elevated blood pressure on three separate occasions. The patient is typically asymptomatic, unless damage to the eyes, heart, brain, or kidney is present. Increased prevalence of nosebleeds, unsteadiness, headaches, blurred vision, depression, and nocturia (urination at night) in untreated individuals without organ damage has also been described. Physical examination and laboratory analysis are done to assess organ damage and detect the infrequent secondary causes of hypertension. Blood and urine analyses, a chest x-ray, and an electrocardiogram (ECG) are routinely included in evaluating a newly diagnosed individual. Kidney tests, x-rays, and more elaborate blood or urine tests are reserved for individuals suspected of having one of the secondary causes of hypertension.

Occurrence

Thirty-one percent of the U.S. population aged 25 to 74 has hypertension. Prevalence increases with age and is higher among blacks than among whites.

Increased blood-lead levels have been associated with high blood pressure: A doubling of the lead level even within normal limits raises blood pressure 1 to 2 mm Hg. Exposure to the heavy metals lead, mercury, cadmium, and arsenic also may cause kidney disease, which in turn may cause hypertension. When kidney disease and/or gout is present, it is more likely that lead may be the cause of hypertension.

Exposure to carbon disulfide, which is associated with ischemic heart disease, may also cause hypertension. Exposure to high noise levels and to psychological stress is associated acutely with an increase in blood pressure. The evidence for an association between hypertension and chronic noise exposure is stronger than that for an association between hypertension and chronic stress exposure.

Causes

Most hypertension is defined as "essential" (the mechanism for loss of control is not known). Correctable causes, which explain fewer than 10% of all cases of hypertension, include renovascular disorders, primary aldosteronism, pheochromocytoma, Cushing's syndrome, and coarctation of the aorta. Kidney disease, depending on the type and extent of kidney insufficiency, is associated with hypertension. Kidney disease can affect production of renin, a potent mediator of the vasoconstrictor angiotensin, as well as the salt and water balance of the body. Heavy metals can cause kidney disease. Lead has been shown to affect renin production. Acute noise exposure causes an increase in catecholamine excretion and plasma levels of cholesterol, triglycerides, free fatty acids, and 11-hydroxycortisol levels; and several studies suggest a probable relationship between workplace noise exposure and elevated blood pressure.

Pathophysiology

Elevated blood pressure, once established, is maintained by increased peripheral vascular resistance, mainly in the small arteries and arterioles, where smooth muscle cell hypertrophy or contraction may account for much of this increased resistance. Stress may lead to structural changes that cause increased vascular resistance. Activation of the sympathetic nervous system, induced by stress, may lead to retention of sodium in the kidney, causing hypertension. Other mechanisms for the initiation and maintenance of hypertension are hypothesized.

Prevention

Reduction in exposure to substances associated with hypertension is necessary for primary prevention. A wellness program that combines reduction in workplace exposures with assistance in nutrition modification to reduce weight, salt intake, and alcohol ingestion would focus on several factors that cause or contribute to hypertension. Modifications of stress factors, such as heavy workloads, a nonsupportive supervisor, limited job mobility, and autonomy, may contribute to prevention. The workplace can be an ideal setting for screening for, treating, and monitoring hypertension. There are many effective approaches to the treatment of hypertension; it is beyond the scope of this book to cover them.

Other Issues

Work in a hot environment (readings over 79°F for men and 76°F for women, with a wet bulb thermometer) or exposure to solvents or nitrates may increase the prevalence and severity of side effects of medications used to treat hypertension. NIOSH has recommended that individuals taking diuretics ("water pills") not work in hot environments.

BSL

Further Reading

Batuman V, Landy E, Maesaka JR, et al. Contribution of lead to hypertension with renal impairment. *N Engl J Med.* 1983;309:17-21.

Jenkins CD. Psychosocial risk factors for coronary heart disease. *Acta Med Scand.* 1982;660(suppl):123-136.

Kaplan NK. Arterial hypertension. In: JH Stein, ed. *Internal Medicine.* 3rd ed. Boston, Mass: Little, Brown; 1990;235-252.

Landrigan PJ, Goyer RA, Clarkson TW, et al. The work-relatedness of renal disease. *Arch Environ Health*. 1984;39:225-230.

Symposium on lead-blood pressure relationships. *Environ Health Perspect*. 1988;78:1-155.

Talbott E, Helmkan PJ, Matthews K, et al. Occupational noise exposure, noise-induced hearing loss, and the epidemiology of high blood pressure. *Am J Epidemiol*. 1985;121:501-514.

Hypothermia—*See* Cold-Related Disorders

Immersion Foot—*See* Cold-Related Disorders

Indoor Air Pollution, Adverse Effects—*See* Building-Related Illness

Infertility and Sexual Dysfunction

ICD-9 610, 628, 302
SHE/O

Identification

The usual clinical definition of an infertile couple is one that has not conceived after 1 year of unprotected sexual intercourse. The World Health Organization has adopted the following definitions:

Primary Infertility: The woman has never conceived despite cohabitation and exposure to pregnancy for a period of two years.

Secondary Infertility: The woman has previously conceived but is subsequently unable to conceive despite cohabitation and exposure to pregnancy for a period of two years.

In many cases, the infertility of the couple can be attributed to either the man or the woman, but in a high percentage of cases, the infertility is due to an interaction of some characteristics of the two. Thus, the term *infertile couple* is more appropriate than *infertile man* or *infertile woman*. For example, a submaximal sperm count in association with some abnormalities of cervical mucus may lead to infertility when neither condition alone would do so.

Sexual dysfunction is defined as a decreased libido, a decreased interest in sexual activity, menstrual disorders in women, or erectile dysfunction in men.

Occurrence

Approximately 10% of all couples are infertile. It is estimated that male factors account for 40% of infertility cases, failure of ovulation account for 10% to 15%, tubal factors for 20% to 30%, cervical factors for 5%, and 10% to 20% of infertility cases have no known cause. There are no good estimates of the percentage of cases of infertility that are due to occupational factors.

Causes

Sexual dysfunction can interfere with fertility, and several occupational causes of sexual dysfunction exist. Decreased libido and impotence have been reported in male workers in association with exposure to chloroprene, manganese, organic and inorganic lead, inorganic mercury, toluene diisocyanate (TDI), and vinyl chloride, as well as with shift work. Menstrual disorders have been reported in female workers in association with exposure to aniline, benzene, chloroprene, carbon disulfide, inorganic mercury, polychlorinated biphenyls (PCBs), styrene, and toluene.

Research about occupational causes of infertility has concentrated primarily on men; therefore, the remainder of this discussion will be limited to male infertility. Early fetal loss, which may be manifest clinically as infertility, is discussed under Pregnancy Outcomes, Adverse.

Two steps are required to demonstrate an occupational cause of male infertility. First, there must be a gonadal disorder as evidenced by at least two of the following abnormal semen analyses: inadequate number, abnormal morphology, poor motility, or a decreased ability of the sperm to penetrate the egg. Second, nonoccupational causes must be evaluated; nothing about an abnormal semen analysis per se suggests an occupational etiology. Table 1 lists the nonoccupational causes of male infertility, and Table 2 lists occupational testicular toxins that have been associated with male infertility. Known toxins are those that have been shown to have an effect in humans, often

Table 1.—Nonoccupational Causes of Male Infertility

1. Primary endocrine disorder
2. Prior nonoccupational testicular injury or testicular surgery
3. Postadolescent mumps
4. Gonadotoxic drugs (eg, chemotherapy with cytotoxic drugs, estrogens)
5. Urologic abnormalities:
   Retrograde ejaculation secondary to diabetes, neurological disease, or prostatectomy
   Ductal obstruction secondary to tuberculosis, sexually transmitted disease, or vasectomy
   Varicocele

at high doses. Suspected toxins are substances that have been studied in humans with suggestive but inconclusive results, or that have produced positive study results in animals. Table 2 includes a scale for measuring the certainty of the causal association between the occupational exposure and the reproductive effect. The scale does not incorporate details of dose; an association is listed as a strong one even if the data exist only for high doses in animals. In applying the details of these tables in an individual exposure situation, dose-response must be considered in more detail.

Table 2.—Occupational Testicular Toxins

Agent	Human Findings	Animal Data	Measure of Association*
Boron	Decreased count Decreased motility	Testicular damage	2,3
Benzene	None	Testicular damage	1
Benzo(a)pyrene	None	Testicular damage	1
Cadmium	Reduced fertility	Testicular damage	2,3
Carbon disulfide	Decreased count, decreased motility	Testicular damage	1,4,7
Carbon monoxide	None	Testicular damage	1
Carbon tetrachloride	None	Testicular damage	1
Carbaryl	Abnormal morphology		3
Chlordecone	Decreased count, decreased motility		2,4
Chloroprene	Decreased motility, abnormal morphology, decreased libido	Testicular damage	1,3,4

Agent	Human Findings	Animal Data	Measure of Association*
Dibromochloro-propane (DBCP)	Decreased count	Testicular damage	2,4
DDVP	None	Decreased count	2
Epichlorohydrin	None	Testicular damage	2,5
Estrogens	Decreased count	Decreased count	2,4
Ethylene oxide	None	Testicular damage	1
Ethylene dibromide (EDB)	Abnormal motility	Testicular damage	2,3,5
Ethylene glycol ethers	Decreased count	Testicular damage	2,3
Heat	Decreased count	Decreased count	2,4
Ionizing radiation	Decreased count		2,4
Lead	Decreased count	Testicular damage, decreased count and motility, abnormal morphology	2,4
Manganese	None	Testicular damage	2
Polybrominated biphenyls(PBBs)	None	Testicular damage	1,5
Polychlorinated biphenyls(PCBs)	None	Testicular damage	1

*1 = limited positive animal data 2 = strong positive animal data
3 = limited positive human data 4 = strong positive human data
5 = limited negative animal data 6 = strong negative animal data
7 = limited negative human data 8 = strong negative human data

Pathophysiology

Reproductive function and sexual interest in both men and women depend, at least in part, on the presence of an intact neuroendocrine system. In men, lutenizing hormone (LH) from the pituitary and testosterone from the Leydig cells of the testes are needed for spermatogenesis and for interest in sexual activity. Pituitary lutenizing hormone and follicle-stimulating hormone (FSH), ovarian and adrenal estrogen, and progesterone are needed for female reproductive function.

This axis can be interrupted by agents that act like steroids or by the neurological input created by stress. Disorders of circadian rhythms, as can occur with some types of rotating work schedules, can also affect the endocrine cycle. The clinical results are menstrual disorders in women and libidinal disorders in both sexes.

Infertility that can be attributed to the male results from some abnormality of semen: an inadequate number of sperm, decreased motility, abnormal morphology, or decreased function. Fertility is a complex process, however, and semen parameters are considered only a surrogate measure. In recent years new assays have been developed to find a measure that is more accurate than the number or motility of sperm, such as the ability of a human sperm to penetrate a hamster egg. No one test is the best measure, and an abnormal sperm count does not always indicate infertility. Most identified occupational causes of male infertility affect the production or function of sperm in the testes, rather than interfering with the hormonal milieu supporting sperm production.

Sperm begin from diploid cells—the primary spermatocytes—and divide over several cell divisions into haploid cells—the spermatids. They then undergo maturation and develop motility in their transit from the testes through the epididymis. The site of action in the testes has not been identified for most of the known agents and can differ from one agent to another. Ethylene glycol ethers are known to affect cell division at the level of the primary spermatocyte, early in sperm production. Ethylene dibromide (EDB) is hypothesized to act on the maturation of spermatids into mature sperm in the epididymis.

Prevention

Primary prevention. Reduction of exposure or substitution of products will prevent the development of sexual dysfunction or infertility in the occupational setting. The specific intervention or level of exposure for each of the agents mentioned above differs.

Secondary prevention. Sperm parameters or fertility status can be monitored in a group exposed to a known or suspected spermatotoxin, with the goal of removing the affected men and reducing exposure if an effect is found. Questionnaire methods of assessing fertility based on the number of live births to exposed and unexposed men have been used; these have not been compared in the same study population to semen analysis, so the relative sensitivities are not known. Repeated semen analyses on the same individuals over time would be a sensitive screening tool; however, voluntary participation could be quite low, while a mandatory company program could be considered an invasion of privacy. Although follicle-stimulating hormone levels rise as sperm counts fall, this is a very insensitive measure in an individual or a small group; oligospermia would need to be severe before it would be detected with FSH as a screening tool.

Tertiary prevention. There is evidence that male infertility secondary to occupational exposure is reversible if the toxic insult is recognized early in the course of the disease. Abnormal sperm counts secondary to dibromochloropropane (DBCP) exposure have reversed at least partially after the agent was removed unless azospermia (absence of sperm) was present; recovery took as long as 18 months in men with oligospermia. Data from patients treated with therapeutic radiation suggest that even azospermia is reversible in some cases, but reversal was not seen until 4 to 5 years had passed.

For this reason, it is desirable to remove an individual with occupational infertility from exposure. Once the diagnosis is made, two semen analyses should be performed, using standardized motility and morphology techniques, before the individual is removed from exposure. The job should be changed, the product substituted, or the exposure eliminated with job controls and protective equipment. If removal is being used as a diagnostic test, it should continue for at least 18 months before the trial can be considered a failure. If monitoring for body burden of the toxin is possible, as it is with lead, these 18 months should begin when the body burden returns to the normal range.

LSW

Further Reading

Barlow SM, Sullivan FM. *Reproductive Hazards of Industrial Chemicals.* New York, NY: Academic Press; 1982.

Mattison DR, ed. *Reproductive Toxicology.* Progress in Clinical and Biological Research. Vol 117. New York, NY: Alan R. Liss; 1983.

Office of Technology Assessment. *Reproductive Health Hazards in the Workplace.* Washington, DC: U.S. Government Printing Office, 1975.

Overstreet JW, Blazak WF. The biology of human reproduction: an overview. In: Mattison DR, ed., *Reproductive Toxicology Progress in Clinical and Biological Research.* Vol 117. New York, NY: Alan R. Liss; 1983:3-5.

Steeno OP, Panskihila A. Occupational effects of male fertility and sterility. *Andrologia.* 1984;16:5-22, 93-101.

Welch LS. Decision-making about reproductive hazards. *Sem Occup Med.* 1986;1:97-106.

Wyrobek AJ, Gordon LA, Bukhart JG, et al. An evaluation of human sperm as indicators of chemically induced alternations of spermatogenic function: a report of the US Environmental Protection Agency Gene-Tox program. *Mutat Res.* 1983;114:77-148.

Infrared Radiation—*See* Radiation, Nonionizing, Adverse Effects

Injuries, Fatal

Identification

Workplace fatalities result primarily from traumatic injuries that occur at work. Unintentional traumatic fatal injuries are discussed here. Homicides that result from assaults are discussed under Homicide and Assault (*see also* Motor Vehicle Injuries).

Occurrence

There is no accurate count of occupational fatalities in the U.S. Estimates vary, depending on the source. The Bureau of Labor Statistics (BLS) estimates the average number of fatalities at 3590 for the years 1985 to 1987, based on its annual survey of a sample of private sector employers of 11 or more employees. NIOSH counted an annual average of 6767 fatalities for the years 1980 to 1985, based on death certificates. The National Safety Council estimated 11 500 fatalities in 1984, using multiple sources; the National Center for Health Statistics estimated 4960 in the same year.

Four major industrial groups—mining, transportation/communication/public utilities, construction, and agriculture (including forestry and fishing)—regularly have fatality rates more than five times greater than those in manufacturing, services, wholesale and retail trade, and finance (Fig. 1). The

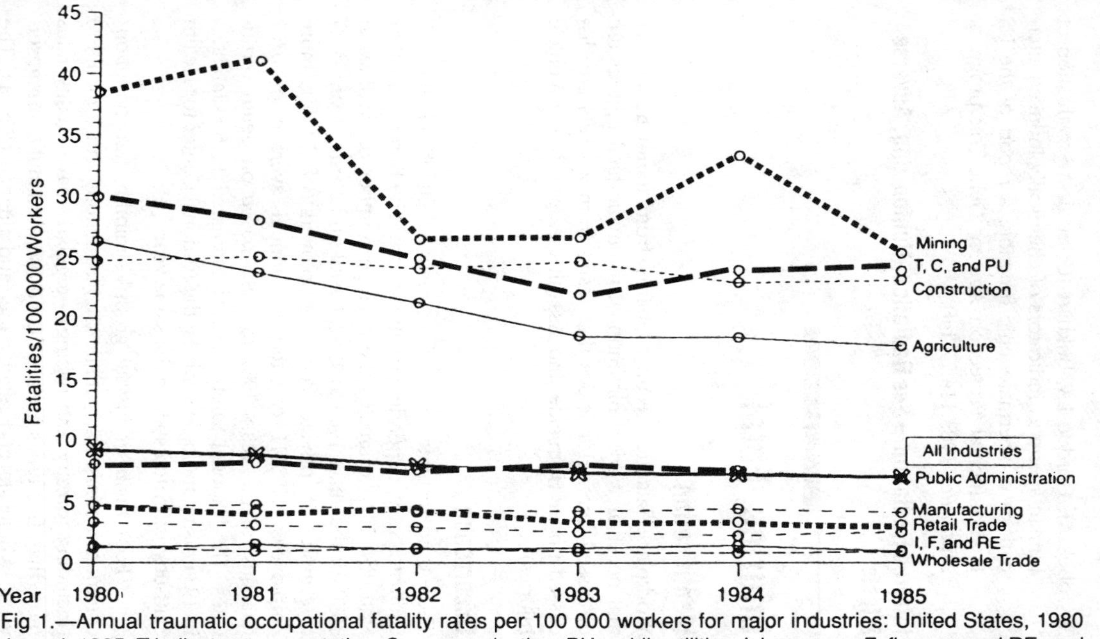

Fig 1.—Annual traumatic occupational fatality rates per 100 000 workers for major industries: United States, 1980 through 1985. T indicates transportation; C, communication; PU, public utilities; I, insurance; F, finance; and RE, real estate.

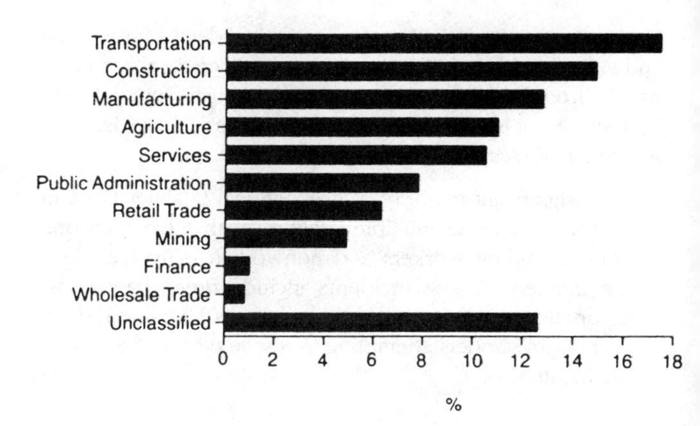

Fig 2.—Distribution of traumatic occupational fatalities by major industry division: United States, 1980 through 1985. Due to lack of comparable employment data, rates for unclassified occupations could not be calculated.

durable goods subsector of manufacturing also has a high fatality rate. Occupations with the highest fatality rates include unskilled laborers, miners, transportation workers, and loggers. The number of fatalities is highest in transportation, construction, manufacturing, agriculture, and service industries, in that order (Fig. 2).

The ratio of male to female workers killed on the job is greater than 12:1. The ratio of nonwhite to white (including Hispanic) workers killed on the job is about 1.2:1. These sex and race ratios reflect that more men than women and more nonwhites than whites hold jobs with high fatality rates. Age-specific fatality rates are greatest among workers over age 60, but the greatest number of fatalities occurs among workers in their 20s and 30s. Age differences vary among industries.

The vast majority (81%) of traumatic deaths are unintentional; among the rest, however, 13% are homicides and the balance are attributed to either suicides or unknown causes.

Homicide accounts for 39% of occupational deaths for women and 11% of such deaths for men. Natural sudden deaths at work, usually from heart disease, outnumber traumatic injury deaths by about 2:1; it is not known what proportion of these deaths is work related (*see* Arrythmias/Sudden Death).

A significant minority of occupational fatalities (25% in one study) occurs as multiple fatalities, with more than one person—including workers and nonworkers—involved in a single incident. These incidents include fires, explosions, transportation accidents, and building failures. Untrained fellow workers or bystanders attempting to rescue others often die in the incidents also.

Causes

Causes differ from those of nonfatal injuries (Table 1). According to BLS data for 1988, the most frequent cause is motor vehicle crashes, involving both on- and off-the-road vehicles.

Other causes vary by industry. In the construction industry, rates are highest in heavy nonbuilding construction and lower in special trades and building construction. According to a study in the state of Washington, the five leading causes of death, accounting for 77% of all fatalities, were falls from elevations, vehicle accidents at the work site, workers being struck by objects, excavation cave-ins, and electrocutions, with rates ranging from 8.0 to 2.3 per 100 000 workers per year. Risk of fatal injury, regardless of cause, increases with decreasing size of the work force per employer. Adverse weather conditions are a factor in 14% of all fatalities and in 20% of electrocutions in the construction industry.

In the transportation industry, most fatalities are due to motor vehicle or airplane crashes. Double-trailer trucks are at greater risk of crash than other heavy trucks. Crashes are also associated with younger drivers, long hours of driving, and operation of empty trucks.

Table 1.—Percentage Distribution of Fatalities by Cause: Reported Occupational Injury and Illness Fatalities for Employers with 11 Employees or More, 1987-88 Average [1]
(in percent)

Cause [2]	Total private sector [3]	Agriculture, forestry, and fishing	Mining–oil and gas extraction only	Construction	Manufacturing	Transportation and public utilities [4]	Wholesale and retail trade	Finance, insurance, and real estate	Services
Total, all causes	100	100	100	100	100	100	100	100	100
Highway vehicles	28	31	23	18	19	42	26	64	51
Falls	12	5	3	21	9	2	19	0	11
Electrocutions	9	5	4	15	5	14	5	0	5
Industrial vehicles or equipment	9	21	18	11	10	6	10	4	1
Heart attacks	8	5	8	9	10	8	3	23	11
Struck by objects other than vehicles or equipment	8	9	32	6	13	3	7	0	2
Assaults	5	6	0	([5])	1	1	21	8	8
Aircraft crashes	4	4	0	1	5	15	1	1	3

Table 1.—Percentage Distribution of Fatalities by Cause: Reported Occupational Injury and Illness Fatalities for Employers with 11 Employees or More, 1987-88 Average [1] (cont.)

(in percent)

Cause [2]	Total private sector [3]	Agriculture, forestry, and fishing	Mining–oil and gas extraction only	Construction	Manufacturing	Transportation and public utilities [4]	Wholesale and retail trade	Finance, insurance, and real estate	Services
Caught in, under, or between objects other than vehicles or equipment	4	2	0	10	4	1	1	0	(5)
Explosions	3	0	3	3	4	1	3	0	1
Gas inhalation	3	8	4	2	5	3	2	2	(5)
Fires	2	0	1	2	3	1	1	0	1
Plant machinery operations	2	1	1	1	9	0	1	1	0
All other [6]	3	4	2	2	3	3	1	0	3

[1] Results are the average of the 2 years because sampling errors for data by cause of fatality are too large to provide reliable annual estimates at the industry division level.

[2] Cause is defined as the object or event associated with the fatality.

[3] Excludes coal, metal, and nonmetal mining and railroads, for which data are not available.

[4] Excludes railroads.

[5] Between 0.1 and 0.5 percent.

[6] The "All other" category includes, for example, contact with carcinogenic or toxic substances, drowning, train accidents, and various occupational illnesses.

NOTE: Because of rounding, components may not add to totals.

In the mining industry, excluding oil and gas extraction, roof falls and cave-ins are the leading causes of fatal injuries, followed closely by accidents involving vehicles. Earlier in this century, fires and explosions, especially catastrophic explosions, were the leading cause of fatalities in coal mining. Since then, however, improved ventilation, the monitoring of methane concentrations, and restrictions on ignition sources have succeeded in reducing the risk. The fatality rate in small underground coal mines is three to four times higher than that in large mines.

Fatalities in agriculture are most often caused by farm vehicles (tractor rollovers being most common) and other machines. Some children under age 10 have died as a result of accidents in farming activities, but nearly half of all farm fatalities occur among people over age 60. Farm fatalities are seasonal, being most common during planting and harvesting.

Fatalities in the logging industry are usually related to machinery, such as skidders and bulldozers (about 17%), and to incidents, such as being struck by a falling or rolling tree (11%).

Pathophysiology

Most occupational fatalities result from severe trauma to the head and neck, massive multiple injuries, or laceration of a major blood vessel or the heart. Other causes include electrocution and asphyxiation (*see* Asphyxiation).

The time from traumatic injury to death occurs in a trimodal distribution following injury. More than half of such deaths occur immediately or very soon (usually less than an hour) after injury. For these people, even prompt state-of-the-art medical intervention would probably not be able to prevent death. Early deaths, accounting for about one fourth of all deaths from traumatic injury, occur within a few hours after injury; for these victims, prompt, high-quality medical attention can prevent death. Emergency medical services and transportation therefore play a critical role in preventing a significant proportion of traumatic fatalities. The remainder of deaths (late deaths) occur weeks after the initial injury, usually from infection or multiple organ failure. The quality of medical care is usually the critical factor in these cases.

Prevention

Preventive measures for fatal occupational vehicle crashes are similar to or the same as those for nonfatal occupational vehicle crashes (*see* Motor Vehicle Injuries). Safety should be improved for off-the-road vehicles. Safety devices on passenger cars, such as seat and shoulder belts, air bags, reinforced sides of vehicles, and improved placement of rear lights should be considered for off-the-road vehicles and trucks. Construction machinery is required to have rollover and falling-object protection canopies and backup alarms. The workplace offers unique opportunities for additional prevention and control measures, including purchaser specifications for vehicles, regular inspection and maintenance, and operator training.

Regulation has helped to reduce the risk of fatal injury in air transportation and mining, suggesting that, if well developed and vigorously implemented, regulation could reduce the risk of fatal injury in other high-hazard industries also. In high-hazard occupations or processes, such as working in confined spaces, hazardous work should be limited, standard procedures should be followed, and workers should be trained in proper rescue and recovery operations.

Prompt and thorough investigation of fatal incidents is essential to identify causes of the incidents and develop preventive measures. These measures include possible legal actions; in some instances, criminal prosecution of negligent employers is warranted and can serve as a deterrent.

Other Issues

There is no consensus on a method for surveillance of occupational fatalities. The OSHA/BLS system identifies only about 60% of occupational fatalities. Many fatalities are not reported to or investigated by OSHA. Yet the OSHA/BLS system has the advantage of integrating fatality data with pertinent denominator data (on numbers of workers at risk) essential for calculating industry- or occupation-specific rates.

Death certificates can identify more fatalities, but the coding of pertinent information, such as whether the fatality occurred at work or elsewhere as a result of work, is inconsistent. Moreover, some states do not code the work-relatedness of death on death certificates, and in states that do, the information is often incomplete or missing.

Workers' compensation systems are designed to determine eligibility for claims, not to conduct surveillance. Consequently, they have records only for occupational fatalities for which claims are filed.

Even though motor vehicles are the leading cause of occupational fatalities, the National Highway Transportation Safety Administration, which has responsibility for surveillance of highway fatalities, does not collect information on whether vehicular fatalities are occupational. Some states and local jurisdictions maintain more comprehensive surveillance systems, but these systems vary. The best ones are registries that include both fatal and nonfatal traumatic injury case reports and distinguish occupational from nonoccupational injuries.

Setting priorities for intervention includes consideration of the high-hazard industries within a particular jurisdiction, the fatal-injury incidence rates by industry and occupation, and the number of fatal occupational injuries. High-quality data systems facilitate setting priorities.

Intoxication by alcohol or other drugs is significantly less of a factor in occupational fatal injuries than in nonoccupational ones. From 4% to 9% of all workers fatally injured at work have blood alcohol concentrations (BACs) above the most common legal limit (0.10%). Considering only victims of vehicle-related occupational fatal incidents, 4% to 13% have BACs above this level. Two percent of drivers of heavy trucks involved in fatal crashes have BACs above this level. In comparison, 25% of drivers of cars involved in fatal crashes have BACs above this level.

JLW

Further Reading

Bell CA, Stout NA, Bender TR, Conroy CS, Crouse WE, Myers JR. Fatal occupational injuries in the United States: 1980 thru 1985. *JAMA*. 1990;263:3047-3050.

Buskin SE, Paulozzi LJ. Fatal injuries in the construction industry in Washington State. *Am J Ind Med*. 1987;11:253-460.

Davis H, Honchar PA, Suarez L. Fatal occupational injuries of women, Texas 1975-84. *Am J Public Health*. 1987;77(12):1524-1527.

Kizer KW. *California Occupational Mortality, 1979-1981*. Sacramento, Calif: California Department of Health Services; 1987.

Kraus JF. Fatal and nonfatal injuries in occupational settings: review. *Annu Rev Public Health*. 1985;6:403-418.

Kraus JF. Homicide while at work: persons, industries, and occupations at high risk. *Am J Public Health*. 1987; 77(1):1285-1289.

Leigh JP. Estimates of the probability of job-related death in 347 occupations. *J Occup Med*. 1987;29(6):510-519.

Massachusetts Department of Public Health. *Dying for the Job: Traumatic Occupational Deaths in Massachusetts, 1989.* Boston, Mass: Massachusetts Department of Public Health; 1989.

National Highway Transportation Safety Administration. *Fatal Accident Reporting System, 1987.* Washington, DC: U.S. Government Printing Office; 1988.

Robinson CC, Kuller LH, Perper J. An epidemiologic study of sudden death at work in an industrial county, 1979-1982. *Am J Epidemiol.* 1988;128(4):806-820.

Salmi LR, Weiss HB, Peterson PL, Spengler RF, Sattin RW, Anderson HA. Fatal farm injuries among young children. *Pediatrics.* 1989;83(2):267-271.

Smith GS, Kraus JF. Alcohol and residential, recreational, and occupational injuries: a review of the epidemiologic evidence. *Annu Rev Public Health.* 1988;9:99-121.

Stout-Weigard N. Fatal occupational injuries in US industries, 1984: comparisons of two national surveillance systems. *Am J Public Health.* 1988;78(9):1215-1217.

Trunkey DD. Trauma. *Sci Am.* 1983;249(3):28-35.

Injuries, Nonfatal

Identification

Nonfatal traumatic occupational injuries include strains, sprains, fractures, dislocations, contusions, abrasions, cuts and lacerations, and crushing injuries that occur at work. Work-related injuries associated with motor vehicles, burns, eye and back injuries, and musculoskeletal disorders associated with cumulative trauma and vibration are treated in more detail in separate entries.

Occurrence

Estimates of the occurrence of occupational injuries vary by data source. National data sources for nonfatal occupational injuries include the Bureau of Labor Statistics (BLS) and the National Center for Health Statistics (NCHS). Other sources include workers' compensation awards and records of emergency department visits. None of these sources provides a complete count of all occupational injuries.

BLS has a statutory mandate to collect occupational injury statistics reported by employers under OSHA jurisdiction; it also collects data reported by mine operators to MSHA. Based on its annual national sample of employer reports, BLS estimated a total of 6.2 million nonfatal occupational injuries in the private sector in the U.S. in 1988. Of these, an estimated 2.9 million injuries resulted in an average of 19 lost workdays each, resulting in a total of 55 million lost workdays due to injury. Average annual injury incidence rates in 1988 were 8.3 per 100 full-time workers for all injuries and 3.8 for lost workday injuries. All these measures of the occurrence of occupational injuries have increased in recent years.

For major industrial divisions (two-digit SIC codes), annual incidence rates for lost workdays and all injuries ranged from a high of 6.8 lost-workday injuries and 14.4 total injuries per 100 full-time workers in the construction industry to a low of 0.9 and 2.0 injuries per 100 workers, in finance, insurance, and real estate. The average number of workdays lost for injuries ranged from 30 days lost for each injury in the mining industry to 17 days lost in wholesale and retail trade. The largest number of injuries occurred in the largest sectors—manufacturing and wholesale and retail trade, each with nearly 20 million workers (Table 1a). Industries with the highest reported annual injury incidence rates are almost always in the manufacturing sector (*see* Table 2), with incidence rates as high as 17.9 for lost workday cases in the sawmill industry and 16.7 in shipbuilding and ship repairing.

Table 1a.—Number of Injuries, Incidence Rates, and Average Severity, U.S. 1988 (BLS)

	Incidence Rate (per 100 workers)			
Industry Group (SIC codes)	Lost-workday Injury Rate	Total	Average Severity (days)	Number of Injuries (x 1000)
Agriculture, Forestry, Fishing (01-09)	5.5	10.4	18	97.3
Mining (10-14)	5.1	8.5	30	62.5
Construction (15-17)	6.8	14.4	21	649.6
Manufacturing (20-39)	5.3	12.1	19	2,288.0
Durable Goods	5.6	13.2	19	1,483.8
Nondurable Goods	5.0	10.6	19	804.2
Transportation, Communication, Public Utilities	5.0	8.8	23	455.8
Transportation (40-47)	6.7	11.5	25	353.9
Communication (48)	1.5	2.9	19	34.6
Public Utilities (49)	3.8	7.3	19	67.3
Wholesale, Retail Trade (50-59)	3.5	7.8	17	1,519.5
Finance, Insurance, Real Estate (60-67)	0.9	2.0	18	116.3
Services	2.5	5.3	18	1,008.6
Average (total) private sector	3.7	8.0	18	

The industry-specific injury incidence rates estimated by NCHS are almost all higher than the BLS estimates. NCHS estimates are derived from the National Health Interview Survey (NHIS, 1983-85), which was based on interviews of a stratified random sample of noninstitutionalized civilians, by household, living in the U.S. An occupational injury was counted if it resulted in restriction of normal activity for more than half a day. Both public and private sector workers were included, regardless of the size of the workplace. Rates were calculated among all persons employed at any time in the 2 weeks prior to the interview. Thus, NCHS and BLS estimates are not strictly comparable.

Table 1b.—Number of Injuries, Annual Average Incidence Rates, and Number of Injuries, 1983-85 (NCHS)

Industry Group (SIC codes)	Incidence Rate (per 100 workers)	Number of Injuries (x 1000)
Agriculture, Forestry, Fishing (01-09)	19.1	616
Mining (10-14)	13.8	140
Construction (15-17)	26.9	1,803
Manufacturing (20-39)	14.5	3,023
Transportation, Communication, Public Utilities	10.1	765
Wholesale, Retail Trade (50-59)	7.6	1,517
Finance, Insurance, Real Estate (60-67)	4.3	287
Services	6.5	2,038
Public Administration	6.8	332
Average or Total		
Private sector	10.5	10,189
Private and public sector	10.2	10,521

Between 1983 and 1985, NCHS estimated an average of 10.5 million occupational injuries annually. Industries with the highest annual incidence rates were construction (26.9 injuries per 100 workers); agriculture, forestry, and fishing (19.1); manufacturing (14.5); and mining (13.8). These rates and those for other industries are shown in Table 1b.

Table 2.—Industries with the Highest Annual Injury Incidence Rates, 1987 and 1986

Rank	Industry (SIC code)	Incidence Rates (injuries per 100 workers)	
		1987	1986
1.	Shipbuilding, Repair (3731)	37.0	20.0
2.	Prefabricated Wood Buildings (2452)	31.3	25.8
3.	Meat-Packing Plants (2011)	30.4	27.0
4.	Mobile Home Manufacturing (2451)	28.7	29.5
5.	Primary Aluminum (3334)	26.5	9.0
6.	Special Product Sawmills (2429)	26.1	24.6
7.	Structural Wood Members (2439)	25.3	27.0
8.	Reclaimed Rubber (3030)	25.3	25.2
9.	Malt Beverages (2083)	25.1	19.0
10.	Primary Lead (3332)	24.7	17.9

In addition to industrial group, injury incidence rates vary by occupation within each industry, work experience, worker's age, rate of business expansion, race, sex, and size of the work force. Younger, less experienced workers usually have higher injury rates than older, more experienced workers. As the economy expands and employment increases, injury rates in manufacturing and construction tend to rise also. In part, this rise is due to an influx of younger, less experienced workers into the work force, but it may also result from new and unfamiliar plants and equipment being put into service. Injury rates among nonwhites and men are higher than among whites and women, but most of the variation is explained either by occupation or by age and experience. In manufacturing, mining, construction, and transportation, mid-sized employers (from 50 to 250 employees) have higher reported injury rates than smaller or larger employers, based on BLS data. In agriculture, the largest employers (employing over 1000 workers) have the highest estimated injury rates. In manufacturing industries, 80% of variation in injury frequency is explained by seven variables, including—in order of the strength of the association—the amount of energy consumed per worker, the rate of new hires, productivity, average hourly wage, and the percentage of the work force unionized.

Injuries that result in compensation awards are classified in the BLS Supplementary Data System (SDS) by industry division, occupation, nature of injury, body part affected, source of injury, and type of accident. In 1986, when BLS estimated 5.6 million lost-time injuries in the private sector, the SDS data set included information on 1.03 million compensation awards in the private sector and 0.15 million in the public sector. This is not a representative sample because only 23 states reported to BLS, and each state has different regulations concerning eligibility. Nevertheless, it is the only data set that is nearly national in scope, classifies injuries by cause, and can be used to show the proportional distribution of awards (Table 3).

Table 3.—Percentage of Compensation Awards by Nature of Injury, Body Part Affected, Injury Source, and Type of Accident (Based on awards in 23 states, 1986)

Nature of Injury (percent)	
Sprains, strains	(41)
Cuts, lacerations, punctures	(13)
Contusions, crushes	(10)
Fractures	(9)
Occupational illnesses	(6)
Burns	(3)
Amputation, enucleation	(1)
Other	(17)

Body Part Affected (percent)	
Back	(24)
Lower extremities	(20)
Upper extremities (excluding fingers)	(14)
Fingers	(12)
Trunk (excluding back)	(11)
Multiple parts	(8)
Head, neck (excluding eyes)	(5)
Eyes	(3)
Body system	(2)
Other	(1)

Source of Injury (percent)	
Working surfaces	(15)
Boxes, containers	(13)
Metal items	(10)
Vehicles	(8)
Machines	(7)
Hand tools	(7)
Wood items	(4)
Chemicals	(2)
Other	(34)

Type of Accident (percent)	
Overexertion	(30)
Struck by or against	(25)
Fall	(17)
Bodily reaction	(7)
Caught in, under, or between	(6)
Contact with radiation or caustics	(3)
Motor vehicle accidents	(3)
Rubbed, abraded	(2)
Temperature extremes	(2)
Other	(4)

Causes

The etiology of occupational injuries is multifactoral, involving specific working conditions and characteristics of individuals. For purposes of prevention, there are four necessary and sufficient conditions, each of which must be present to cause an injury: (a) there must be a source of energy (mechanical, electrical, thermal, or chemical); (b) it must released; (c) there must be a vehicle to transmit the energy to the victim; and (d) the energy must be absorbed at a rate that exceeds the ability of the body to avoid damage. By controlling any one of these causes, it is possible to prevent injury. Classifying causes in this manner lays the foundation for successful engineering control of injury hazards. (This description of causes is discussed and illustrated in more detail in part 1, Injury Prevention.)

Pathophysiology

The pathophysiology of injuries varies by the nature of the injury and the source of the energy and vehicle. Mechanical energy usually results in physical trauma to soft tissue or bone. Such injuries fall into several classes, including lacerations, punctures, abrasions, fractures, dislocations, contusions, strains, and sprains. Electrocutions can result in localized burns and neurological disorders. The pathophysiology of burns, eye injuries, low back pain, cumulative trauma disorders, and fatal traumatic injuries is discussed in more detail in specific entries.

Prevention

Injury control is the systematic analysis of injury risk and the development and application of control measures for the primary and secondary prevention of injuries. A general strategy for injury control, illustrated with several examples, is described in part 1.

Other Issues

National data sources are currently inadequate for effective prevention of occupational injuries. They do not provide enough information about the causes of injuries or about the at-risk population, nor do they provide information needed for targeting intervention. Information collected by MSHA is an exception: In the mining industry, all injuries, some "near-miss" accidents, and hours worked must be reported, and this information is available for every mine operating in the U.S.

The proportion of nonfatal occupational injuries associated with intoxication by alcohol or other drugs is not well documented. Based on very limited data, workers acutely intoxicated by alcohol or other drugs (medications or illegal drugs) face about twice the risk of nonfatal injury as other workers. However, the proportion of workers who may be intoxicated with alcohol or other drugs while on the job is unknown, so the proportion of all injuries for which intoxication is a factor is unknown. Therefore, on-the-job testing for drug or alcohol use has unknown value for preventing occupational injuries.

Some occupational safety specialists, as they develop safety programs, consider accident proneness—the concept that there are some individuals who are more likely than others to have injuries because of their psychological or physical makeup. Individuals are occasionally labeled "accident prone" after having been injured more than once. However, the number of such accident repeaters is often not significantly different from what would be expected by chance alone, based on existing injury rates. Also, the proportion of all injuries that occur as repeat occurrences is small, and thus the potential benefit of prospective identification of accident-prone workers is also small. Attempts to identify such individuals and to intervene prospectively to reduce injury frequency have not been successful. Thus, for practical purposes, the concept of accident proneness is not useful for developing injury prevention programs.

JLW

Further Reading

Arno PS. *The Political Economy of Industrial Injuries*. New York, NY: New School for Social Research; 1984. Dissertation.

Collins JG. *Health Characteristics by Occupation and Industry: United States, 1983-85*. Washington, DC: National Center for Health Statistics; 1988. Vital Health Stat 10(170).

Haddon W Jr. The changing approach to the epidemiology, prevention, and amelioration of trauma: the transition to approaches etiologically rather than descriptively based. *Am J Public Health*. 1968;58(8):1431-1438.

Haddon W Jr. A logical framework for categorizing highway safety phenomena and activity. *J Trauma*. 1973;12:193-207.

Kriebel D. Occupational injuries: factors associated with frequency and severity. *Int Arch Occup Environ Health*. 1982;50(3):209-218.

Pollack ES, Keimig DG, eds. *Counting Injuries and Illnesses in the Workplace: Proposals for a Better System*. Washington, DC: National Academy Press; 1987.

Robertson LS. *Injuries: Causes, Control Strategies, and Public Policy*. Lexington, Mass: Lexington Books; 1983.

Robinson JC, Shor GM. Business-cycle influences on work-related disability in construction and manufacturing. *Milbank Q*. 1989;67(suppl 2, pt 1):92-113.

Smith GS, Kraus JF. Alcohol and residential, recreational, and occupational injuries: a review of the epidemiologic evidence. *Annu Rev Public Health*. 1988;9:99-121.

U.S. Department of Labor, Bureau of Labor Statistics. *Annual Survey of Occupational Injury and Illnesses, 1988.* Washington, DC: U.S. Government Printing Office; 1990.

Waller J. *Injury Control: A Guide to the Causes and Prevention of Trauma.* Lexington, Mass: Lexington Books; 1985:467-471.

Zwerling C, Ryan J, Orav EJ. The efficacy of preemployment drug screening for marijuana and cocaine in predicting employment outcome. *JAMA.* 1990;264:2639-2643.

Ischemic Heart Disease

ICD-9 410-414

Identification

The World Health Organization defines ischemic heart disease as "myocardial impairment due to imbalance between coronary blood flow and myocardial requirements caused by changes in the coronary circulation." The clinical presentation can include repeated chest pain (angina), dyspnea (due to congestive heart failure), palpitations, syncope, sudden death (due to arrhythmias), and acute chest pain (myocardial infarction) (*see* Congestive Heart Failure and Arrhythmias/Sudden Death). Patients with angina typically develop their pain with exertion and are relieved by rest or nitroglycerin. Different presentations of pain may occur with unstable angina, where there is a changing pattern of pain or pain at rest, or with Prinzmetal's (variant) angina secondary to coronary vasospasm, where myocardial ischemia also occurs at rest. Results of physical examination, chest x-rays, and electrocardiogram (ECG) are typically normal, unless some other manifestation of ischemic heart disease is present. Exercise stress testing, which may include a radionuclide scan, and cardiac catheterization may be performed, particularly if cardiac bypass surgery is being considered.

Myocardial infarction typically presents with severe acute chest pain, although silent infarcts and infarcts not brought to the attention of the medical system reportedly account for up to 20% of all heart attacks. Physical examination will show normal results unless other manifestations of ischemic heart disease are present. The ECG will show distinct changes indicating the location and extent of heart muscle destruction. During the acute phase, certain enzymes (LDH, SGOT, CPK [MB isoenzyme]) in the blood are typically elevated.

Occurrence

Approximately 6 million individuals in the U.S. are estimated to have ischemic heart disease; 800 000 individuals a year have new heart attacks, and another 450 000 individuals have a second or further heart attack. Despite the marked reduction in mortality from heart disease since the 1950s, ischemic heart disease (including sudden death) is still the most common cause of death, with more than 500 000 people dying from it each year. The disease is 1.5 to 3.2 times more common in men than in women, with the ratio decreasing with increasing age. Incidence increases with age (approximately twice as frequent among individuals more than 65 years old). Twenty percent of men will develop the disease before the age of 60.

Causes

Much is known about the risk factors of ischemic heart disease. The major risk factors are increasing plasma cholesterol level, cigarette smoking, and hypertension. Other important risk factors include obesity, diabetes, lack of sufficient physical activity, Type A personality, heavy alcohol use, and use of oral contraceptive drugs. The etiology is multifactorial, and there is an increase in risk of heart disease among people with multiple risk factors. Some authors have estimated that the known risk factors explain only 50% of the disease.

There is good evidence of a causal association between ischemic heart disease and exposure to carbon monoxide, carbon disulfide, or nitrates. There is limited evidence that stress, noise, or arsenic causes ischemic heart disease. There are hypothetical and anecdotal reports on an association between ischemic heart disease and hot or cold temperature extremes, radiation, fibrogenic dusts, cadmium, and ethanol and phenol used as solvents. There are no data to quantify the importance of occupational exposures. Carbon monoxide is ubiquitous and found wherever there is incomplete combustion (fuel-powered engines). If a strong association were to be found with noise and/or stress, the contribution of occupation to the etiology of ischemic heart disease would be very significant. However, given the prevalence of ischemic heart disease in the U.S., even if occupational factors were important in only a small percentage of cases, mitigation of such factors would have important public health benefits.

Pathophysiology

The predominant cause of changes in the coronary circulation is atherosclerosis of the coronary heart muscle. The three major pathogenic hypotheses are the lipogenic theory (lipids initiate and are essential to progression of disease), the thrombogenic theory (thrombi initiate plaques, and formation of thrombi on plaques causes progression of disease), and the monoclonal theory (plaques develop from single precursor cells and progress like tumors). Each theory includes factors known to be important in atherosclerosis. The final explanation for the formation of plaques (atherosclerosis) in blood vessels will probably include, at a minimum, elements of all three theories.

Acute exposure to carbon monoxide reduces the amount of oxygen that is transported and released to myocardial tissues and other organs, and it also inhibits mitochondrial enzymes such as cytochrome oxidase. Sufficient exposure causes a reduction in cardiac output. Hemoglobin has an affinity for carbon monoxide that is 200 times that of oxygen, whereas myoglobin has an affinity for carbon monoxide 50 times that of oxygen. The percentage of carbon monoxide bound to hemoglobin (percent carboxyhemoglobin) equals the percentage of carbon monoxide in air multiplied by the time in minutes multiplied by a constant K (K = 3 at rest, 5 for light physical work, 8 for moderate physical work, and 11 for heavy physical work). The time in the equation cannot be greater than the time it takes for carbon monoxide to reach equilibrium (8 hours at rest, less for increased activity). Carboxyhemoglobin levels as low as 2.8% have been associated with acute symptoms.

Chronic exposure to carbon monoxide may be associated with increased atherosclerosis, although data suggesting this are mixed. Increased vascular permeability and increased platelet adhesiveness have been found in animals exposed to carbon monoxide.

Suggested mechanisms for the atherosclerotic effect of carbon disulfide include a role in causing hypertension, hypercholesterolemia, and/or an antifibrinolytic effect. Nitrates cause an acute rebound coronary vasospasm on removal from exposure but may also be associated with increased atherosclerosis after chronic exposure.

Prevention

Reduction in exposure to substances that cause acute changes (such as carbon monoxide) or chronic atherosclerotic changes (such as carbon disulfide) is necessary for primary prevention. Whether adequate protection is afforded by current occupational standards is not known.

Early diagnosis with removal of symptomatic individuals exposed to carbon disulfide has been effective in reducing the morbidity and mortality from ischemic heart disease. This is even more likely to be true for people exposed to carbon monoxide and nitrates, where the evidence for an acute effect is stronger than the evidence for a chronic atherosclerotic effect.

Because of the poor predictive value of stress tests that do not include radionuclide scans, brief questioning using a standardized questionnaire (the Rose Questionnaire) and a resting ECG remain the best screening tests for clinically significant ischemic heart disease in a working population.

Because ischemic heart disease has a multifactorial etiology, any reduction in known personal life-style risk factors will reduce the risk of developing symptomatic ischemic heart disease.

Other Issues

Symptoms of ischemia (angina and claudication) occur with increased frequency and duration as levels of carboxyhemoglobin increase, even under the legal exposure limits for carbon monoxide. Other evidence suggests that acute heart attacks and death secondary to ventricular fibrillation are more likely to occur in people with underlying ischemic heart disease who are exposed to carbon monoxide within current legal limits. Therefore, individuals with underlying ischemic heart disease may be particularly susceptible. Certainly such individuals are more likely to have an acute cardiac event during stress or strenuous activity in cold or hot conditions.

Fire fighters and police are accorded special compensation rights for ischemic heart disease in approximately 20 states.

KR

Further Reading

Harlan WR, Sharret AR, Weill H, et al. Impact of the environment on cardiovascular disease. Report of the American Heart Association Task Force on Environment and the Cardiovascular System. *Circulation.* 1981;63:243A-271A.

Leading work-related diseases and injuries: United States. *Cardiovascular Diseases.* 1985;34:219-222.

National Academy of Sciences. Carbon Monoxide. Washington, DC: National Academy of Sciences; 1977.

Nurminen M, Hernberg S. Effects of intervention on the cardiovascular mortality of workers exposed to carbon disulphide: a 15-year follow-up. *Br J Ind Med.* 1985;42:32-35.

Rosenman KD. Environmentally related disorders of the cardiovascular system. *Med Clin North Am.* 1990;74:361-375.

Kidney Disease—*See* Glomerulonephritis; Renal Disease, Chronic: Tubular and Interstitial; and Renal Failure, Acute

Laryngeal Cancer

ICD-9 161
SHE/O

Identification

Cancer of the larynx may be asymptomatic initially. With supraglottic (above the glottis) cancers, hoarseness occurs late, and the first symptom may be vague pain or a lump in the neck. Cancers of the glottis lead to hoarseness early. Subglottic cancers rarely cause hoarseness but may lead to shortness of breath due to partial airways obstruction. Direct examination of the larynx by laryngoscopy and biopsy yields the diagnosis.

Occurrence

Worldwide occurrence is much greater in males than in females. In the mid-1970s, the SEER (Surveillance, Epidemiology, and End Results) Program of the National Cancer Institute, which covered about 10% of the U.S. population, found incidence rates of 8.5 (males) and 1.3 (females) per 100 000 people. Often, but not always, laryngeal cancer incidence varies proportionally with lung cancer incidence from one population to another. In most parts of the world, laryngeal cancer is increasing among males and, in the more developed countries, also among females. The median age of diagnosis of laryngeal cancer is usually in the 50s or 60s. Age-adjusted death rates are highest among the poor.

Causes

Cigarette smoking and alcohol consumption have both been shown to have a strong association with laryngeal cancer. Studies have revealed a dose-response relationship for each of these factors taken alone, as well as a synergistic, or multiplicative, effect of these two factors taken together. Asbestos exposure is a major risk factor. The relationship between cancer of the larynx and occupations that do not involve asbestos exposure is less well substantiated, although some studies have shown mustard gas and nickel exposure to be major risk factors. Incidence and mortality studies of laryngeal cancer have also raised the possibility that the following occupations, among others, are risk factors: woodworkers; workers in the chemical and printing industries; and brewers, wine makers, bar workers, and other workers with access or exposure to alcohol in their work.

Pathophysiology

Laryngeal cancer refers almost exclusively to squamous cell carcinomas. It is believed that direct contact with the carcinogenic agent is responsible for initiating development of this malignancy in most cases. Supraglottic cancers with late symptoms are the type most likely to metastasize.

Prevention

Prevention rests on reducing exposure to the well-known responsible agents and on early detection and effective treatment of people with this disease. Early detection largely depends on early recognition of symptoms among people with this cancer and on their seeking medical attention.

Other Issues

Smokers with leukoplakia (of the mouth) may be at greater risk of developing laryngeal carcinoma than smokers without leukoplakia.

HF

Further Reading

Austin DF. Larynx. In: Schottenfeld D, Fraumeni JF, eds. *Cancer Epidemiology and Prevention.* Philadelphia, Pa: WB Saunders; 1982:554-563.

Flanders WD, Rothman KJ. Occupational risk for laryngeal cancer. *Am J Public Health.* 1982;72:369-372.

Rothman KJ, Cann CI, Flanders D, et al. Epidemiology of laryngeal cancer. *Epidemiol Rev.* 1980;2:195-209.

Laser Radiation—*See* Radiation, Nonionizing, Adverse Effects

Lead Poisoning

ICD-9 984

Identification

Lead poisoning is a syndrome of intoxication caused by absorption of metallic lead, inorganic lead compounds, or, in rare instances, organic lead compounds. Centuries after the dangers of lead intoxication were recognized, lead poisoning remains one of the most common environmental diseases. Because lead poisoning can masquerade with a wide variety of clinical presentations, diagnosis depends on a high degree of clinical suspicion. Patients may present with predominantly neurological, gastrointestinal, or rheumatologic complaints. Diagnosis can be confirmed by demonstration of elevated levels of lead in body fluids or tissue.

Occurrence

The useful properties of lead, its widespread availability, and the ease with which it can be manipulated by even simple technology contribute to the widespread occurrence of lead poisoning. Because lead can be melted at relatively low temperatures (327.4°C), is malleable, is water resistant, and makes brightly colored pigments, it has many uses throughout the world. The village tinker in Palestine can use lead to steady a wobbly millstone; the moonshiner in the United States can readily obtain lead pipe and solder to build a still. Lead pigments turn up as folk remedies in Mexico and as cosmetics in India.

Nor does sophisticated technology provide a margin of safety from this "old" disease. State-of-the-art "mass-burn" solid-waste incinerators produce ash with a high content of lead that poses a potential threat to operators. Sophisticated air pollution control devices and complex chemical plants rely on the time-honored skills of lead burning for their maintenance.

About 9 million metric tons of lead are produced each year worldwide, and many more millions of tons are recycled. In the U.S., approximately 1 million tons are consumed annually, half from new production of lead and half from recycling scrap. Two thirds of the U.S. consumption is used in manufacturing lead-storage batteries. Because batteries have a limited life span, about 80% of this lead is returned to scrap to be recycled. Sheet lead is used for waterproofing, noise and vibration reduction, and lining chemical reaction vessels.

Large amounts of lead compounds are used in paints for color, for rust prevention, and as drying agents. Although the use of indoor paint containing significant amounts of lead has been banned in the U.S., red lead paint is still used extensively as a primer on structural steel. Lead is also used in the plastics and ceramics industries, and in both bullets and primer for firearms.

Organic lead compounds, mainly tetraethyl lead, have been used since the 1920s as anti-knock additives in fuels. Substantial atmospheric pollution can result from the inorganic lead compounds produced by the combustion of these fuels.

Epidemiological study of lead poisoning focuses not only on the incidence and prevalence of overt cases of symptomatic lead poisoning, but also on the relatively high levels of lead absorption among the general population in many industrialized countries. Both cross-sectional and longitudinal studies have shown convincingly that current population norms for lead content reflect substantial nonoccupational absorption. Recent studies that demonstrate how blood lead levels previously thought to be well below the safe "threshold" affect the cognitive development of children have focused attention on this issue. Absorption can occur from exposure to lead in urban air, soil, and house dust (contaminated by lead from automobile exhausts and paint); in and on food (particularly lead solder in cans); and in drinking water (particularly where corrosive water is delivered through pipes with relatively new lead solder). However, the dramatic reduction of blood lead in the U.S. population that followed the reduced use of leaded gasoline showed that the "normal background level" is not immutable. Because workers carry the burden of nonoccupational exposures into any workplace encounter with lead, overall reductions in population exposures will reduce the impact of occupational exposures.

Estimates of prevalence of occupational lead poisoning have been hampered by disagreement on case definition. As recently as two decades ago, 80 μg/dL was considered an acceptable limit for lead in blood, and many exposed workers had levels just below this. Industry-sponsored mortality studies have since shown that these levels of exposure caused increased mortality from renal and cardiovascular disease. The OSHA lead standard requires medical evaluation at a blood level of 40 μg/dL and requires that physicians conducting surveillance be provided with a copy of guidelines describing signs and symptoms of intoxication. Studies of "subclinical" and "asymptomatic" lead poisoning have shown that workers with lead levels in the 40 to 60 μg/dL range not only show neurological abnormalities, but also have symptoms when asked.

States that have implemented surveillance systems increasingly use the CDC definition of abnormal blood lead (>24 μg/dL) as a trigger for laboratory reporting and subsequent investigation of cases. In Maryland, an industrialized state with a population of approximately 4.5 million, 114 cases of adult lead poisoning from 27 workplaces were identified during 1987 using this criterion. (While far from a common disorder, this incidence of 2.5/100 000 makes lead poisoning about as common as multiple myeloma or ulcerative colitis.) In the same year there were 305 new reported cases of lead poisoning (lead levels >30 μg/dL) among children in Maryland. Only cases in which the diagnosis was suspected and the blood lead was measured are counted in this surveillance system. Because an accurate diagnosis is difficult to obtain, many cases undoubtedly go unrecognized and uncounted.

Causes

Occupational lead poisoning can occur in a variety of settings. Perhaps the most common is where work is done with lead as a paint or coating. Although exposure can occur while paint is being applied, the worst exposures occur when existing paint is disrupted. Workers removing lead paint from homes during renovation or remodeling can produce much fine lead dust, especially during burning or powered sanding operations. Such processes may not only poison workers but also render a home much more dangerous to occupants.Similarly, workers removing old paint from steel structures by grinding or abrasive blasting prior to repainting may also be exposed. Demolition of such structures by the use of oxyacetylene torches, an extremely common practice, has resulted in many documented epidemics of lead poisoning.

Because lead poisoning has long been recognized in the battery industry, hygienic conditions in many plants have been improved. Nevertheless, dangers still exist in production and especially recycling old batteries.

Lead used for waterproofing can be a hazard both for those installing it and for those involved in repair or demolition. Several episodes of lead poisoning have occurred in small plants reclaiming telephone cable that has a protective lead sheath, and overexposure of cable splicers has been documented. The use of lead as a cleaning and coating agent in the manufacture of steel and wire may also be a hazard.

The use of lead solder in manufacturing and repairing automobile and truck radiators exposes many individuals, usually in very small and poorly ventilated shops. Additionally, lead in bullets and primer presents a significant hazard on indoor target ranges. Discharge of pistols and rifles produces much lead fume and particulate, a significant hazard to firing-range employees and regular users.

Lead poisoning may also occur in the arts with the use of lead-containing pigments and glazes for painting and ceramic work. Stoneware made with lead glazes may allow lead to leach into acidic foods and beverages. Lead is also used in stained-glass work and in making metal objects, figurines, and fishing sinkers.

A number of folk remedies from Mexico and other countries have been shown to contain large quantities of lead. Some calcium supplements from health food stores are produced from animal bones and may be heavily contaminated with lead. Hair coloring and some cosmetics may also contain lead.

Pathophysiology

Lead as a fume or very fine particulate is readily deposited in and absorbed through the lungs. Inhalation is probably the most significant route of absorption in most occupational cases. Larger particles that are inhaled will usually be trapped by the mucociliary defensive system; unfortunately, most lead that is cleared from the tracheobronchial tree is swallowed. Lead dust that contaminates hands, clothing, food, and tobacco products may also be ingested. Absorption from the gastrointestinal tract is particularly efficient in children. Although organic lead compounds, such as tetraethyl lead, can be absorbed through the skin, inorganic lead cannot.

Once in blood, lead is largely bound in red cells. With circulation, it is distributed to soft tissues and bone. Soft-tissue concentrations are highest in the liver and kidneys. Lead is largely excreted in the urine and, to a small extent, in the feces. Excretion is slow, with estimates of a half-life as long as 10 years. Lead also crosses the blood-brain barrier and the placenta.

There is relatively slow equilibration of lead among body compartments, so that levels of lead in blood may decrease after an acute exposure as lead is distributed to soft tissue and bone. After equilibration, about 90% of total body lead is present in bone. Although bone may act acutely as a metabolic "sink," trapping lead that would otherwise be available to cause greater harm elsewhere, the gradual accumulation of lead in bone that occurs with chronic exposure may maintain soft-tissue concentrations years after occupational exposure ceases. Under circumstances that cause calcium mobilization, lead that has been stored for some time in bone may be remobilized and may elevate blood lead levels. The mobilization of stored lead with pregnancy, menopause, and aging is a cause of concern.

Lead is a potent central nervous system poison. Recent studies have demonstrated subtle CNS effects at blood lead levels that had previously been thought to be safe. Severe episodes of intoxication may leave the affected worker with permanent cognitive and emotional changes. The earliest symptoms of intoxication are usually moodiness, irritability, change in sleep patterns, and difficulty with concentration. Patients typically develop headaches, a general decrease in well-being and energy, and a loss of libido. In many cases, these symptoms develop gradually, and patients and their family members tend to attribute them to an emotional or psychological cause rather than to a physical one. Medical attention may not be sought for some time, and if the patient is unable to alert the physician to the possibility of lead poisoning, diagnosis is often delayed. The insidious development and slow recognition of CNS manifestations of lead poisoning can cause great family and social disruption, perhaps the most devastating aspect of this illness.

Lead attacks the peripheral nerves and may cause a peripheral neuropathy, usually manifested by motor weakness and, in severe cases, by wrist drop or foot drop. Lead poisoning also predisposes to nerve entrapment syndromes, especially in workers who use their arms vigorously and repetitively. Although it may be present at time of diagnosis, entrapment at the wrist or elbow sometimes does not become clinically apparent until the patient is recovering from lead poisoning and has been returned to work at a physically demanding job.

Gastrointestinal symptoms are also prominent. Frequently, it is the development of abdominal pain that finally causes the affected person to seek medical attention. Some patients may develop the severe spasmodic pain of "lead colic," but most patients have symptoms that are difficult to distinguish from peptic ulcer disease or irritable bowel syndrome. Often patients develop constipation. It is believed that these symptoms are due to autonomic dysfunction.

The renal effects of lead poisoning rarely cause symptoms, but are a major cause of long-term morbidity and mortality. Lead principally attacks the renal tubular cells. Aminoaciduria may develop, and abnormalities of uric acid excretion may lead to episodes of gout. With severe poisoning, a picture of end-stage renal disease may supervene.

The relationship between lead exposure and hypertension is of great research interest. Studies have shown that individuals with prior lead exposure may have persistent renal dysfunction and hypertension. Mortality studies of battery workers show increased deaths from hypertensive cardiovascular disease, renal disease, and stroke.

The effects of lead on reproduction have engendered major public policy debate. In men, lead is associated with a reduction in sperm counts and some increase in the number of abnormal sperm. In women, lead exposure has been associated with spontaneous miscarriage and stillbirth. Studies of the relationship between increased umbilical cord blood lead and delayed cognitive development suggest that lead can be a significant transgenerational toxin.

The hematopoietic effects of lead have been extensively studied. The impact of lead on heme synthesis is typically measured in blood, but heme is critical not only to hemoglobin but also to cytochromes and other enzyme systems in many other tissues. Measurement of the various porphyrin markers in blood serves as a marker for the biochemical disruptiveness of lead in other organs.

Lead causes a normochromic normocytic anemia, in which red cells may show characteristic basophilic stippling. Even when patients are not frankly anemic, their hemoglobin values are usually depressed and typically rise slightly during recovery. When patients are frankly anemic, it is often due, in part, to hemolysis related to lead effects on the red cell membrane.

Endocrine dysfunction in patients with lead poisoning has been described, but its extent and significance are not understood.

Prevention

Primary prevention can be achieved by substitution of other materials, by elimination of exposure through process changes, and by good work practices.

Substitution has led to elimination of tetraethyl lead as an anti-knock additive, elimination of much of the lead soldering in steel-can manufacture, use of epoxy paints in place of red lead paint on some exterior metal surfaces, use of fluidized silica beds to replace traditional lead baths in wire manufacture, and changes in automobile design to remove the hazard posed by solder grinding.

Where lead is still used, processes can be changed to reduce fumes dramatically or to reduce the concentration and increase the size of lead particulate. Removing lead paint from the interior of homes traditionally involves burning and sanding surfaces, methods that can leave a house more dangerous to inhabitants than before and can poison those doing the work. State regulations banning the use of powered sanders and open flames and requiring postabatement measurement of surface dust lead levels can dramatically reduce this risk. To comply with these regulations, contractors and property owners can replace woodwork, perform off-site stripping, and chemically remove paint on site.

Other process changes can reduce the risks in painting exterior steel surfaces. Careful inspection and use of sanders with HEPA vacuum attachments on problem areas, or the use of wet methods, can reduce the enormous amount of leaded dust that is generated when surfaces painted with old lead paint or primer are prepared by dry abrasive blasting. Such methods can also reduce environmental pollution.

Jacketed ammunition and low-lead primers have been used to reduce the burden on ventilation systems at indoor pistol ranges.

When exposure cannot be eliminated by changes in process, enclosure and ventilation can be highly effective. To avoid turning an occupational exposure into a community exposure, filtration and extraction of lead fume and particulate from exhaust air may be necessary.

In industry, and even on construction and demolition sites, critical thinking about process and industrial hygiene can prove cost-effective by reducing the burden placed on personal protective equipment as a final line of defense. The use of wet methods of cleaning, HEPA vacuums, and particle accumulators or precipitators can reduce dustiness. Longer torches can enable demolition or scrapping workers to be farther back from the lead-containing plume.

Wherever dust is being generated or fume may be settling out, surfaces must be cleaned frequently. Providing facilities for frequent washing at work and forbidding eating, drinking, and smoking at the workplace can reduce the likelihood of lead ingestion. The use of protective clothing, coupled with a shower before leaving work, can eliminate the danger of lead particles being transported to workers' homes.

Good respiratory protection is a major part of primary prevention in settings where other methods have not been sufficient. Both workers directly involved in the generation of fume or dust and those working nearby need respiratory protection. Positive-pressure respirators are preferred; the OSHA lead standard provides that they be made available at the employee's request. Especially in construction and demolition work, where exposures can vary greatly from day to day and can be extremely high at times, negative-pressure respirators may be inadequate.

Because of the insidious nature of lead poisoning, worker education is critical to primary prevention. Worker training should include specific information about the danger of lead to workers and their families, symptoms of lead poisoning, importance of engineering controls and good work practices, and the provisions of the OSHA standard and other applicable regulations.

The OSHA industrial lead standard (CFR 1910.1025), promulgated in 1978, remains the cornerstone of regulatory prevention of occupational lead poisoning. It established a permissible exposure limit of 50 $\mu g/m^3$ of lead in air and an action level of 30 $\mu g/m^3$, and it prescribed that engineering controls be used whenever feasible to meet these goals. The standard requires removal of workers from exposure when blood lead levels exceed 60 μg/dL (or average over 50 μg/dL). Medical evaluation is required whenever a worker has a blood lead level over 40 μg/dL, complains of possible symptoms, or expresses concern about adverse reproductive effects. The standard also directs affected workers' physicians to design suitable restrictions and remedies, relying on their medical judgment.

In interpreting surveillance data, physicians should be sensitive to early symptoms of lead intoxication and not rely on the numerical requirements of the OSHA lead standard. Although the idea of defining specific levels that call for immediate action by employers is sound, blood lead levels that mandate worker removal in the current OSHA standard represent outdated thinking about "safe" levels. Beyond absolute values, a rising blood lead level is a signal that preventive measures have broken down and that the individual is at risk.

Although the standard was based on the now untenable idea that it was appropriate to allow working populations to endure an average blood lead level of 40 μg/dL, enforcement of the standard has achieved much better-than-anticipated results in many industries. The elaborate requirements for sampling, surveillance, employer-paid second medical opinions, protective equipment, and income maintenance while on medical removal have stimulated many employers to implement preventive measures to achieve exposures well below the action level.

In addition to OSHA regulation, some states and cities have adopted regulations linked to efforts to abate lead in housing. Some of these are more proscriptive and prescriptive than OSHA regulations and treat all facets of the problem, from identification of children at risk to disposal of lead-contaminated waste.

Construction and demolition work were exempted from the federal OSHA standard. In most of the U.S., construction workers have little regulatory protection from lead exposure. Maryland extended a somewhat modified version of the standard to construction in 1984 and has demonstrated that such regulations can be effectively enforced in the construction industry.

While never a substitute for the primary preventive measures described above, careful medical surveillance can detect evidence of lead absorption and intoxication in time to identify gaps in primary prevention and prevent permanent damage. Health professionals conducting medical surveillance must be alert for the subtle symptoms of lead poisoning elicited by careful history taking. Equally important is repeated reinforcement of the health education the worker has received about lead intoxication. Without these two elements, monitoring of laboratory tests may have little lasting benefit.

Measurements of intermediates in the heme synthesis pathway can serve as simple screening tests and can improve interpretation of blood lead measurements. Protoporphyrin IX, bound to zinc in vivo, can be measured directly and instantaneously with a hematofluorometer. Because zinc protoporphyrin (ZPP) can also be elevated by iron deficiency, results must be interpreted carefully, especially in screening children. Iron deficiency is so infrequent in working men, however, that an elevated ZPP nearly always indicates lead intoxication.

Measurement of blood lead itself may be used to confirm an abnormal ZPP or to serve as the primary screening test for lead absorption. Careful choice of an experienced laboratory and good quality control are important. Blood lead determination can be difficult, and confidence in the laboratory being used is critical. Even with CDC efforts at laboratory certification, much variability and inaccuracy still plague commercial clinical labs. State health departments may provide the most reliable services or give advice about selecting a dependable lab.

Because there are many nonoccupational sources of lead poisoning and because many occupational cases occur in settings where the potential for lead exposure has been ignored, sporadic cases are likely to continue despite the best efforts at organized control. Public health workers have begun to address this problem in some states by educating primary medical care practitioners and by requiring clinical laboratories to report elevated lead levels. Surveillance systems allow health departments to respond to each identified case by providing information and support to the patient and physician and, when appropriate, by involving OSHA.

Whenever a case of lead poisoning occurs, the nature of the patient's exposure should be characterized, family members or coworkers who may have also been exposed should be identified, and interventions should be made to stop exposure to lead.

The principal therapy of occupational lead poisoning is removing the worker from exposure and investigating the source. In cases in which blood lead levels or body burdens are high, where patients are symptomatic, or there is evidence of end-organ damage, chelation therapy is indicated. Therapy usually begins with ethylene diamine tetra acetate (EDTA) given intravenously, sometimes together with intramuscular dimercaprol (BAL). Dimercaptosuccinic acid (DMSA), an effective oral agent, has recently been licensed for use in children; it is likely to supplant the use of penicillamine. Chelation therapy should never be given prophylactically or administered to a patient when lead absorption may be continuing. Indications for chelation therapy remain controversial, and few controlled trials have been done.

JK

Further Reading

Agency for Toxic Substances and Disease Registry. *The Nature and Extent of Lead Poisoning in Children in the United States: A Report to Congress*. Washington, DC: U.S. Government Printing Office; 1988.

Annest JL, Pirkle JL, Makuc D, et al. Chronological trend in blood lead levels between 1976 and 1980. *N Engl J Med*. 1983;308:1373-1377.

Baker EL, White RF, Pothier LJ, et al. Occupational lead neurotoxicity: improvement in behavioral effects after reduction of exposure. *Br J Ind Med*. 1985;42:507-516.

Batuman V, Landy E, Maesaka JK, Wedeen RP. Contribution of lead to hypertension with renal impairment. *N Engl J Med.* 1983;309:17-21.

Batuman V, Maesaka JK, Haddad B, et al. The role of lead in gout nephropathy. *N Engl J Med.* 1981;304:520-523.

Cooper WC, Wong O, Kheifets L. Mortality among employees of lead battery plants and lead producing plants,1947-1980. *Scand J Work Environ Health.* 1985;11:331-345.

Cullen MR, Robins JM, Eskenazi B. Adult inorganic lead intoxication. *Medicine.* 1983;62:221-247.

Mahaffey KR, Annest JL, Roberts J, Murphy, RS. National estimates of blood lead levels: United States, 1976-1980. *N Engl J Med.* 1982;307:573-579.

Mantere P, Hanninen H, Hernberg S, Lukkonen R. A prospective follow-up study on psychological effects in workers exposed to low levels of lead. *Scand J Work Environ Health.* 1984;10:43-50.

Wedeen RP. *Poison in the Pot: The Legacy of Lead.* Carbondale, Ill: Southern Illinois University Press; 1984.

Leukemia

ICD-9 204-208
SHE/O

Identification

Leukemias are malignant neoplasms of the blood-forming system. They include myeloproliferative and lymphoproliferative disorders. The myeloproliferative disorders are clonal neoplasms arising in a pluripotent hematopoietic stem cell and characterized either by an excessive production of phenotypically normal mature cells (chronic myeloproliferative disorders) or by an impaired or aberrant maturation of hematopoietic precursor cells (acute myeloproliferative and myelodysplastic disorders). They include chronic myelogenous leukemia (CML) and acute myelogenous leukemia (AML), the latter also referred to as acute nonlymphocytic leukemia (ANLL).

Acute myelogenous leukemia is a hematopoietic stem cell disorder resulting in lethal overgrowth of incompletely differentiated bone marrow precursor cells. Various subtypes are identified by predominant cell type (myelomonocytic, monoblastic, promyelocytic, and megakaryocytic leukemias, and erythroleukemia). Chronic myelogenous leukemia is also a clonal stem cell disorder characterized clinically by overproduction of mildly defective granulocytes. Ninety percent of CML patients have the Philadelphia marker chromosome. In contrast to the myelogenous leukemias, acute and chronic lymphocytic leukemias (ALL and CLL) are lymphoproliferative disorders of committed lymphopoietic stem cells, originating in lymphoid precursors of marrow, thymus, and lymph nodes.

Symptoms include fatigue, weakness, weight loss, repeated infections, enlarged lymph nodes, bruising, and bleeding. Diagnosis is based on symptoms, on signs such as an enlarged spleen and enlarged lymph nodes, and on laboratory tests, including complete blood counts and smears (which reveal abnormally low or high blood cell counts and abnormal cells) and examination of the bone marrow.

Occurrence

In 1989, there were approximately 27 300 new cases of leukemia in the U.S., divided about equally between acute and chronic cases. Of these, 25 000 were estimated to occur in adults. Leukemia ranks as the fifth cause of new cancers both in men and women. Overall, it accounts for 3% of cancer incidence in the U.S., with acute leukemia accounting for 1%. The incidence of acute leukemia remained steady between 1975 and 1985. Age-adjusted incidence is somewhat higher among whites than among blacks.

Leukemia ranks as the third leading cause of death among cancers in men and fourth among cancers in women. In 1985, total deaths from leukemia overall were 18 100. In that same year, age-adjusted mortality rates of leukemia in the U.S. were 8.4 per 100 000 for males and 5.0 per 100 000 for females; these rates were 50% higher for men aged 25 and over than for women. Death rates from acute leukemia rise sharply with age among men, with a rate of 1.0 deaths per 100 000 among those 25 to 34 years old, 11.6 among those 65 to 74 years old, and 27.7 for those aged 85 and older. Geographical variation of death rates from leukemia in the U.S. has been observed, with elevated death rates occurring in several central states (Nebraska, Minnesota, Oklahoma, Colorado, Montana, Kansas, and Iowa). These are among the top 10 states for employment in agriculture. Maine, Oregon, and California also have relatively high death rates from leukemia.

Causes

Occupational causes include the following:

Benzene. Based on numerous case reports and various cohort studies (with standardized mortality ratios [SMRs] rising to 6500 among the most heavily exposed), benzene is clearly causally related to AML (especially erythroleukemia), less strongly to CML, and suggestively to Hodgkin's and non-Hodgkin's lymphomas, ALL, CLL, and multiple myeloma. There is a dose-response relationship between cumulative benzene exposure and leukemia risk, although other dose aspects are incompletely explored. The dependence of benzene-induced AML on a preceding aplastic anemia or pancytopenia (high-dose phenomena) has been asserted but cannot be considered established.

Several retrospective cohort studies have indicated excess mortality from leukemia. In one study of rubber plant workers, a fivefold excess mortality from leukemia was observed; in another, an SMR of 330 was observed among white males. A factorywide retrospective study of workers exposed to benzene in China found an SMR of 574, with mortality from leukemia occurring more among workers employed in organic synthesis, painting, and rubber production. Other occupations at risk for developing leukemia include pharmaceutical and plastics workers exposed to solvents, gas station attendants, petroleum production or products workers, and shoe and leather workers.

The use of benzene is also associated with hobbies; benzene is a component of paint strippers; dry-cleaning and spot removers; and furniture-refinishing products, paints, and waxes.

Ionizing Radiation. Based on studies of radiologists, Japanese atomic bomb survivors, and patients receiving therapeutic x-irradiation, increases in acute and chronic leukemia (but not CLL) are definitely linked to ionizing radiation exposure in excess of 50 rads in a dose-dependent fashion. Epidemiological studies of lower-dosage occupations and situations such as nuclear facility workers, nuclear shipyard workers, and soldiers and civilians exposed to weapons-test fallout have yielded mixed results, although radiation is considered the prototype for a nonthreshold carcinogen. Alpha radiation, associated with radon gas, does not appear to be a leukemogen, based on studies of miners.

Exposure to ionizing radiation resulted in a 20-fold increase in the incidence of ANLL and CML among Japanese atomic bomb survivors; survivors who also were exposed to benzene and to medical x-rays experienced an almost threefold risk for ANLL and an almost double risk for CML. Therapeutic radiation has been associated with the occurrence of leukemia; further epidemiological research is required.

Ethylene Oxide (EtO). About 75 000 U.S. health care workers had potential exposure to this potent alkylating gas in the early 1980s. Based on excess leukemias in production workers (relative risk = 10) and supporting studies of cytogenetic damage and other adverse health effects (stomach cancers, reproductive effects, animal carcinogenicity), OSHA lowered the TWA from 50 ppm to 1 ppm.

Excess mortality from and incidence of leukemia have been reported in Sweden both among workers exposed to EtO during a sterilizing process and among EtO operators and maintenance workers involved in its production. A larger study of Swedish EtO-exposed workers found a significant increase in leukemia. The International Agency for Research on Cancer (IARC) has concluded that there is evidence, albeit limited, that EtO is carcinogenic to humans, and NIOSH and OSHA have concluded that EtO is an occupational leukemogen. Further epidemiological research is required, however. Workers at risk of occupational exposure to EtO include hospital and health care workers; EtO is used to sterilize medical equipment.

Agricultural Industries. Agricultural workers and inhabitants of agricultural regions are at increased risk for leukemia, although specific exposures have not been clearly identified. Increased risk exists for agricultural workers in general, as well as for meat workers, cannery workers, lumber mill workers, pesticide appliers, and farmers, including livestock and poultry farmers (who have repeatedly been demonstrated to be at excess risk for acute leukemias, especially lymphatic, and also have had increased risk for lymphomas and myeloma). Herbicide exposure was recently reported to produce a dose-dependent increase of risk for non-Hodgkin's lymphoma.

Antineoplastic Agents. Health care workers who prepare and administer antineoplastic agents, such as the alkylating agents melphalan, cyclophosphamide, chlorambucil, and busulfan, may be at risk for developing leukemia, although this has not yet been demonstrated. Mutagenicity has been detected in the urine of nurses and pharmacists.

Electromagnetic Fields. Excess mortality from leukemia has been reported among workers who were exposed to strong electromagnetic fields. Such workers include electrical and electronic technicians, radio and telegraph operators, telephone and power line workers, power station operators, radio and television repair workers, welders, electricians, aluminum-reduction workers, and movie projectionists. Excess mortality from leukemia, particularly from myeloid leukemia, has been reported among electrical workers and coal miners exposed to underground power lines. Further epidemiological studies are required, however. The relationship remains speculative because of other negative studies, absence of dose responses, and possible confounding by benzene or ionizing radiation.

Chlordane and Heptachlor. Case reports have accumulated associating these termiticides with cases of leukemia (and aplastic anemia), and these pesticides have been suggested as the causal agents for some of the increased hematological neoplasms in farmers. Cohort studies of production workers and applicators have not confirmed leukemogenicity. However, certain of these compounds are carcinogenic in rodent liver.

Nonoccupational causes of leukemia include alkylating agents, particularly melphalan, cyclophosphamide, chlorambucil, and busulfan. Hodgkin's disease patients receiving such antineoplastic drugs have a 5- to 10-fold increase in risk for AML, and multiple myeloma patients have a 100-fold increase in risk for this disorder. Chlorambucil has been associated with an excess incidence of ANLL. Chloramphenicol-induced aplastic anemia patients have subsequently developed leukemia. Excess incidence of ANLL has been reported among patients with ovarian cancer, multiple myeloma, and breast cancer following therapy with alkylating agents.

Individuals with Down's syndrome are reported to have a higher risk of myelogenous leukemia. Genetic factors, which include the occurrence of chromosomal damage or unstable chromosomal patterns, are associated with ANLL.

Pathophysiology

The leukemias are all felt to be clonal disorders and are often characterized by chromosomal translocations and other aberrations. The known leukemogens, ionizing radiation and benzene, are potent clastogens, and EtO has been shown to cause sister chromatid exchange in a dose-dependent fashion in exposed workers, suggesting a possible unifying mechanism. Chemotherapeutic agents are also clastogenic. Induction of leukemia by radiation or chemotherapy peaks at about 5 to 10 years after exposure, compared with the 20- to 30-year latency for radiation-induced solid tumors.

Prevention

The reduction of OSHA's benzene exposure standard from 10 ppm to 1 ppm reflects concern for reducing increased leukemia risk from an estimated 95 deaths per 1000 workers to fewer than 10 deaths for a lifetime working exposure, based on the risk assessment favored by OSHA. The absence of effective secondary or tertiary prevention makes reduction of exposure critically important. Screening populations or individuals who are exposed to leukemogens for hematologic abnormalities is not likely to indicate significant deviation from exposure standards or to provide clinically meaningful early detection of neoplasia.

HK, BG

Further Reading

Goldstein BD. Clinical hematotoxicity of benzene. In: *Carcinogenicity and Toxicity of Benzene.* Vol 4. Princeton, NJ: Princeton Scientific Publishing; 1983:51-61.

Landrigan PJ. Occupational leukemia. In: Brandt-Rauf PW, ed. *Occupational Cancer and Carcinogenesis. State of the Art Reviews in Occupational Medicine.* 1987;2:179-188.

Linet MS. *The Leukemias: Epidemiologic Aspects.* New York, NY: Oxford University Press; 1985.

Savitz DA, Pearce HA. Occupational leukemias and lymphomas. *Semin Occup Med.* 1987;2:283-289.

Liver Cancer

ICD-9 155.0

(See also Hepatic Angiosarcoma)

Identification

Symptoms of liver cancer include right upper quadrant pain, fever, nausea, weight loss, and anorexia. Physical examination is generally unremarkable early. In rare instances, hepatic masses are noted, and ascites is infrequent on presentation. Results of liver injury tests are often abnormal but may not fit a specific diagnostic pattern. Alpha fetoprotein and carcinoembryonic antigen (CEA) levels are generally abnormal. A CEA level over 1:1000 is over 95% specific for liver cancer. Radionuclide scanning, abdominal computed tomography scanning, and sonography are useful diagnostic procedures.

Occurrence

Liver cancer represents approximately 1.5% of all cancer in the U.S., although the proportion reaches 30% in Southeast Asia. More than 90% of all liver cancers are hepatic adenomas. The 5-year survival rate is less than 4%. A rare liver tumor, hepatic angiosarcoma, has been associated with work exposure to vinyl chloride and to arsenic.

Causes

A large number of substances are documented animal carcinogens leading to liver cancer. These substances include aflatoxins, dioxins, dibenzofurans, halo-azo compounds, aliphatic halogenated hydrocarbons, pyrrolizidines, and alkaloids. Liver cancer can be a complication of cirrhosis of the liver, whether from infectious (hepatitis B) or noninfectious (toxic) causes. Several occupations have been associated with an increased risk, even in the absence of elevated liver cirrhosis rates; these occupations include farming, possibly due to the associated pesticide exposure, and road work, possibly due to exposure to coal tar pitch products. Workers in the manufacture of polychlorinated biphenyls (PCBs) may be at increased risk for liver cancer, primarily for cholangiocarcinomas. Chemists and possibly rubber workers and dry cleaners may also be at increased risk.

Pathophysiology

The precise mechanism is unknown. Agents may have intrinsic carcinogenic properties or may be activated in the liver to carcinogens. Inadequate degradation of these activated compounds may be the direct mechanism.

Prevention

Primary prevention involves exposure control. No dose-response relationships have been developed, so no clear guidelines may be established (for hepatic angiosarcoma). Screening has not been advocated as a secondary prevention measure, because no effective therapy is currently available. Concerning tertiary prevention, the effectiveness of liver transplantation, considered curative in the past, has recently been questioned. This costly procedure is not widely available.

MH

Further Reading

Dusheiko G. Molecular biology and animal models. *Gastroenterol Clin North Am.* 1987;16(4):575-590.

Jones DB, Koorey DJ. Screening and markers. *Gastroenterol Clin North Am.* 1987;16(4):563-573.

Kassianides C, Kew MC. The clinical manifestations and natural history of hepatocellular carcinoma. *Gastroenterol Clin North Am.* 1987;16(4):553-562.

Neugut AI, Wylie P, Brandt-Rauf PW. Occupational cancers of the gastrointestinal tract. In: Brandt-Rauf PW, ed. *Occupational Cancer and Carcinogenesis. State of the Art Reviews in Occupational Medicine.* 1987;2:137-151.

Nicholson WJ. *Review of the Health Effects and Carcinogenic Risk Potential of PCBs. Report to the Special Panel on Occupational Risk of Mortality from Cancer from Exposure to Polychlorinated Biphenyls.* New York, NY: Mount Sinai School of Medicine, Division of Environmental and Occupational Medicine; 1987.

Rustgi VK. Epidemiology of hepatocellular carcinoma. *Gastroenterol Clin North Am.* 1987;16(4):545-551.

Low Back Pain Syndrome

ICD-9 724.2

Identification

Low back pain syndrome is a common presentation for a large number of independent pathological conditions that affect the bones, tendons, nerves, ligaments, and intervertebral disks of the lumbar spine. It is classically divided into cases of regional low back pain, with or without pain radiating into the legs (ie, sciatica), and back pain due to systemic disease (eg, cancer, osteomyelitis, spondyloarthropathies, Paget's disease). The latter type is very uncommon in the working-age population, but it should be considered in workers over age 60 and in workers with backache at rest or with backache and fever, regardless of age.

Regional low back pain may result from a specific incident, injury, or fall; from chronic or repetitive trauma or strain; or from no definable precedent. Localized pain, decreased range of spinal motion, and often muscle spasm are the predominant features. Standard x-rays of the lower spine will either be normal or show minimal degenerative changes consistent with the patient's age.

Sciatic-type low back pain includes mild to severe pain with radiation down one or both legs. Normal walking is difficult, and mobility of the lumbar spine is limited. Pain is aggravated by jarring movements, such as riding in a car, coughing, or sudden changes in position. Computed tomography (CT) scanning and electromyography may be useful for a more complete diagnosis of sciatica.

The prognosis for an isolated episode of acute regional low back pain is excellent: 80% of those affected will be better in 2 weeks, and 90% will resolve in 1 month with conservative therapy (eg, rest and exercise) alone.

Occurrence

Low back pain occurs in approximately 80% of people in the course of their lives. It is relatively uncommon before age 20. First episodes most often occur between the ages of 20 and 40. After age 65, the prevalence of low back pain decreases. Men and women are equally likely to be affected, but men are much more likely to receive workers' compensation. Racial differences have not been well studied. However, low back pain is more commonly reported among those in lower, rather than upper, social classes. Sciatica occurs in about 1% of cases.

Estimates of the cost of low back pain are as high as $16 billion annually in the U.S. The cost of compensable low back pain is estimated to be $11 billion annually. About 75% of workers who leave work because of low back pain return to work within 1 month. From 5% to 10% are disabled for more than 6 months and account for 70% of lost workdays and workers' compensation costs due to low back pain.

Low back pain is more common among farm and nonfarm laborers, service workers, operatives, and craftspeople than it is among managers, professionals, salespeople, and clerical workers. Truck drivers have the highest rate of compensated back injuries. Truck drivers incur their injuries most often while loading or unloading trucks rather than while driving, although driving may be an important antecedent. Materials handlers, nurses, and nurses' aides also have high rates of compensated back injuries.

Understanding the work factors associated with reported low back pain is confounded by the common requirement for proof of work-relatedness to be eligible for workers' compensation benefits. Nevertheless, studies have consistently found low back pain to be associated with heavy manual labor. Lifting, twisting, and awkwardly moving on the job; prolonged sitting in one position; and prolonged driving of motor vehicles and other causes of whole-body vibration are thought to contribute to low back pain and herniation of the intervertebral disks. Low back pain is more common in workers who have muscle strength less than that routinely required by their jobs. People who are least fit physically may be more at risk for acute back pain, but those who are more fit may have more expensive injuries. Low back pain is also associated with boring, repetitious, and dissatisfying work and with occupational stress. Psychological factors affect reporting and recovery from low back pain, but there is little evidence to support a primary etiologic role.

Nonoccupational associations that have been cited by some studies but contradicted by others include height, increased weight, full-term pregnancy, and cigarette smoking. The association with recreational activities is weak except for interior linemen on football teams and gymnasts with spondylolisthesis. The best predictor of future back pain, both disabling and nondisabling, is a prior history of back pain; the rate of recurrence is approximately 90% within 10 years following an initial episode.

Causes

The causes of low back pain are different for different jobs and may be acute or chronic or both. Among drivers, causes may include vibration, acceleration or deceleration, inability to alter position, lack of back support, and leg position. Among materials handlers and nurses, causes include lifting objects or patients in awkward positions, with the load's center of gravity separated from the person doing the lifting. Causes also include slips and falls associated with poor walking or working surfaces.

Pathophysiology

The pathophysiology of regional low back pain is thought to be due to injury or microtrauma to the supporting structures of the lumbar spine or to the spine itself. The injury may arise from direct trauma (such as a fall), a single overexertion (such as lifting a particularly heavy or bulky object), or repetitive loading of tissues. The pathophysiology of sciatic-type back pain is thought to be through either nerve root compression or other mechanisms of nerve root irritation. Nerve root compression can be due to a herniated disk or to arthritic changes in the spine in the region of the neural foramen, where the nerve root exits the spinal canal. Irritative causes of sciatic-type back pain include arachnoiditis, active rheumatoid arthritis, and the so-called facet syndrome. The pathology of low back pain in any individual is almost always indeterminate.

Prevention

Primary preventive measures include the following:

- Designing the job to reduce known stresses associated with increased risk. Examples of effective ergonomic design to reduce low back pain are described below.
- Training workers to perform job tasks safely and effectively so the risk can be minimized.

Improved ergonomic design can best be accomplished by mechanical aids and good workstation design. Effective mechanical aids will reduce the load on the supportive structures of the lumbar spine during manual handling of heavy loads. Good workstation design reduces unnecessary bending, twisting, and reaching. Prolonged sitting and standing can be modified by providing flexible workstations that allow workers to perform their tasks in a variety of positions. Excessive loads can often be reduced by purchasing supplies in packages of appropriate size and weight. For drivers, good design can include improved vehicle suspension and cab design and improved loading and unloading facilities at truck terminals. For materials handlers and nurses, it can include mechanical lifting devices. In the unlikely event a job cannot be redesigned, strength testing of workers, combined with job placement for workers who are likely to be subjected to high-risk jobs or tasks, may be appropriate.

The NIOSH *Work Practices Guide to Manual Lifting* summarizes the research on acceptable lifting practices and has determined lifting conditions and weights that will be acceptable to 75% of the industrial population. For lifts above these "acceptable" limits, this document recommends that administrative controls (eg, worker selection and training) or ergonomic modifications (such as described above) be implemented.

Job placement (worker selection) may be appropriate for certain jobs that are difficult to design and control, such as fire fighting, police work, and certain warehouse and construction jobs. Selection techniques can be grouped into medical examination, strength and fitness testing, and job-rating programs. No published studies have yet shown the effectiveness of preplacement medical examinations in identifying workers who are susceptible to back pain or in preventing future back pain. Preemployment low back x-rays have been found to be both ineffective as a preventive tool *and* potentially harmful to workers due to unnecessary x-ray exposure. Although several studies have shown a relationship between strength and fitness and the incidence of low back pain, no studies have shown the effectiveness of using strength and fitness criteria as a preplacement technique in reducing the incidence of future low back pain in a particular job. In otherwise asymptomatic people, all available methods for screening workers for low back pain have serious technical, ethical, and legal limitations and should not be used in lieu of ergonomic controls as a way of preventing low back pain.

Job-rating programs are structured attempts to evaluate the job as well as the worker to obtain a good match between the two. Rather than screening out applicants, job-rating programs identify appropriate work for people with varying abilities. These programs appear promising, but to date, no studies have demonstrated the success of any job-rating program in modifying the incidence or severity of work-related low back pain syndrome.

Educating and training workers are the oldest and most commonly used approaches to preventing back problems in industry. Equally important is educating management, unions, and clinicians who deal with back pain in the workplace. Worker training programs have concentrated on safe lifting and on strength and fitness. Some uncontrolled studies have shown reductions in back disabilities as a result of training in safe lifting techniques, but no controlled studies have documented this finding yet. Although back strength and overall fitness of workers appear to be related to the incidence of back pain, it has not yet been shown that improving strength and fitness through a supervised training program in the workplace will reduce the occurrence of back pain. A variety of back schools have been used as coordinated ways of training workers either before or after an injury. The results look promising; however, there have been few controlled studies to document the effectiveness of back schools in the asymptomatic worker. Some studies have shown the value of back schools in rehabilitating injured workers so they can return to work earlier than they would otherwise.

Secondary preventive measures are aimed at identifying workers who are at increased risk for developing low back pain but who are not yet impaired or disabled by their symptoms. Medical surveillance, retraining, and job modification can play a role in the secondary prevention of low back pain and diminish its progression to low back impairment.

Tertiary measures, which overlap secondary preventive measures, include medical treatment and rehabilitation, work hardening, and, most important, job modification. Job modifications for the impaired worker include changing the tools, position, or manner in which a worker does a job, thus allowing the worker to accomplish the essential tasks despite physical impairment. Modified duty (sometimes called "light duty") is a related concept in which the worker is assigned to a new job with requirements that will not cause aggravation of pain or progression of disease. In the ideal situation, the job modifications will be individualized based on the degree and nature of the individual's impairment. For the worker with low back pain, the modifications usually include decreasing the frequency and degree of spinal flexion, decreasing weight loads lifted by the worker, and providing for flexible sitting or standing postures. The same ergonomic changes that have been associated with decreased incidence of low back pain are effective in enabling workers with mild low back pain to continue to work.

Modified duty may not be available in many workplaces. In these situations, workers and their physicians may be told that a worker needs to be "100% better" before being allowed to return to work. The purpose of work-hardening programs is to allow such workers an opportunity to return gradually to full working capacity in a supervised setting that is similar to the workplace but provides for modifications and worker autonomy while the worker completes the healing process. Recent studies that integrate work-hardening techniques with intensive physical and psychological rehabilitation programs have shown promising results in patients with chronic disabling back pain.

In summary, prevention and control of low back pain syndrome in the workplace require a combination of approaches that may be grouped into job design (ergonomics), job placement (worker selection), and education/training.

Other Issues

One of the most difficult problems in placing a worker with a history of low back pain is determining a safe job assignment that will protect the worker from serious future injury without limiting his or her work and economic opportunity. Since low back pain is such a common condition in the general population, it is clearly inappropriate to put limitations on a worker merely because of a history of prior back pain. If a worker has a history of recurrent, severe, and disabling back pain in relation to job tasks and requirements that are similar or identical to those anticipated in the job placement in question, however, modifications of the job or restrictions from known exacerbating factors would clearly be indicated. Absolute refusal of job opportunities based on the history, routine physical exam, or x-ray is rarely justified, given the lack of evidence that such restrictions are effective in preventing future episodes of back pain.

JH

Further Reading

Dayo RA, ed. Occupational back pain. *Spine: State of the Art Reviews.* 1987;2(1).

Frymoyer JW. Back pain and sciatica. *N Engl J Med.* 1989;318:291-300.

Himmelstein JH, Andersson G. Low back pain: fitness and risk evaluations. *State of the Art Reviews in Occupational Medicine.* 1988;3(1):255-269.

Kelsey JL. *Epidemiology of Musculoskeletal Disorders.* New York, NY: Oxford University Press; 1982:3. Monographs in Epidemiology and Biostatistics.

Leigh JP, Sheetz RM. Prevalence of back pain among full-time US workers. *Br J Ind Med.* 1989;46:651-657.

National Institute for Occupational Safety and Health. *Work Practices Guide to Manual Lifting.* Cincinnati, Ohio: National Institute for Occupational Safety and Health, Division of Biomedical and Behavioral Science; 1981. Technical Report 81-122.

National Institute for Occupational Safety and Health. *Low Back Atlas of Standardized Tests/Measures.* Morgantown, WVa: National Institute for Occupational Safety and Health, Division of Safety Research; 1988.

Pope MH, et al., eds. *Occupational Low Back Pain.* St. Louis, Mo: Mosby Yearbook; 1991.

Webster BS, Snook S. The cost of compensable low back pain. *J Occup Med.* 1990;32:13-15.

Lung Cancer

ICD-9 162
SHE/O

Identification

Occupational lung cancer includes all types of lung cancer that appear in the general population.

Occurrence

Lung cancer has been on the increase for the past 50 years, first among males and now among females as well. There are more than 125 000 deaths from this cancer each year. Most of these deaths occur in men; however, at the current rate of increase, women are expected to have equivalent death rates by early in the next century. It has been difficult to determine how much of lung cancer is directly attributable to occupation, because lung cancer is strongly associated with cigarette smoking. Latency between initial exposure and cancer onset is, on average, around 20 years, which makes identification of the association with historical exposure also difficult.

Causes

The best-recognized cause of lung cancer is cigarette smoking. There is some evidence that air pollution and involuntary inhalation of cigarette smoke are associated with excess risk. Experimental studies suggest a genetic predisposition to lung cancer in those with altered ability to metabolize hydrocarbons (with inducible aryl hydrocarbon hydroxylase). The causes of lung cancer that are due to occupation are listed below, accompanied by a brief review of the current understanding of the association between each agent or environment and lung cancer.

Arsenic. The best-described association between occupational arsenic exposure and lung cancer has been documented among smelter workers in studies from Japan, Sweden, and the U.S.; whether other agents in this environment (particulates and sulfur dioxide) may play a role in this risk is under study. Excess risk of lung cancer among arsenical pesticide manufacturers and applicators has also been documented.

Asbestos. One of the earliest and best-described workplace agents associated with lung cancer is asbestos, now suspected of causing increased risk of lung cancer, regardless of fiber type. It has been difficult to compare risk by fiber type, because unmixed exposure is rare. Elevated lung cancer has been found to be highest among insulators and textile workers, intermediate for asbestos product manufacturers, and lowest among asbestos miners. Asbestos-related lung cancer may, on average, be more peripherally located and more common in the lower lobe than is non-asbestos-related lung cancer. Moreover, asbestos-related lung cancer does not have a different cell-type distribution. With high exposure, even exposures lasting a few months have been shown to increase lung cancer risk. Of particular importance is the well-documented multiplicative (synergistic) effect of asbestos and cigarette smoking on lung cancer risk. Asbestos workers who smoke cigarettes have a 60-fold increased risk of lung cancer compared with workers who are not exposed to asbestos and do not smoke, while cigarette smokers without asbestos exposure experience a 10-fold increased risk. Excess risk is also present for asbestos workers who do not smoke.

Chloromethyl Ethers. Among the many alkylating agents used in modern industry, bis-chloromethyl-ether (BCME), a carcinogen in animals, is a frequent contaminant of chloromethyl ether (CME). As such, BCME is strongly associated with lung cancer in workers who encounter it as a contaminant of CME. Oat cell carcinoma occurs more frequently among such workers than in the general population, but other cell types have also been described.

Chromium. Chromium salts have been associated with lung cancer in several different types of employment. The accepted associations have occurred among those involved in chromate production and, more recently, among those using chromate pigments. The risk for chrome platers is still under study, but there is suggestive evidence of excess lung cancer in this employment group as well. There is general agreement that the hexavalent chromium salts are human carcinogens, with uncertainty, as yet, about the trivalent salts.

Coke Oven Emissions. A striking excess risk of lung cancer has been documented among coke oven workers, especially those who were exposed on the top of coke ovens. New environmental controls have dramatically reduced exposure to polyaromatic hydrocarbons (the suspected carcinogen group), so future risk is expected to be reduced.

Ionizing Radiation. Workers exposed to ionizing radiation, particularly alpha-emitting radioactive isotopes in the form of radon progeny, have shown an excess risk of lung cancer. This was first documented among uranium miners but now is also suspected to occur in other mining groups, such as hematite (iron ore) and other metal ore miners. There is an apparent excess of oat cell–type tumors among these workers and, among uranium miners, a multiplicative association with cigarette smoking. Excess risk, however, has been documented in non-smokers.

Nickel. A substantial excess risk of lung cancer has been documented among nickel production workers. Studies suggest that the risk was associated in the past with early stages of the refinery processes and that the risk may be expected to be controlled in modern production operations.

Vinyl Chloride. Vinyl chloride monomer has been determined to be a lung carcinogen in addition to its other cancer risks. The excess risks appear to be limited to environments where exposures have been substantial. For the most part, these are locations where polyvinyl chloride (PVC) polymerization is being carried out, rather than where vinyl chloride is being manufactured or where polyvinyl chloride is being put to final use.

Miscellaneous Materials and Work Environments

A number of other industrial exposures have been associated with lung cancer excess in working populations. The studies in support of these associations have not been frequently or widely replicated. These materials should be considered as suspect lung carcinogens.

Acrylonitrile. Acrylonitrile has been studied experimentally and found to be mutagenic and an animal carcinogen. Several human epidemiology studies have shown lung cancer excess, but only in the aggregate is the evidence strong enough to be convincing.

Aluminum Industry. Four mortality studies of aluminum workers, when combined, suggest that there is an excess of lung cancer among primary aluminum industry workers. A specific agent has not been identified.

Beryllium. Scientific debate continues about the lung cancer risk associated with beryllium exposure. For the present, the evidence that beryllium causes lung cancer in humans is limited.

Butchers. A number of studies have associated lung cancer with employment as a butcher or slaughterhouse worker. However, negative studies have also been reported, and the role of cigarette smoking in explaining the excesses has been suggested repeatedly.

Foundry Workers. Epidemiological studies consistently suggest that foundry workers are at a greater risk of dying from lung cancer than is the general population and that this risk varies with job performed, calendar time employed, duration of exposure, and type of foundry. Current studies should assist in refining the location and source of excess risk.

Man-Made Mineral Fibers (MMMF). Suspicion of lung cancer risk as a result of MMMF exposure results from evidence that the physical dimensions of asbestos fibers are key to the mechanism of their carcinogenicity. Studies of MMMF, however, have been confounded by a continuing development of smaller and thinner fibers and by the fact that many employees have previous exposure to asbestos. IARC has aggregated studies from Europe, and the evidence, when added to that found in U.S. studies, suggests that some fibers are associated with excess lung cancer risk.

Oil Mists (Cutting or Machining Fluids). Studies of oil mist exposures in various settings have determined these agents to be carcinogenic. Whether there is a specific risk for lung cancer, however, is still under study. When the major studies of machining operators and metal industry workers are taken together, the evidence supports such an association. Because industrial processes require a variety of machining fluids (straight mineral oil, semisynthetic oils, and synthetic oils) and the composition of any of these fluids varies from lot to lot and over time, it has been difficult to specify the agent(s) that might be causative. Cigarette smoking may also be confounding the understanding of this risk.

Printing Industry. An excess of lung cancer risk has been shown repeatedly in studies of this industry. The excess risk is not high, and the agent(s) has not been identified. It is possible that oil mists, common in the industry, are a source of the risk.

Silica. Recently, there has been a great deal of interest in whether silica is a carcinogen or at least a promoter. Various work environments that include silica exposures have been reviewed (eg, mines, quarries, granite industry, foundries, ceramics, stone work), and there appears to be a clear excess of lung cancer in workers exposed to silica. However, it is not clear whether silica is the causative agent or a marker for other carcinogen exposures. In addition, the fibrogenic property of silica may cause it to act as a promoter.

Transportation Workers. Case-control and cohort studies of truck drivers and other transportation workers have, with reasonable consistency, shown an excess risk of lung cancer. Because the work requires prolonged driving time, it is difficult to separate cigarette smoking from exhaust emissions as agents that might explain the risk. When controlled for smoking, recent studies have suggested an increased risk from exposure to diesel particulates.

Pathophysiology

There are three stages in the natural history of lung cancer. The development stage occurs with exposure to a carcinogen. Basal cell hyperplasia, stratification, and squamous metaplasia develop in areas of high exposure. As the tumor enlarges, the production of signs and symptoms signals the second stage. Tumor spread to contiguous areas marks the end of this stage, and the third stage occurs coincident with metastases. Tumor growth rates vary by cell type. Oat cell cancers grow most rapidly, and adenocarcinoma grows least rapidly. Residence time between estimated onset and manifestation of symptoms is 2 to 3 years for oat cell, 7 to 8 years for squamous cell (epidermoid), and 15 to 17 years for adenocarcinoma. In about half of patients with lung cancer, the initial symptoms are respiratory, but many lung cancer patients have coincident pulmonary disease (chronic bronchitis or emphysema), which makes it difficult to identify the new symptoms of cancer.

Prevention

Lung cancer has a high case-fatality rate and low 5-year survival. There is no accepted means of medical surveillance to identify tumors at an early enough stage to affect the ultimate course of the cancer. Extensive trials of surveillance by chest radiography and sputum cytology have not been shown to affect the mortality of those at risk, even in cases where the risk is high (eg, cigarette smokers and coke oven workers).

The most important approach to prevention, therefore, must be through primary prevention. This includes reducing or eliminating exposures to any of the above-noted occupational carcinogens as well as cessation of cigarette smoking. Further study of the interaction between cigarette smoking and occupational exposures, along with a better understanding of genetic, nutritional, and immunologic factors, is needed before host-factor susceptibility is likely to be identified or alternative therapies developed.

Other Issues

There are no means to distinguish occupational lung cancer from lung cancer from other causes. Cigarette smoking interacts with asbestos and radon progeny to multiply the risk of lung cancer in exposed workers. Similar effects have been inconsistently reported for workers exposed to arsenic and CMEs.

DHW

Further Reading

Alderson M. *Occupational Cancer.* Kent, England: Butterworths; 1986.

Cone JE. Occupational lung cancer. *State of the Art Reviews in Occupational Medicine.* 1987;2(2):273-295.

Frank AL. Occupational cancers of the respiratory tract. *State of the Art Reviews in Occupational Medicine.* 1987;2(1):71-83.

International Agency for Research on Cancer. *Overall Evaluations of Carcinogenicity: An Updating of IARC Monographs Volumes 1 to 42.* Lyon, France: International Agency for Research on Cancer; 1987. IARC Monographs on the Evaluation of Carcinogenic Risks to Humans, Supplement 7.

Schottenfeld D, Fraumeni JF Jr, eds. *Cancer Epidemiology and Prevention.* Philadelphia, Pa: WB Saunders; 1982.

US Public Health Service. *The Health Consequences of Involuntary Smoking: A Report of the Surgeon General.* Washington, DC: U.S. Government Printing Office; 1986.

Mass Psychogenic Illness—*See* Collective Stress Disorder

Median Nerve Entrapment Syndrome—*See* Carpal Tunnel Syndrome

Memory Impairment

ICD-9 310.1, 780.9

Identification

Memory impairment is often a symptom of chronic exposure to neurotoxins. Memory depends on the specific physiological and psychological processes of perception of a stimulus and of recall or reproduction of this perception. Anterograde memory impairment is the inability to assimilate new material. Retrograde memory impairment is the inability to recall previously learned material. Clinically, a distinction is made between deficits in long-term memory (in which retrieval is usually hours, days, or years after presentation), short-term memory (in which retrieval is several minutes to an hour after presentation), and immediate memory (in which retrieval is seconds to a few minutes after presentation). Distinguishing retrograde from anterograde memory disorders is easier if there is a specific time of injury, as there is with head injury or an acute, one-time high chemical exposure. It is more difficult if memory loss is due to chronic toxic exposure.

Memory impairment may lead to an inability to adapt to new situations requiring that new information be learned; therefore, for efficient work practices to continue, it is critical that occupational health professionals pay attention to signs of memory impairment. When questioned about recent events, memory-impaired workers may confabulate, either because they are embarrassed about their inability to remember or because they believe their words to be true. Generally, workers suffering from memory disorders do not identify themselves as such. It is, therefore, difficult to describe memory loss objectively as a symptom. Interviewing friends and family members about the patient's recall ability is helpful.

Deficits in short- and long-term memory may be more rigorously assessed by means of a neuropsychological evaluation that includes subtests of the Wechsler Adult Intelligence Scale and Wechsler Memory Scale (Information, Orientation, Mental Control, Memory Passages, Digit Span for immediate recall, Visual Reproductions, and Associated Language). The naming test, in which subjects are asked to identify famous people's faces (Albert's Test), may be another testing method to assess memory deficits. Neuropsychological testing may quantify deficits in memory by comparing results with preexposure test data, which implies that there is a worker surveillance system in place. When preexposure information is not available, tests of vocabulary ability that measure old, established verbal skills are of use in estimating baseline ability. Depressive symptoms concomitant with memory difficulties may be a response to the cognitive impairment experienced by the worker in exposure cases. Other neuropsychological signs that are often seen in workers with memory impairment include impaired initiative, speed, intellect, and concentration; easy fatigability; mood changes; anxiety; nervousness; emotional lability; and irritability.

Occurrence

The occurrence of memory impairment caused by occupational exposure is unknown. It is infrequently seen in isolation and is often a component of the symptom complex associated with toxic encephalopathy. It occurs as one of several symptoms of head injury or systemic poisoning by heavy metals, organic solvents, or pesticides. Occurrence data are usually classified by these specific etiologic agents. (*See also* the Occurrence section of Peripheral Neuropathy.)

Causes

Acute or chronic exposure to heavy metals (eg, arsenic, lead, manganese, mercury), solvents (eg, carbon disulfide, perchloroethylene, toluene, trichloroethylene), or organophosphates can lead to impaired memory. The reversibility of the symptom depends on the dose and, thus, extent of damage. Short-term memory deficits, especially of visual memory, in the presence of tremor and psychomotor disturbances are indicative of inorganic mercury intoxication. Workers exposed to lead demonstrate short-term memory problems, depression, and poor problem-solving abilities; long-term memory recall is relatively intact. As with metal exposure, solvent exposure causes mainly short-term and immediate recall impairment, along with impaired visuospatial functioning and affective problems such as depression. With immediate recall problems, deficits in learning ability are evident.

Nonoccupational causes of memory impairment include alcoholism, senility, depressive pseudodementia, Alzheimer's disease, and cerebral concussion. Korsakoff's syndrome consists of anterograde amnesia, variable retrograde amnesia with intact early memories, and confabulation. This syndrome has been associated with trauma, vascular lesions, surgically induced lesions, metabolic or nutritional deficiencies, central nervous system infections, or exposure to toxic agents. Korsakoff's disease or psychosis is the name of the alcohol-related syndrome of amnesia, and, in many cases, it presents in patients with Wernicke's encephalopathy. Wernicke's encephalopathy is associated with thiamine deficiency; people with this condition are in a confusional state and have ocular motor dysfunctions, memory disturbances, perceptual and conceptual function disturbances, ataxia, and, commonly, peripheral neuropathy. Cortical dementias, such as in Alzheimer's disease, involve loss of both anterograde and retrograde memory, impaired language function, and diminished ability to execute motor tasks due to cognitive deficits.

Pathophysiology

Evidence for the exact pathological target responsible for memory ability is not conclusive. Memory deficits in the presence of impaired verbal fluency, reduced psychomotor speed, and depression suggest bilateral frontotemporal dysfunction with lesions involving the amygdala and hippocampus. This is the suspected target in lead exposure. Short-term memory deficits in solvent-exposed workers have been attributed to diffuse brain damage. Memory is also thought to be a complex interaction of acetylcholine, neurotransmitters, and neuropeptides; thus, toxic chemical exposures may affect memory through interference in neurotransmitter function.

Prevention

Exposure to neurotoxins should be minimized. Effective industrial hygiene monitoring and control of respiratory, gastrointestinal, and dermal routes of chemical entry are discussed in part 1. Biological monitoring of workers for specific chemical exposures may be used as a guide to check on individual exposure levels, but careful attention must be paid to the interpretation of individual results. For more information, the American Conference of Governmental Industrial Hygienists (ACGIH) has issued biological exposure indices (BEIs) for a number of chemicals.

Periodic neuropsychological testing to assess memory problems in workers may be initiated, but it requires trained neuropsychological personnel.

Other Issues

People affected by alcoholism often develop memory disorders, which may confound the diagnosis of occupational toxic exposure. Other confounding influences, fatigue and other attentional deficits, and learning disabilities can affect the diagnosis of memory impairment as a result of toxic chemical exposure.

RF, SP

Further Reading

Baker EL, White RF. The use of neuropsychological testing in the evaluation of neurotoxic effects of organic solvents. In: Joint/WHO Nordic Council of Ministers Working Group. *Chronic Effects of Organic Solvents on the Central Nervous System and Diagnostic Criteria.* Copenhagen, Denmark: World Health Organization; 1984;219-242.

DeJong RN. *The Neurologic Examination.* 4th ed. Hagerstown, Md: Harper & Row; 1979.

Feldman RG, Travers PH. Environmental and occupational neurology. In: Feldman RG, ed. *Neurology: The Physician's Guide.* New York, NY: Thieme-Stratton; 1984:191-212.

Juntunen J. Alcoholism in occupational neurology: diagnostic difficulties with special reference to the neurological syndrome caused by exposure to organic solvents. *Acta Neurol Scand.* 1982;66(suppl 92):89-108.

Lindstrom K. Behavioral effects of long-term exposure to organic solvents. *Acta Neurol Scand.* 1982;66(suppl 92):131-141.

Lindstrom K. Criteria for the psychological diagnosis of toxic encephalopathy. In: Joint WHO/Nordic Council of Ministers Working Group. *Chronic Effects of Organic Solvents on the Central Nervous System and Diagnostic Criteria.* Copenhagen, Denmark: World Health Organization; 1984:243-262.

Mikkelsen S, Browne E, Jorgensen M, Gylehensted C. Association of symptoms of dementia with neuropsychological diagnosis of dementia and cerebral atrophy. In: Joint WHO/Nordic Council of Ministers Working Group. *Chronic Effects of Organic Solvents on the Central Nervous System and Diagnostic Criteria.* Copenhagen, Denmark: World Health Organization; 1984:166-184.

White RF. Differential diagnosis of probable Alzheimer's disease and solvent encephalopathy in older workers. *Clinical Neuropsychologist.* 1987;1(2):153-160.

Mercury Poisoning

ICD-9 985.0

Identification

Mercury poisoning can occur from either inorganic (including elemental) mercury or organic mercury, especially alkyl mercury compounds (methyl and ethyl mercury). Diagnosis is based on the combination of symptoms, the signs on physical examination, and a history of exposure. Laboratory tests to determine the concentration of mercury in blood and urine may be helpful but usually do not correlate well with toxic effects. Signs and symptoms vary by type of mercury or mercurial compound and by type of exposure (acute or chronic).

Acute poisoning by elemental mercury or inorganic mercurial compounds is often associated with respiratory tract or skin irritation, kidney damage, stomatitis (inflammation of the mouth), and, if the mercury has been ingested, gastrointestinal disturbances. Chronic poisoning (more typical for occupational mercury poisoning) is associated with a classic triad of tremor, psychological irritability (erethism), and stomatitis, among other symptoms.

Acute poisoning by organic mercurial compounds is associated, if there has been skin contact, with irritation and even burns. Chronic poisoning is associated with tremor, paresthesias (tingling sensations), dysarthria (slurred speech), ataxia (loss of balance), decreased visual fields, and mental disturbances.

Occurrence

Reliable data on the actual incidence of mercury poisoning do not exist. More than 65 occupations have been identified as having exposure to mercury, including the manufacture of certain pharmaceuticals, paints, catalysts, control devices, textiles, paper, and jewelry; the manufacture and use of fungicides; gold mining; chlor-alkali mercury cell operations for production of chlorine gas; and work in certain laboratories and dental offices.

Causes

Because occupational exposure to alkyl mercury compounds is uncommon in the U.S. at this time, most occupational mercury poisoning in the U.S. now is due to elemental mercury or inorganic mercurial compounds. Environmental exposures involve both inorganic and organic mercury.

Pathophysiology

Mercury is primarily absorbed through the lung, although absorption through the skin and gastrointestinal tract also occurs. A large fraction of absorbed mercury is deposited in the kidney, but it is also deposited in the brain, liver, red blood cells (especially alkyl mercury compounds), and other tissues and organs. Through its affinity for sulfhydryl groups, mercury exerts much of its toxic effect by inhibiting enzymes (at least 30 common enzymes are so affected) and by binding to cell membranes. The type of damage varies by organ. In the kidney, damage ranges from isolated tubular injury to the nephrotic syndrome with large amounts of protein in the urine. Tissue concentrations of mercury do not correlate with toxic effects; for example, methyl mercury can reach high tissue concentrations in the kidney without significant toxicity, but it can be present at lower concentrations in the brain with significant neurological damage.

Prevention

Prevention is accomplished by reducing exposures, monitoring both workplaces where mercury is used and workers who are potentially exposed, investigating outbreaks of mercury poisoning, and ensuring that control measures are developed and implemented. Engineering control, often with enclosure or ventilation of processes that use mercury or mercurial compounds, is often the most effective means of reducing exposure. Education of workers regarding the hazards of mercury and ways of reducing exposure plays an important part in prevention. Protective clothing for exposed workers is often important because mercury can be readily absorbed through the skin. Preplacement and periodic medical examinations, with emphasis on the skin and mouth, respiratory tract, neurological system, mental status, kidney function, and urinalysis, are recommended for exposed workers. Although blood mercury levels correlate reasonably well with acute toxicity due to recent exposure, urine mercury levels corrected for creatinine can be helpful for monitoring exposure over time. Methods and timing of collection are important for the interpretation of urine levels.

Other Issues

Clothing contaminated with mercury and brought home by workers can be a source of mercury poisoning for family members. Mercurial compounds discharged into the air from industrial plants can cause adverse health effects in people living nearby. Many foods, especially seafoods, contain mercury. Mercury poisoning can occur by ingestion of food contaminated with mercurial fungicides. An outbreak of over 100 cases and 41 deaths occurred in residents around Minamata Bay in Japan, who consumed fish contaminated by methyl mercury from waste dumped into the water by an industrial plant.

BSL

Further Reading

Battigelli MC. Mercury. In: Rom WD, ed. *Environmental and Occupational Medicine.* Boston, Mass: Little, Brown; 1983;449-463.

Finkel AJ, ed. *Hamilton and Hardy's Industrial Toxicology.* 4th ed. Reading, Mass: PSG Publishing; 1983:93-106.

Proctor NH, Hughes JP, Fischman ML, eds. *Chemical Hazards of the Workplace.* 2nd ed. Philadelphia, Pa: JB Lippincott; 1988.

Mesothelioma

ICD-9 158, 163
SHE/O

Identification

Mesothelioma is a malignant tumor of the pleura, peritoneum, or pericardium. The diagnosis is difficult because of nonspecific signs, symptoms, and laboratory findings, such as chest or abdominal pain, weight loss, ascites, or pleural effusion. Mesothelioma is often misdiagnosed initially as a metastatic abdominal tumor. Diagnosis usually requires a large tissue section demonstrating mucin staining. Although closed-needle biopsy or effusion cytology may be sufficient for diagnosis, open biopsy or thoracotomy is usually necessary to obtain sufficient tissue. Even then, mesothelioma may be misdiagnosed as metastatic adenocarcinoma. Tissue should be sent to a pathologist experienced with the diagnosis or to a mesothelioma registry. Treatment protocols have been ineffective in altering the median survival time of 11 months from diagnosis to death.

Occurrence

In the general population, mesothelioma is rare. Mortality rates among U.S. men are 4.4 to 11 deaths per million and among women, 1.2 to 3.8 deaths per million. About two thirds of men who die from mesothelioma are over 65 years old. Projected estimates of mortality from mesothelioma range from 850 to 2000 deaths per year into the next century.

The highest estimated incidence rates are reported in South Africa, with 32.9 cases per million among white adult males, over 100 cases per million for white males aged 55 and over, and 24.8 per million for nonwhite males. Differences for racial groups may be due to differences in diagnostic accuracy resulting from differences in access to health services.

Mesothelioma is more frequent in the abdominal cavity than in the chest cavity by about 2:1. Its occurrence in the pericardium is rare.

Death rates vary by geographic location, with high rates in regions with a history of shipbuilding and of manufacturing asbestos products. Thus, states with the highest reported rates are Washington and New Jersey. A high death rate in Florida is believed to be attributable to a large number of retirees migrating from the northeastern U.S.

Causes

Occupational and nonoccupational exposure to asbestos is found in most cases of mesothelioma. Exposure may have been very brief, and latency can be 40 years or longer. Other possible causes include exposure to ionizing radiation, thorotrast, and natural and artificial fibrous materials. In case series, the proportion attributable to asbestos has ranged from 20% in retrospective chart reviews to over 95% when living subjects were interviewed about their occupational histories.

It has been suggested that amphibole fibers are significantly more carcinogenic than chrysotile fibers for producing mesothelioma, but the evidence for this conclusion is indirect. In some epidemiological studies, the occurrence of mesothelioma is more frequent among workers exposed to asbestos consisting mainly of amphibole fibers than chrysotile. Amphibole fibers are more common in the lungs of people autopsied for mesothelioma. However, in other investigations, mesothelioma also occurs among workers exposed only to chrysotile, although the risk appears less. In laboratory animals, both fiber types produce malignant mesothelioma following inhalation and implantation.

Mesotheliomas have been caused experimentally by exposing laboratory rodents to fibrous materials other than asbestos.

Pathophysiology

The precise carcinogenic mechanism whereby either chrysotile or amphibole fibers produce mesothelioma is unknown. Based on investigations in animals, carcinogenesis may depend on fiber length and width. Thus, long (8 micrometers and longer) fibers with a high aspect ratio (length to width) tend to be more carcinogenic than shorter (5 micrometers or less), thicker fibers. An alternative mechanism attributes carcinogenicity to fiber surface characteristics and immunologic mechanisms.

Prevention

Disease control methods are discussed in the entry on asbestos-related diseases.

Other Issues

Smoking does not appear to alter the risk of mesothelioma.

MH, DHW

Further Reading

Churg A. Chrysotile, tremolite, and malignant mesothelioma in man. *Chest.* 1988;93(3):621-628.

Lilienfeld DE, Mandel JS, Coin P, Schuman LM. Projection of asbestos related diseases in the United States, 1985-2009, I: Cancer. *Br J Ind Med.* 1988;45:283-291.

Lillis R. Mesothelioma. In: Merchant JA, ed. *Occupational Respiratory Diseases.* Washington, DC: U.S. Government Printing Office; 1986:671-688. DHHS (NIOSH) publication 86-102.

Neugut AI, Wylie P. Occupational cancers of the gastrointestinal tract. In: Brandt-Rauf PW, ed. *Occupational Cancer and Carcinogenesis. State of the Art Reviews in Occupational Medicine.* 1987;2:109-135.

Zwi AB, Reid G, Landau SP, et al. Mesothelioma in South Africa, 1976-84: Incidence and case characteristics. *Int J Epidemiol.* 1989;18:320-329.

Metal Fume Fever

ICD-9 506.0-506.3
SHE/O

Identification

Metal fume fever, or metal fever, is an acute and usually self-limited syndrome that follows exposure to metal fumes or very fine metal dust particles. It is characterized by a complex of symptoms that includes fever, chills, excess sweating, nausea, weakness, fatigue, throat irritation, cough, headache, myalgias, and arthralgias. The first symptoms are often thirst and a metallic taste in the mouth. Onset usually occurs after a delay of several hours from time of exposure to the offending agent(s). Peripheral leukocytosis often occurs.

The diagnosis is made by taking a careful history and making special note of recent exposures. Repeated episodes may occur. There is no chronic form of the disease, although the occurrence of wheezing has reportedly been associated with a prior history of metal fume fever.

When, in the absence of an accurate occupational history, attention is focused on the most recent exposures, metal fume fever is often incorrectly identified as the flu.

Occurrence

No population-based studies are known to have been conducted to establish the frequency of this condition adequately. However, given the common occurrence of metal fume exposure in the workplace, the number of workers at risk is large, and the syndrome is considered common.

Causes

Metal fume fever is caused by exposure to metal fumes (ie, very small condensation products of metal vapor) and by fine particles of metal oxides. The most common metals causing this syndrome are zinc, copper, and magnesium. Other metals that have been shown to cause it include aluminum, antimony, cadmium, copper, iron, manganese, nickel, selenium, silver, and tin. It commonly occurs following welding (eg, on galvanized steel), brazing, soldering, or other exposure to molten metal. Exposure in a poorly ventilated workplace may be a contributing factor.

Exposure to cadmium or manganese may also result in pneumonitis (*see* Hypersensitivity Pneumonitis), and exposure to nickel may also result in occupational asthma (*see* Asthma). There are also case reports of asthma following exposure to zinc fumes.

Pathophysiology

A full understanding of the pathophysiology of this syndrome has not been achieved. Metal fume fever can be reproduced in laboratory animals. It has been postulated that the pathophysiological mechanism is as follows: minute particles (0.05- to 0.5-μm particles in 5- to 10-μm aggregates) of metal fume or dust penetrate deeply into the respiratory tract and easily reach alveoli; they then activate cells (eg, macrophages) with subsequent release of pyrogenic mediators. Unless accompanied by bronchiolitis obliterans or pneumonitis, they cause no permanent structural damage. Curiously, tolerance is acquired after continuous exposure but lost after several days. This suggests some form of unknown immunologic mechanism.

Prevention

Primary prevention involves reducing exposure to offending agents through engineering controls or local exhaust ventilation. Respirators are available but should be selected to filter particles in the submicron-size range. Early identification and diagnosis of this syndrome, documentation of its occurrence, and follow-up evaluation are also important. Workers potentially exposed to molten metal (eg, welders, solderers, foundry and smelter workers) or to very fine metallic aerosols should be alerted to this condition and urged to take precautionary measures such as use of ventilation. Often, experienced workers know the syndrome and are a useful source of information for less experienced workers.

DCC

Further Reading

Cotes JE, Feinmann EL, Male VJ, Rennie FS, Wickham CA. Respiratory symptoms and impairment in shipyard welders and caulkers/burners. *Br J Ind Med.* 1989;46:292-301.

Nemery B. Metal toxicity and the respiratory tract. *Eur Respir J.* 1990;3:202-219.

Pierce JO. Metal fume fever. In: Parmeggiani L, ed. *Encyclopedia of Occupational Health and Safety.* 3rd ed., rev. Geneva, Switzerland: International Labour Organization; 1983.

Ross DS. Welders' metal fume fever. *J Soc Occup Med.* 1974;24:125-129.

Microwave/Radiofrequency Radiation—*See* Radiation, Nonionizing, Adverse Effects

Mononeuritis Multiplex—*See* Carpal Tunnel Syndrome

Motor Vehicle Injuries

ICD-9 E810-E825

Identification

Occupational injuries from motor vehicle crashes occur when people are performing occupational duties by driving a motor vehicle. These injuries can involve both highway and off-the-road vehicles. Injuries vary from superficial contusions to massive crushing injuries affecting several regions of the body. A relatively injury-free external surface can be deceptive; closed head, chest, and abdominal organ injury should always be considered.

Occurrence

The leading cause of fatal occupational injuries is motor vehicle crashes, accounting for 28% of all fatal occupational injuries in the U.S. in 1988. The principal data source for on-road fatal motor vehicle injuries is the Fatal Accident Reporting System (FARS), published annually by the National Highway Transportation Safety Administration (NHTSA) and based on local police reports. A classification for occupationally related injuries was recently added to FARS data, but reporting is incomplete. It is inferred that, most often, drivers of medium and heavy trucks and buses who incur injuries were performing occupational duties.

According to this source, in 1987 there were 5451 fatal crashes involving medium and heavy trucks and buses, resulting in 5878 fatalities. The vast majority of fatalities (85%) occurred to nonoccupants or to occupants of other vehicles.

According to the BLS, highway vehicles were involved in 28% of all occupational fatalities in 1988 and were the leading cause of fatalities in each major industrial sector except mining. Most fatalities involving highway vehicles occurred in the transportation and public utilities sectors, followed by construction and manufacturing. Industrial vehicles were involved in 9% of all fatalities, most often in construction and manufacturing.

Causes

The causes of injuries from motor vehicle crashes can be classified by characteristics of the vehicle operators, of the vehicles involved, and of environmental factors. The severity of injuries is a function of energy exchange in the crash: energy = $mv^2/2$, where m = mass and v = velocity. Energy increases with the square of speed, but vehicles with greater energy-absorbing properties are less likely to injure severely at the same speeds.

Risk factors for drivers include gender, age, length of driving immediately prior to the crash, and alcohol consumption. Risk factors for vehicles include dimensions, weight, maintenance, center of gravity, and energy management in the event of a crash (eg, seat belts, air bags, and reinforcing and energy-absorbing structures). Environmental factors include conditions of roadways, shoulders, and berms; weather; and time of day.

Motor vehicle injury is twice as common among men per mile driven. The rate of injuries per mile declines exponentially with age and increases exponentially with blood alcohol concentrati n. Alcohol is involved most often in motor vehicle injuries among males 20 to 45 years old. The risk of injury increases per number of convictions for violations in the prior 3 years. Truckers driving more than 6 hours increase crash risk by about one third.

Fatal crash involvement of tractor-trailer trucks is 16 times that of cars per vehicle registered, and double-trailer trucks are in crashes 2 to 3 times more often than single-trailer trucks on the same roads. Deaths are usually in vehicles that collide with the trucks because a truck's mass is greater.

Brake defects were found about 1.5 times as often in trucks involved in crashes as in trucks not involved in crashes, and steering defects were found more than twice as often.

Motorcycle occupant fatalities per vehicle registered are three times those of cars. Occupant deaths in cars with wheelbases less than 100 inches are twice those in cars with wheelbases of 120 inches or more per vehicle registered due to reduced distance for occupant deceleration.

Pathophysiology

Death may occur within minutes or hours, most often from nervous system damage, internal hemorrhage, or asphyxiation from damage to the larynx. Later deaths are related to complications of infection and immobility.

Prevention

Prevention strategies follow from the known risk factors. Teenagers should not be employed in jobs that require driving.

People with convictions for moving violations in the prior 3 years should not be employed in jobs that involve driving.

High-risk vehicles, such as motorcycles, small utility vehicles, small cars, and double-trailer trucks, should be avoided. When utility vehicles are necessary, the stability coefficient should be 1.2 or more. (The stability coefficient is the ratio of the distance between the center of the tires to twice the height of center of gravity.) Antilock brakes should be installed on all vehicles, and headlamps should be kept on at all times. Reflective tape should be placed around the outline of the truck to enhance visibility at night. Speed limits should be enforced.

Underride barriers on the rear of truck trailers should be replaced with lower, energy-absorbing structures. Seat belts and air bags should be installed in all vehicles, and only vehicles with air bags should be purchased. Vehicles with sharp points or edges on their fronts that increase severity of injury to pedestrians should be avoided.

Other Issues

Many occupational injury programs are based on education (*see* part 1). However, there is no good experimental evidence that driver training reduces crash risk. Studies of high school driver education, defensive driving courses, and professional driver courses indicate that they do not reduce the risk of crash or injury. Most studies that claim such courses are effective do not control for selectivity of lower-risk drivers into the courses. Available studies do not preclude the possibility, however, that an effective program could be devised.

LSR

Further Reading

Baker SP, O'Neill B, Karpf RS. *The Injury Fact Book.* Lexington, Mass: DC Heath; 1984.

National Highway Transportation Safety Administration. *Fatal Accident Reporting System*, 1987. Washington, DC: U.S. Government Printing Office; 1989.

Robertson LS. *Injuries: Causes, Control Strategies and Public Policy.* Lexington, Mass: DC Heath; 1983.

Robertson LS. Risk of fatal rollover in utility vehicles relative to static stability. *Am J Public Health.* 1989;79:300-303.

Stein HS, Jones IS. Crash involvement of large trucks by configuration: a case-control study. *Am J Public Health.* 1988;78:491-498.

Stein HS, Jones IS. Defective equipment and tractor-trailer crash involvement. *Accid Anal Prev.* 1989;21:469-481.

U.S. Department of Labor. Bureau of Labor Statistics. *Occupational Injuries and Illnesses in the United States by Industry, 1988.* Washington, DC: U.S. Government Printing Office; 1990.

Multiple Chemical Sensitivities

Identification

Multiple chemical sensitivities (MCS) is an incompletely understood syndrome whose hallmark is symptoms affecting multiple organs and occurring after exposure to small amounts of diverse chemicals. Although specific causes are unknown, MCS is sufficiently distinctive and common to have stimulated public and scientific discussion. A general consensus has emerged on its name and working definition. The cardinal features of MCS include each of the following:

1. Symptoms first appear in a previously healthy person after a typical occupational or environmental illness, such as intoxication or injury due to a high-level ex-

posure to a known toxin. This precipitating illness may be isolated or recurrent, mild or severe, and usually becomes known to a physician in the course of taking a patient's history.

2. Reexposure to decreasing amounts of the same or a similar toxin causes symptoms resembling those of the precipitating illness.
3. Symptoms become generalized to other organ systems, almost invariably including the central nervous system.
4. The causes become generalized. Increasingly diverse and chemically unrelated classes of substances at decreasing concentrations evoke symptoms. Responses typically occur at exposures that are orders of magnitude below accepted exposure limits.
5. Common tests of organ function cannot explain the constellation of symptoms.
6. Psychosis or major medical conditions are absent.

Multiple chemical sensitivities may be a complication of a well-characterized occupational or environmental disease, but it is distinct from any toxic or allergic reaction yet described because of the diversity of symptoms and causes and because of very low exposure. The inability of organ function tests to explain symptoms and the absence of major medical conditions or overt psychosis may suggest a milder psychiatric disorder or neurosis. More severely disabled people may, in fact, become anxious or depressed. However, the features of full-blown MCS are sufficiently distinct to warrant this diagnosis until its pathogenesis and means of control are understood.

This distinctiveness is less clear when not all the above cardinal features are met. Current knowledge is far too limited even to predict the likelihood that people with only some of the features have the same underlying disturbance. In such cases, other diagnoses should be considered, especially when longstanding symptoms predate known occupational or environmental disease or when symptoms vary without relation to environmental reexposures, however small.

Occurrence

Because no uniform definition has been accepted, data on occurrence are limited. The following summary is based on anecdotal reporting and several small surveys.

Although no reliable prevalence of the data exists, there is a general impression that incidence of MCS has increased during the 1980s. This may represent an artifact of recognition, but stable occupational medicine practices with defined referral bases appear to share this perception.

Demographic factors associated with higher risk include female gender, higher socioeconomic and education levels, and Caucasian race. Very few cases have been reported among impoverished or marginal workers, especially nonwhites. However, the extent to which any of these risk factors is truly predictive—as opposed to being a reflection of illness perception, of access to occupational or environmental physicians, or of reporting biases—is unknown. Multiple chemical sensitivities seems to have a higher incidence among younger individuals—mostly in their 30s and 40s—than among older people, although cases have been described even among retirees.

Cases have usually occurred in isolation from each other, suggesting some host idiosyncrasy, although host factors predisposing to illness remain obscure. Multiple cases have occurred after some outbreaks of occupational disease, especially "sick building syndrome" (*see* Building-Related Illness). There have also been reports of clusters occurring in families, suggesting either genetic or acquired common risk factors, but these clusters have not been analyzed in detail.

Causes

Although the cause or causes are not known, enough experience with MCS has now accrued to allow discussion of some environmental and occupational factors associated with it. These exposure factors appear to be a necessary component of the illness sequence; hence, they may appropriately be considered as "causes" even while it remains evident that other contributors, including host factors, must also be causally linked.

Virtually any toxic substance in the workplace or ambient environment, when present in sufficient concentration to cause predictable and acute toxicological effects, appears capable of precipitating MCS in some individuals. Overwhelmingly, most important among these substances are organic solvents, pesticides, and respiratory tract irritants. It is unknown whether these large classes of compounds have unique features or whether these agents are prominent merely because of their ubiquitous presence in the environment, coupled with their high potential to cause perceptible, albeit often mild, clinical effects.

A very important environmental precipitant is the constellation of agents that collectively make up the air of "sick buildings." Among individuals adversely affected in large outbreaks of sick building syndrome, some develop MCS, often with little hiatus or clear demarcation between the responses that are shared by large proportions of exposed workers and the insidious development of more generalized responses in a few. Sick building syndrome and MCS must be distinguished, although the latter may complicate the former in some individuals. Unlike most of their co-workers, those who develop MCS will not predictably improve when control measures are adopted to improve indoor air quality.

Once the illness is established, patients with MCS will react adversely to an extraordinary array of environmental agents, both toxic and nontoxic. Substances with low thresholds for irritation and/or odor, as well as those with various properties similar to those of the original offending agent, seem to be the most potent stimuli. This may reflect the ease with which the patient recognizes such substances rather than any special importance the substances have in precipitating symptoms as compared with other widely distributed chemicals. Exhaust fumes, aerosols, and indoor contaminants such as formaldehyde and tobacco smoke often appear on the long lists of intolerable substances.

Pathophysiology

The mechanism(s) by which a toxic exposure and subsequent disorder precipitates MCS in a susceptible worker is unknown. The pathways that lead from low-level exposure to the provocation of often disabling symptoms are also obscure. Moreover, it is unclear whether MCS is due to a single disease process.

One or more of several theories may explain MCS. Most important from a societal point of view is that MCS and its variations represent a form of global, cumulative poisoning of the immune system resulting in immune dysregulation. This proposal, espoused for many years by a group of practitioners known as clinical ecologists or environmental physicians, explains MCS as a result of an excessive total body burden of toxic compounds—largely man-made chemicals of petrochemical origin. According to this view, collective toxicity of these prevalent chemicals in hosts who are perhaps predisposed because of nutritional, infectious, or other stressors creates an effect that exceeds the potential of any toxin alone. The result is the development of "allergy" to a broad range of substances, which can be mitigated only by extreme chemical avoidance coupled with modification of other stressors. Although there is little scientific information to support this theory, it has been widely disseminated, many MCS patients probably have learned of it, and some have clung to it as a plausible basis for an otherwise inexplicable illness.

The three major alternative theories entail psychological mechanisms: (a) that MCS is a somatization disorder adapted to the culture of postdevelopment society; (b) that MCS is a variant of posttraumatic stress disorder (PTSD), in which the precipitating illness fulfills the role of the trauma with subsequent exposures serving to recapitulate the trauma, expressed as somatic symptoms rather than as nightmares or flashbacks that are more typical of PTSD; and (c) that MCS is a form of "psychological sensitization" to workplace and/or environmental chemicals, with symptoms serving as conditioned responses to stimuli that evoke an initial unpleasant experience. None of these views has been substantiated.

Other hypotheses include the possibility that MCS may represent an atypical host response to neurotoxins or airway irritants, which are the most common initial precipitants. It is also plausible that MCS may result from multiple or complex mechanisms or from differing mechanisms in different settings.

Only open-minded scientific inquiry appears likely to resolve these issues.

Prevention

No established primary prevention strategy can be offered to reduce the occurrence of MCS. Because acute occupational diseases, such as intoxication by solvents and pesticides or injury due to airway irritants, appear to predispose to MCS, reduction in the incidence of these illnesses would likely be somewhat effective. On the other hand, global reductions in chemical use in our culture, recommended by those who postulate an immune basis for MCS due to "total body burden" of xenobiotics, would have uncertain benefits based on present confirmed knowledge of the condition.

Because victims of acute occupational and environmental disorders appear to be at risk for MCS, it would be prudent to focus secondary preventive efforts on these individuals. Unfortunately, it remains unclear what to do for these individuals to alter the likelihood of MCS occurring as a complication. If psychological formulations prove to be correct, interventions aimed at uncoupling the trauma of the acute illness from the experience of less noxious exposure to chemicals might be justified, although no effort of this kind has been systematically undertaken or evaluated. In any event, it certainly makes sense to follow patients who have experienced acute illnesses more closely and expectantly than the level of reversible injury might otherwise demand. In particular, sensitive and early follow-up of all cases of acute occupational disease can be justified, because this intervention alone may be of value; at a minimum, early recognition and treatment of MCS could ensue, avoiding the crises in care reported by most patients.

Regarding tertiary prevention, once MCS is established, the goal of therapy must be to maximize patient function and minimize suffering. For most patients, this begins with establishing a therapeutic relationship with a clinician who accepts the illness, even though the nature of the underlying disorder is obscure, and requires abandoning the tireless search for the cause that will vindicate their suffering and prove someone has harmed or poisoned them. For clinicians, this relationship entails receiving the patient reports with full seriousness, even though test results do not reveal pathology.

In practical terms, the biggest issues become (a) the degree to which avoidance of chemicals should be undertaken; (b) determination of appropriate benefits, such as workers' compensation; and (c) the use of specific treatment modalities, including the experimental ones now available based on immunologic theory. It seems best to minimize avoidance by emphasizing to the patient that there is little basis for the "total body burden" view; it is proposed that the harm of each offending exposure is limited to the symptomatic consequences, however unpleasant. If this is accepted, patients are encouraged to remain in contact with offending environments to the extent that these environments are important for continued functioning. Although the patient should not be directly returned to the work environment in which the precipitating illness occurred, a rapid return to appropriately modified work, with job retraining if necessary, is strongly encouraged.

To allow the usually obligatory job modification, legitimate claims for compensation benefits should be supported, viewing MCS as a complication of occupational disease. Without such benefits, it is inevitable that clinical symptoms will be severely aggravated by material loss, although rarely is such loss sufficient to press the patient back into inappropriate work. Although therapy should encourage prompt return to some work, rarely will the financial benefit of compensation provide a strong disincentive to do so.

Radical therapies may be difficult to endorse. Often, however, the patient has placed considerable emotional investment in one or more of these, so the value of criticism is unlikely to outweigh the cost. On the other hand, since many experimental therapies are very expensive and are frequently coupled with admonitions to avoid all chemicals, it is often necessary to discuss the harm of these modalities.

The mainstay of therapy, given present knowledge, is support. The goals of therapy should be realistic, focusing on treatable co-pathology and short-term, achievable life objectives.

MRC

Further Reading

Bascom R. *Chemical Hypersensitivity Syndrome.* Baltimore, Md: Department of the Environment, State of Maryland (2500 Broening Highway, Baltimore, MD 21224); 1989.

Cullen MR, ed. Workers with multiple chemical sensitivities. *Occupational Medicine: State of the Art Reviews.* 1987;2(4).

Nasal (or Sinonasal) Cancer

ICD-9 160.0
SHE/O

Identification

Cancer of the nose and paranasal sinuses (often referred to as sinonasal cancer) includes malignancies of the internal nose; the middle and inner ear; and the maxillary, ethmoid, frontal, and sphenoid sinuses. Presenting features can include pain, bleeding, sores that do not heal, chronic sinusitis, and cranial nerve abnormalities. Diagnosis is confirmed by biopsy and microscopic examination. An occupational association relies heavily on a carefully obtained occupational history.

Occurrence

In the U.S., the age-adjusted incidence rates are 0.8 for males and 0.5 for females per 100 000 annually. Incidence increases with age, with the highest rates occurring for the group aged 75 years and older. Particularly high rates have been found in Japan, Uganda, and an industrial city in Zimbabwe. Five-year survival rates in the U.S. are more than 40% for whites but less than 20% for blacks.

Causes

The following occupational causes have been identified: (a) nickel compounds in refining processes (nickel subsulfide, oxide, or carbonyl) as well as metallic nickel, with the risk greatest among workers involved in furnace operations; (b) wood and other organic dusts in the furniture-making industry, with relative risks up to 70 reported; (c) boot and shoe dust; (d) other organic dusts, as suggested by studies on textile industry workers, bakers, and flour mill workers; (e) chromium, based on studies of workers manufacturing chromate pigments; (f) radioisotopes, based on studies of radium dial painters; and (g) work in the petroleum and chemical industries. Inhalation of snuff has also been implicated.

Pathophysiology

More than half these cancers are squamous cell carcinomas. It is presumed that direct inhalation accounts for exposure of the target tissues. Known carcinogenic agents may be trapped in the nose and sinuses because of large particle size. Delayed mucociliary transport may be another factor. Specific histological types of cancer are associated with specific agents. For example, the incidence of ethmoid sinus and nasal cavity adenocarcinomas in woodworkers and boot- and shoemakers leads to speculation that a common etiologic agent, such as aflatoxins from fungi, may be involved.

Prevention

Prevention relies heavily on reducing or eliminating worker exposure to the carcinogenic agents. As discussed in part 1, exposure can be reduced by engineering measures, changed work practices, substitution, and personal protective equipment. Education and training of workers and managers is also important. It may be useful to educate workers as well as physicians and dentists to recognize symptoms early, which would lead to earlier and more effective treatment. Screening of asymptomatic individuals is not of proven value.

BSL

Further Reading

Hernberg S, Collan Y, Degerth R, et al. Nasal cancer and occupational exposures. *Scand J Work Environ Health.* 1983;9:208-213.

Redmond CK, Sass RE, Roush GC. Nasal cavity and paranasal sinuses. In: Schottenfeld D, Fraumeni JF Jr, ed. *Cancer Epidemiology and Prevention.* Philadelphia, Pa: WB Saunders; 1982:519-535.

Roush GC, Meigs JW, Kelly J, et al. Sinonasal cancer and occupation: a case-control study. *Am J Epidemiol.* 1980;111:183-193.

Noise-Induced Hearing Loss—*See* Hearing Loss, Noise-Induced

Organic Dust Toxic Syndrome

Identification

Organic dust toxic syndrome (ODTS) is characterized by acute onset of symptoms following exposure to concentrations of organic dust during agricultural work or the handling of agricultural products. Symptoms include a flu-like complex of fever, chills, arthralgias, myalgias, cough, shortness of breath, and sometimes chest constriction. Notable laboratory features include normal findings on both chest x-ray and usual pulmonary function tests (ie, no demonstrable bronchial constriction). Occasionally, the diffusion capacity is reduced. The syndrome is distinct from silo-filler's disease, which is bronchiolitis obliterans after exposure to oxides of nitrogen. Recovery from ODTS is similar to that from metal fever and polymer fever and is usually complete within 24 hours. Interestingly, after repeated exposures there is tolerance to at least the febrile effects of exposure.

Occurrence

The prevalence of ODTS is not known, but the syndrome has been described among a number of populations. The varieties of ODTS have been reported using descriptors that refer to the particular work environment. These varieties include mill fever in cotton textile workers; grain fever in grain handlers; humidifier fever among workers exposed to humidified air; pulmonary mycotoxicosis among farmers exposed to moldy silage; and fowl fever among workers in poultry confinement buildings. Because of tolerance to the febrile effects, the syndrome is probably underreported.

Cause

Various organic dusts have been associated with this syndrome. An agent common to most exposures is gram-negative bacterial endotoxin. Threshold exposure and dose-response relationships for endotoxin-induced ODTS have not been defined.

Pathophysiology

Activation of pulmonary macrophages and release of mediators are probably the responsible mechanisms for the responses seen after inhalation of organic dusts. Activated macrophages are capable of releasing a number of mediators that are biologically active, locally and systemically. These include interleukin-1, interleukin-8, prostaglandins, platelet-activating factor, and polymorphonuclear cell chemotactic factors. In addition, some organic dusts have been shown to activate the complement system directly, resulting in inflammation.

Prevention

The syndrome occurs after high levels of exposure and after initial exposures. Thus, primary prevention involves reducing exposures to organic dusts. This is the most feasible way of controlling the problem, because elimination of bacterial endotoxin is impossible in most of these environments. In the case of moldy silage, modification of storage procedures to minimize mold growth would constitute primary prevention.

DCC

Further Reading

Donham KJ. Hazardous agents in agricultural dusts and methods of evaluation. *Am J Ind Med.* 1986;10:205-220.

do Pico GA. Health effects of organic dusts in the farm environment: report on diseases. *Am J Ind Med.* 1986;10:261-265.

Merchant JA. Agricultural exposures to organic dusts. In: Rosenstock L, ed. *State of the Art Reviews in Occupational Pulmonary Disease.* 1987;2:409-425.

Pancreatic Cancer

ICD-9 157

Identification

Early symptoms of pancreatic cancer are generally vague. The classic triad of pain, weight loss, and jaundice implies late disease and, therefore, a very poor prognosis. Diabetes mellitus, acute pancreatitis, and thrombophlebitis are other associated findings that may make the diagnosis more likely. Abdominal CT scanning may demonstrate a pancreatic mass. Fine-needle biopsy or laparotomy allows a histological diagnosis.

Occurrence

An estimated 27 000 new cases of pancreatic cancer and 25 000 deaths from it occurred in the U.S. in 1989. It is the fourth most common cause of death from cancer in men and the fifth most common cause in women. It occurs more often in men than in women and more often in blacks than in whites. Risk increases with age; the disease rarely occurs before age 40 and has the highest incidence rates between ages 65 and 79. No estimate of the proportion of pancreatic cancer attributable to occupational agents is available.

Causes

Studies have implicated a variety of workplace exposures, primarily hydrocarbons (petrochemical products, paints and degreasers, gasoline, vinyl chloride, and epichlorohydrin). Exposed groups thought to be at risk include both production workers (eg, in petroleum refining and coke oven byproducts) and users, such as rubber and furniture workers and chemists.

Radiation has also been implicated in several studies. Dry-cleaning and laundry workers who are exposed to carbon tetrachloride, petroleum solvents, trichloroethylene, and tetrachloroethylene have had increased risk reported. Tanners and tannery workers in Sweden exposed to chromium and chlorophenols have experienced a threefold increase in pancreatic cancer risk. Increased mortality from pancreatic cancer has been reported in furniture workers exposed to resins, glues, and solvents and in pulp and paper mill workers exposed to sulfites. A consistent increase in mortality from pancreatic cancer has been reported in several studies of workers in petrochemical plants and oil refineries in the U.S. Aluminum production and reduction plant workers have had increased risk reported, as have other metal workers. These workers have a diversity of exposures with no single agent or group of substances being clearly identified as causative.

Nonoccupational risk factors include smoking, diabetes mellitus, heavy alcohol use, and chronic pancreatitis. Combinations of risk factors may lead to substantially more disease than might the presence of individual risk factors.

Pathophysiology

The mechanism of carcinogenesis is unknown. As with other agents, a multifactorial model, with inducers and promoters, has been proposed. Animal models of pancreatic cancer have implicated both pathways in the etiology of chemically induced pancreatic cancer. Agents may be carried to the pancreas through blood. On the other hand, reflux of agents from the duodenum and biliary tract has led to cancer in some experiments.

Prevention

Primary prevention includes prevention of exposure to agents with known carcinogenicity. In addition, smoking cessation and reduction of alcohol use may lead to a lower probability of developing disease. As regards secondary prevention, no screening procedures for the early detection of disease have been investigated. There is no evidence that early treatment of disease leads to better long-term prognosis. The tertiary prevention measures surgical intervention and chemotherapy both have low 2-year survival rates.

HF

Further Reading

Lin RS, Kessler IK. A multifactorial model for pancreatic cancer in man. *JAMA*. 1981;245:147-152.

Neugut AI, Wylie P, Brandt-Rauf PW. Occupational cancers of the gastrointestinal tract. In: Brandt-Rauf PW, ed. *Occupational Cancer and Carcinogenesis. State of the Art Reviews in Occupational Medicine*. 1987;2:137-151.

Sleisenger MH, Fordtran JS. *Gastrointestinal Disease*. 2nd ed. Philadelphia, Pa: WB Saunders; 1983.

Parkinsonism—*See* Tremors

Peptic Ulcer Disease

ICD-9 532, 533

Identification

Peptic ulcer disease usually presents with epigastric pain or dyspepsia. The classical presentation—pain within 1 to 3 hours after meals, relieved by antacids and food, with nocturnal awakening—occurs in fewer than 50% of individuals. The diagnosis can be made with either x-ray techniques (85% sensitivity) or upper endoscopy (up to 100% sensitivity).

Occurrence

The estimated prevalence of peptic ulcer disease ranges from 0.3% to 15%, with higher estimates generally occurring in groups with other underlying diseases. Men are twice as likely as women to develop the disease. Blue-collar workers are twice as likely as white-collar workers to develop disease, suggesting the possibility of occupational etiologies. There is some suggestion that specific activities, such as welding, may increase symptoms of ulcer disease.

Causes

Both heredity and reactions to stressful situations have been clearly implicated in the etiology of peptic ulcer disease. For the latter, anxiety has been demonstrated as leading to chronic acid oversecretion and decreased mucosal resistance. Cigarette smoking is a contributor to both onset and recurrence of disease, but coffee is generally not so considered. In some countries, lower socioeconomic status has been considered a risk factor.

Pathophysiology

Development of peptic ulcers generally requires excess production of acid and secretion of pepsin together with decreased mucosal resistance. Increased numbers of parietal cells, increased basal secretion rates, and increased sensitivity of parietal cells to secretory stimuli (secretory drive) may all play a role. Decreased mucosal resistance has been attributed to anatomic predisposition related to underlying muscular structures, extension of antral mucosa from the pylorus, and gastritis or duodenitis.

Prevention

For primary prevention, see entry on Stress. Dyspepsia is commonly treated empirically, without diagnostic procedures, for 6 weeks. If disease initially improves but then recurs, or if it fails to improve, diagnostic procedures are warranted. Smoking cessation and avoidance of drugs that damage the mucosa are integral to long-term success.

Ulcers may heal spontaneously. In addition, individuals with ulcers under treatment may become asymptomatic despite the persistence of ulcers. For this reason, long-term treatment with reduced doses of histamine (H_2) receptor blockers is usually employed.

MH

Further Reading

Arnetz B. Lifestyle and gastro-duodenal ulcers: a critical review of the causes of peptic ulcer disease. *Int J Psychosomat.* 1987;34:35-41.

Sleisenger MH, Fordtran JS. *Gastrointestinal Disease.* 2nd ed. Philadelphia, Pa: WB Saunders; 1983.

Peripheral Nerve Entrapment Syndromes

Identification

Peripheral nerve entrapment is a general term for compression or pinching of any nerve in the arm or leg. The peripheral nerves may be injured by internal sources of pressure, such as muscle contraction or tendon swelling, or by external sources of pressure, such as the sharp edge of a desk or tool handle or the hard surface of a chair or knee pedal.

One of the best-known peripheral nerve entrapments is carpal tunnel syndrome (*see* entry). Other examples include compression of the median nerve under the muscle of the forearm (pronator teres syndrome); the ulnar nerve at the wrist (Guyon's canal syndrome) or at the elbow (cubital tunnel syndrome); the sciatic nerve and its branches, such as the peroneal nerve, in the hip area or just below the knee; the radial nerve in the upper arm; and any or all of the three nerves that travel from the neck through the shoulder and into the arm (thoracic outlet syndrome, which also involves compression of the accompanying blood vessels). Other compression disorders at the level of the shoulder include scalene anticus syndrome, costoclavicular or claviculocostal syndrome, and pectoralis minor syndrome. A single nerve may be compressed at more than one location, as in the "double-crush" syndrome, which involves compression of the median nerve at both the shoulder and the wrist.

The symptoms and signs of peripheral nerve entrapment depend on whether the affected nerve is sensory, motor, autonomic, or all three. Compression of sensory nerve fibers usually results in tingling, numbness, or pain in the supplied areas. The symptoms often begin gradually and intermittently. They may progress to burning, painful numbness, deep dull aching, or a sensation of swelling without objective signs; sometimes they spread up or down the limb. Compression of motor nerve fibers often causes weakness or clumsiness in the supplied muscles, as evidenced by difficulty in holding small objects without dropping them. Compression of an autonomic nerve interferes with normal sweating and blood flow.

When the compression is at the thoracic outlet, the distribution of pain varies and can include the neck, shoulder, arm, or hand; pain in the area supplied by the ulnar nerve is most typical (the little finger and the outside of the fourth finger), but pain may also occur in the area supplied by the median nerve (*see* Carpal Tunnel Syndrome). Symptoms may occur at night or after prolonged sitting. Because there is simultaneous compression of blood vessels, there may be other symptoms, such as blanching, discoloration, coldness, or swelling. Symptoms may be less localized and less intermittent than those in a simple nerve entrapment.

Abnormalities on physical examination may be minimal. Signs of sympathetic nerve fiber compression include extremely dry and shiny skin and blanching associated with poor circulation. Sensory nerve damage may lead to decreased sensitivity to touch, pain, temperature, or vibration; two-point discrimination may be poor. With motor nerve damage, there may be objective loss of strength or dexterity in the affected muscles, with atrophy (muscle wasting) in severe cases. Tendon reflexes may be weak or absent with either sensory or motor nerve involvement, depending on the severity.

Sensory symptoms may be provoked or worsened by specialized tests for specific compression syndromes. Usually such a test consists of either percussion over the site of entrapment (eg, Tinel's sign over the ulnar nerve at the wrist crease for Guyon's canal syndrome) or resisted motion involving muscular contraction at the site of entrapment (eg, resisted pronation with clenched fist and flexed wrist for pronator teres syndrome). For thoracic outlet syndrome, symptoms may be precipitated or exacerbated by Adson's (hyperabduction or costoclavicular) test, which involves raising the arm or extending it behind the head, with or without turning the head to the side.

Various laboratory tests may be used to attempt to confirm the diagnosis. For some compression syndromes, nerve conduction studies may be performed to measure whether the nerve impulse is conducted more slowly (decreased nerve conduction velocity) or less efficiently (decreased amplitude and prolonged residual latency) through the entrapped section of the nerve than through other segments of the same nerve or the corresponding nerve on the other side (if only one limb is affected). Vibrometry (*see* Carpal Tunnel Syndrome), magnetic resonance imaging, and computed tomography have been proposed for diagnostic testing; however, the latter two are expensive and unlikely to be used routinely.

Diagnosis of these conditions tends to be problematic. Neither examination methods nor diagnostic criteria have been fully standardized. The validity of almost all the available tests remains to be demonstrated. In addition, there are no data on how well results of examinations in the early stages of a nerve compression disorder predict progression.

The differential diagnosis includes inflammation of local tendons, compression of the same nerve at another location, and various systemic diseases.

Occurrence

There is a lack of accurate data on the incidence and prevalence of nerve entrapment syndromes, both for the general population and for exposed occupational groups, partly because of the diagnostic difficulties noted above. Many cases have been described in the medical literature, but few of these reports include sufficient information for adequate statistical analysis to determine the role of occupational or nonoccupational factors. The prevalence of any single syndrome is probably low, except perhaps in populations highly exposed to factors that produce nerve compression at a particular body site.

In one study of three occupations, 18% of the workers had thoracic outlet syndrome (TOS) on physical examination.[1] The highest rate (32%) was observed in female cash-register operators who had to keep their right arm elevated continuously, producing high static load on the muscles of the shoulders, neck, and back. The rate was 17% among men and women working on a heavy assembly line and 10% in office workers using video display terminals.

The frequencies of these disorders by age or gender are not well known. In the study of TOS, the prevalence was higher among women than among men in both office work and heavy industry. Gender differences in body size and resulting worker-machine fit may have accounted for some of this discrepancy.

Causes

The specific cause of a nerve entrapment depends on the particular nerve and where it is compressed.[2] As stated above, a peripheral nerve may be entrapped by either an internal or an external source of pressure (*see also* Carpal Tunnel

Syndrome). Internal entrapment may occur at a point where the nerve passes between two anatomic structures, such as muscle and bone, and is compressed by muscular contraction, muscle hypertrophy, tendon inflammation and swelling, or narrowing of the passage in certain body postures. External sources of pressure include mechanical force concentrations created by contact of the body with hard surfaces and sharp edges of tools, workstations, chairs, and equipment, especially where the nerve is close to the surface and vulnerable to external pressure.

For example, with repeated pronation (palm-down rotation) of the forearm, and especially with repetitive or forceful wrist or finger flexion, the pronator muscle becomes hypertrophied and compresses the median nerve as the nerve travels through the forearm (pronator teres syndrome). Repetitive elbow flexion, especially in combination with wrist flexion, may increase pressure within the elbow joint and compress the ulnar nerve as it passes through the elbow (cubital tunnel syndrome). Frequent, prolonged, or forceful exertions with the arm raised to or above shoulder height, or with the arm extended back behind the midline of the body, compresses the nerves and blood vessels for the entire arm at the thoracic outlet between the shoulder muscles, the rib cage, and the collarbone. Compression of these tissues may also be secondary to tendon irritation and swelling (resulting from arm work in one of these postures, or from repetitive or forceful motions that crowd the interior of the shoulder joint).

External mechanical compression of a nerve by hard or sharp edges often occurs at a site where the nerve is close to the surface; an example would be resting the elbow on an unpadded table surface or corner (ulnar nerve compression in cubital tunnel syndrome) or repetitive use of a thigh-operated pedal (compression of the peroneal nerve). Mechanical pressure applied to the palm (using the hand as a hammer or using a tool whose handle presses into the center of the hand) may compress branches of the median or ulnar nerve, producing symptoms similar to those of carpal tunnel syndrome. Molded tool handles with indentations too wide or too narrow for an individual's fingers may compress the small nerve fibers along the sides of the fingers. Carrying a weight on the shoulders, such as a mailbag or other type of knapsack, may cause mechanical compression of the soft tissues at the shoulder. The sciatic nerve may become compressed from prolonged sitting on a small or inadequately padded seat.

Other ergonomic risk factors include highly repetitive or forceful exertions, especially with the joint in a bent or twisted position that brings the nerve closer to the surface or irritates local tendons (*see* Carpal Tunnel Syndrome). Segmental vibration (*see* Hand-Arm Vibration Syndrome) and low temperatures may interfere with the microcirculation of the nerve and may also worsen the effects of other ergonomic stressors.

In addition to ergonomic causes, occupational or nonoccupational exposure to solvents or heavy metals, such as lead or mercury, damages the peripheral nerves and may cause symptoms and signs resembling those of a nerve entrapment. This damage may also make the nerves more susceptible to the effects of mechanical pressure, although there are few epidemiological data on the effects of such combined exposure. Occupational or nonoccupational acute trauma, such as fracture of a bone near the site of the nerve damage, may leave scar tissue or a bone fragment pressing on a nerve.

There are no epidemiological data on the effects of combined exposures to ergonomic stressors and heavy metals or other peripheral neurotoxins, such as lead or solvents. There are also no data on whether smoking might play a possible role as an aggravating factor because of its negative effect on peripheral circulation.

Nonoccupational causes include alcoholism, diabetes mellitus, rheumatoid arthritis, kidney failure, malnutrition, possibly obesity, and other systemic conditions causing swelling that may lead to nerve entrapment. These conditions are relatively common in the general population and should not be permitted to distract attention from the workplace and easily preventable causes, especially when a high rate of one or more specific nerve compression syndromes is observed. Uncommon nonoccupational associations include Paget's disease of bone, neoplasms including multiple myeloma, gout, myxedema, acromegaly, renal disease anemia, Guillain-Barré syndrome, and toxic shock syndrome.

Pathophysiology

The mechanism of damage is mechanical compression or irritation of a peripheral nerve between two body structures or between an internal structure and an outside source of pressure. This compression causes slowed or incomplete transmission of sensory and motor nerve signals through the limb to the part supplied by that nerve. The results are discomfort and loss of nervous and muscular function of the innervated body part. Simultaneous compression of accompanying blood vessels may cause loss of blood supply to the nerve and aggravate the nerve damage.

In thoracic outlet syndrome, there is compression or irritation of the entire neurovascular bundle at the shoulder. The neurovascular bundle consists of the major artery and vein that supply the arm, as well as the nerve fibers that give rise to most of the peripheral nerves of the arm. The symptoms vary depending on whether the nerves, the blood vessels, or both are compressed.

Prevention

Primary prevention involves the use of engineering controls (ergonomic design and selection of tools, tasks, and workstations) to minimize the force and repetition of manual exertions, mechanical compression, sources of excessive cold, and frequency and amplitude of vibration. Ergonomic design of the workstation and tools should eliminate the necessity of working with the body in any awkward or nonneutral posture (such as wrist bent, forearm pronated, or upper arm elevated) (*see* An Overview of Occupational Musculoskeletal Disorders in part 3).

Sharp edges and hard surfaces of workstations and tools should be padded wherever they come into contact with the worker's body. Sources of mechanical compression of the palm should be eliminated by measures such as lengthening and padding tool handles that rest in the palm of the hand. The hand should not be used as a hammer; if such use is completely unavoidable, gloves with well-padded palms should be provided to protect the soft tissues and distribute forces over a larger surface area. Tool handles with overly molded indentations for the fingers should be avoided, because indentations that do not fit the hand well may press on the small nerves in the sides of the fingers.

If analysis of motion patterns demonstrates a true potential for greater variety in manual work by combining various work tasks (job rotation) or alternating among them (job enlargement), these administrative controls may reduce some of the repetitiveness of motions and the duration of exposure to ergonomic stressors.

For secondary prevention, health care providers should be trained in the interview and clinical examination procedures necessary to identify occupational peripheral nerve entrapments at an early stage. Once reported, cases should be treated conservatively, and jobs should be analyzed for ergonomic features that may be modified. Removing a worker from source of exposure is critical to prevent the disorder from progressing to irreversible nerve damage. Follow-up is important to ensure that job modifications have been effective, that "light-duty" jobs have been correctly selected to avoid continuing ergonomic stress, and that symptoms and signs have not progressed.

Surgery to release the compressed nerve is rarely used and should be considered only as a last resort. It may be only temporarily effective if the worker is returned to an ergonomically stressful job that has not been modified; possible loss of muscle strength or build-up of scar tissue following surgery makes it imperative that job assignments be selected carefully to avoid recurrence. Use of transitional workshops for affected workers and graded retraining under the supervision of an experienced physical therapist may also reduce risk of recurrence.

LP

References

1. Sallstrom J, Schmidt H. Cervicobrachial disorders in certain occupations, with special reference to compressions in the thoracic outlet. *Am J Ind Med.* 1984;6:45-52.

2. Feldman RG, Goldman RH, Keyserling WM. Peripheral nerve entrapment syndromes and ergonomic factors. *Am J Ind Med.* 1983;4:661-681.

Further Reading

Armstrong TJ, Silverstein BA. Upper extremity pain in the workplace: role of usage in causality. In: Hadler NM, ed. *Clinical Concepts in Regional Musculoskeletal Illness.* New York, NY: Grune & Stratton; 1987.

Caillet R. *Neck and Arm Pain.* 2nd ed. Philadelphia, Pa: F.A. Davis; 1984.

Caillet R. *Soft Tissue Pain and Disability.* Philadelphia, Pa: F.A. Davis; 1983.

Peripheral Neuropathy

ICD-9 357.7
SHE/O

Identification

Peripheral neuropathy presents with insidious onset of symptoms such as intermittent tingling and numbness and may progress to dysesthesias and an inability to perceive sensation. Muscle weakness and eventual atrophy result from damage to the motor nerve fibers. Based on the structural components of the nerve primarily involved, toxic polyneuropathies may be subdivided into axonopathies, which present as distal sensorimotor loss (most evident in the lower extremities where the axons are the longest); myelinopathies, such as the spotty segmental demyelination; and neuronopathies. The most common pattern seen in metabolic or toxic neuropathies is "dying back" or distal axonopathy with segmental demyelination occurring as a secondary effect. Objectively, insensitivity to pinprick or touch indicates peripheral neuropathy. Two-point discrimination, position, vibration, and temperature sensation may also be impaired.

Depending on the type or severity of neuropathy, electrophysiological examination of nerves can reveal slowed conduction velocities, reduced sensory or motor action potentials or amplitudes, and/or prolonged latencies. Prolonged sensory action potentials serve as a complementary test to confirm clinically observed diminished sensation. Slowed motor or sensory conduction velocity is generally associated with demyelination of the nerve fibers, while preservation of normal values in the presence of muscle atrophy indicates axonal neuropathy. However, exceptions occur when there is progressive loss of motor/sensory nerve fibers in axonal neuropathy that affects the maximal conduction velocity (which results from the dropping out of the large-diameter, fast-conduction fibers) and when there is immature regeneration of nerve fibers. Regenerating fibers are known to conduct slowly, so conduction may be slowed early in the recovery period of axonal neuropathy; this usually occurs only in the distal nerve segments, but occurrence depends on the severity of damage.

Occurrence

Knowledge about occurrence of peripheral neuropathy or any other occupational neurological disease is limited to case series and a partial listing of compensation awards. In 1986, a total of 8723 workers' compensation awards for occupational conditions of the nervous system among both public and private sector employees were reported to BLS from 23 states. The vast majority of these awards (98%) were for "diseases of the nerves and peripheral ganglia." The remainder were for conditions that either were unclassified or affected the central nervous system. Most (63%) were for conditions that occurred in the manufacturing sector. Of related interest, the same data source reported 8300 compensation awards for mental disorders, occurring predominantly among public sector employees (43%), workers in the services industry (16%), and workers in manufacturing (13%). The percentages attributable to either exposure to workplace toxins or other causes are unknown.

Causes

Occupational causes of peripheral neuropathy include neurotoxic exposure to heavy metals, organic solvents, and insecticides (*see* Table 1 at end of entry). Lead accumulation in the body may cause numbness and tingling of the fingers and toes, followed by motor weakness. In lead neuropathy, motor nerve conduction velocity is affected more than sensory nerve conduction velocity, because the underlying process is primarily segmental demyelination and remyelination. Later, the slower fibers are affected as well when axonal changes occur. Lead-exposed workers, at blood lead levels of 50 to 70 mcg/dl (micrograms per deciliter), have abnormal nerve conduction velocity and amplitude. Persistent reductions in nerve conduction velocities after blood lead levels have fallen can be attributed to incomplete remyelination of a sufficient number of fibers, which thereby impairs overall conduction velocity.

Clinical manifestations of arsenical neuropathy initially present as losses of sensation in the feet and hands and as dysesthesias. Neuropathy induced by thallium exposure is progressive, beginning with paresthesias of the extremities; pain increases in severity as weakness and atrophy of muscles develop. The skin of the trunk is painful to touch, and the muscles and nerves are tender to pressure.

Neuropathy caused by organic solvents such as methyl-n-butyl ketone (MBK) and n-hexane is initially manifested as tingling paresthesias in the fingers and toes with loss of sensation to pinprick, temperature discrimination, and touch. Organic solvent intoxication results in slowed motor conduction velocities and prolonged terminal latencies because of secondary changes in the myelin sheath. Cranial neuropathies have been observed following exposure to trichloroethylene; the effects are primarily on the trigeminal nerve, and exposure has been associated with motor and sensory losses in the face.

Organophosphates, such as malathion, mipafox, leptophos, and tri-cresylphosphate, can produce a rare type of delayed onset polyneuropathy: organophosphate ester-induced delayed neuropathy (OPIDN). The delayed neurotoxic effects are due to the inhibition of the enzyme neuropathy target esterase (NTE), which occurs within hours following exposure and is gradually restored; the clinical onset is not for 3 to 4 weeks later, however. Symptoms present primarily as a distal symmetrical motor polyneuropathy characterized by distal weakness in the limbs. Sensory loss may occur, depending on dose and duration. Electrophysiological studies of subjects with organophosphate-induced neuropathy reveal small or absent amplitudes of sensory nerve action potentials.

Repetitive manual motions may damage the peripheral nerves from external compression or entrapment. Other mechanical injuries, such as lacerations, stretch vibration, and repeated trauma, may also lead to peripheral neuropathy.

Nonoccupational causes of peripheral neuropathy may be genetic, nutritional, infective or postinfective, associated with malignant conditions (carcinoma or leukemia), metabolic (diabetes or thiamin deficiency), or physical (exposure to cold or radiation). Occurrence may also be due to alcoholism, uremia, paraproteinemia, amyloidosis, or sarcoidosis. Alcoholic neuropathy is characterized by axonal degeneration and regeneration; electrophysiological testing reveals fibrillation potential, positive sharp waves, increased mean motor unit potential duration, reduced interference pattern, and reduced sensory and motor nerve conduction velocities.

Pathophysiology

Axonal degeneration and segmental demyelination of nerve fibers can occur. Lead-induced peripheral neuropathy is generally of mixed segmental demyelination with some axonal changes. Pathological studies of thallium-exposed persons has revealed axonal degeneration. Most organic solvent intoxication causes mainly axonal neuropathy; trichloroethylene is known to also damage myelin. In organophosphate-induced delayed neuropathy, the longer nerve fibers to the legs undergo axonal degeneration, as do the long spinal cord dorsal columns. Due to involvement of the spinal cord long tracts, organophosphate-induced delayed neuropathy is generally irreversible.

Prevention

Exposure to neurotoxins should be minimized. Effective industrial hygiene monitoring and control of respiratory, gastrointestinal, and dermal routes of chemical entry are discussed in part 1. Biological monitoring of workers for specific chemical exposures may be used as a guide to check on individual exposure levels, but careful attention must be paid to the interpretation of individual results. For more information, the American Conference of Governmental Industrial Hygienists (ACGIH) has issued biological exposure indices (BEIs) for a number of chemicals.

New portable machines are being used to screen workers for some of the early sensory signs and symptoms that are sometimes the only manifestation of early peripheral nervous system dysfunction. One of these machines, the Optacon Tactile Test, has been used in a screening capacity to assess tactile and vibratory thresholds. After some undesirable technical features are worked out, it will potentially be able to detect either worsening or improvements in sensation. Careful workstation and tool design will reduce the risk of physical injury to the nerves. Surgical release of entrapped nerves will correct peripheral neuropathy of physical origin.

Other Issues

Diseases of the spinal cord can cause problems in differential diagnosis. Workers who have been exposed to lead may later develop a clinical picture indistinguishable from the amyotrophic lateral sclerosis-motor neuron disease forms seen in people without lead or other metal exposures. The diagnosis of toxic neuropathy due to workplace hazards must include a comprehensive examination that takes into account confounding presentations. The patient's medications should be evaluated, as some may produce neuropathic symptoms. Individuals with metabolic or hereditary disorders that manifest with peripheral neuropathy present variables that can confound the diagnosis of toxic peripheral neuropathy due to exposure to workplace hazards. Workers with neuropathy due to diabetes, alcohol, previous injury, or postinfectious syndromes are more susceptible to compression neuropathy.

RF, SP

Further Reading

Bleeker M. The Optacon: a new screening device for peripheral neuropathy. In: Gilioli R, ed. *Advances in the Biosciences*. Vol 45. *Neurobehavioral Methods in Occupational Health*. New York, NY: Pergamon Press; 1983:41-46.

Feldman RG, Travers PH. Environmental and occupational neurology. In: Feldman RG, ed. *Neurology: The Physician's Guide.* New York, NY: Thieme-Stratton; 1984:191-212.

Juntunen J, Haltia M. Polyneuropathies in occupational neuropathy: pathogenic and clinical aspects. *Acta Neurol Scand.* 1982;66(suppl 92):59-74.

Seppalainen AM, Hernberg S, Koch B. Relationship between blood lead levels and nerve conduction velocities. *Neurotoxicology.* 1979;1:313-332.

Spencer PS, Schaumburg HH, eds. *Experimental and Clinical Neurotoxicology.* Baltimore, Md: Williams & Wilkins; 1980.

Table 1.—Exposures Associated with Peripheral Neuropathy

Neurotoxin	Major Uses or Sources of Exposure
Metals	
Arsenic	Pesticides Pigments Antifouling paint Electroplating industry Seafood Semiconductors Smelters

Neurotoxin	Major Uses or Sources of Exposure
Lead (Motor neurons only)	Solder
	Lead shot
	Illicit whiskey
	Insecticides
	Auto body shops
	Foundries
	Storage battery manufacturing plants
	Lead-stained glass
	Lead-based paint
	Smelters
	Lead pipes
Mercury	Scientific instruments
	Photography
	Pigments
	Electroplating industry
	Amalgams
	Taxidermy
	Electrical equipment
	Felt making
	Textiles
Thallium	Rodenticides
	Manufacturers of special lenses
	Infrared optical instruments
	Fungicides
	Mercury and silver alloys
	Photoelectric cells

Neurotoxin	Major Uses or Sources of Exposure
Solvents	
Carbon disulfide	Manufacturers of viscose rayon Preservatives Rubber cement Electroplating industry Paints Textiles Varnishes
Methyl-n-butyl ketone	Paints Quick-drying inks Metal-cleaning compound Varnishes Lacquers Paint removers
N-hexane	Lacquers Rubber cement Pharmaceutical industry Stains Printing inks Glues
Perchloroethylene	Paint removers Extraction agent for vegetables and mineral oils Dry-cleaning and textile industries Degreasers

Neurotoxin	Major Uses or Sources of Exposure
Trichloro-ethylene	Degreasers Adhesive in shoe and boot industry Process of extraction of caffeine from coffee Paints Painting industry Varnishes Dry-cleaning industry Rubber solvents Lacquers
Monomers	
Acrylamide	Paper, pulp industry Water, waste treatment facilities Grouting materials Photography Dyes
Insecticides	
Organophosphates	Agricultural industry

Pesticide Poisoning

ICD-9 989.2-989.4
SHE/O

Identification

Exposure to herbicides, insecticides, fungicides, and rodenticides causes both acute and chronic adverse health effects. Acute effects are much more easily associated with pesticide exposure than chronic effects. Identification of acute pesticide poisoning is made on the basis of signs and symptoms (*see* Table 1 at end of entry), a history of exposure, and, for some pesticides, laboratory tests that measure (a) the concentration of the pesticide or its metabolite, or (b) a physiological effect.

(For example, activity of the enzyme cholinesterase is depressed following exposure to organophosphates or carbamates.) Most people with acute symptoms of pesticide poisoning have been exposed to organophosphates. Chronic effects include cancer as well as reproductive and developmental, immunological, and neurological and behavioral effects, summarized below in Pathophysiology.

Occurrence

The World Health Organization (WHO) estimates that there are 3 million severe acute cases of pesticide poisoning annually worldwide. An estimated 2 million of these are the result of suicide attempts. Many cases of acute pesticide poisoning are not recognized by workers or health personnel because symptoms are nonspecific, attributed to other causes, or not present. WHO also estimates that there are 220 000 deaths attributable to acute pesticide poisoning annually and 700 000 cases of specific and 40 000 cases of nonspecific chronic effects each year worldwide.

People who are at greatest risk for pesticide poisoning are those who produce, mix, and apply pesticides on farms and plantations and those who are engaged in structural pest control. Risks are higher in developing countries because they have less pesticide regulation than developed countries, poorer application methods, and lower availability of protective equipment; in addition, developing countries often import and use pesticides banned in developed countries. WHO estimates that 50 million people are at high risk worldwide and another 500 million are exposed less intensively (and thus generally know less about the hazards of pesticides and preventive measures than those who are more intensively exposed). In addition to producers, mixers, and applicators, those at risk for unintentional exposure include (a) agricultural field-workers who may be exposed to pesticide aerosols or to pesticide residues on crops; (b) household contacts of agricultural workers who may be exposed to pesticides brought home on work clothes or may consume food or fluids that have been contaminated by pesticides or stored in containers that previously contained pesticides; and (c) young children, who may accidentally ingest pesticides stored in or near the home. Children and family members of agricultural workers who live near agricultural operations where pesticides are used may also be exposed following application if, for example, they live downwind.

There have been some large outbreaks of acute pesticide poisoning due to unintentional heavy pesticide contamination of fruits, vegetables, and other food items. However, such outbreaks probably represent only a very small fraction of pesticide poisoning cases. Pesticide manufacture may present risks to individuals living near the production facility, as occurred in the disaster in Bhopal, India. In addition, there are many instances in which pesticides have contaminated groundwater, presenting potential risks to those who use the water.

The best data on acute pesticide poisoning incidence in the U.S. come from California, where reporting of poisoning by pesticides, including antimicrobials, is required by law. At least six other states require reporting of occupational pesticide poisoning. In 1988, physicians reported 2118 illnesses and injuries associated with pesticide exposure (including definite, probable, and possible cases). About half were systemic in nature, about one fourth affected the eyes, and about one fourth affected the skin. The most common ways in which individuals were affected were through application (674), exposure to residue (666), mixing and loading (246), "coincidental exposure" (193), exposure to concentrate (72), fumigation (49), and exposure during emergency responses (39). A total of 357 individuals were disabled for 1 or more days, 52 were hospitalized for 1 or more days, and 3 died, 1 as a suicide. Exposure to combinations of pesticides was often reported as accounting for illness or injury.[1]

Surveillance data from elsewhere are generally less reliable. Annual rates of unintentional acute pesticide poisoning range from 0.3 to 18 cases per 100 000 population, with many of these cases due to occupational exposure. In Malaysia and Sri Lanka, approximately 7% of agricultural workers who used pesticides reported having pesticide poisoning (or hospital admission records indicated this diagnosis) during the previous year, and about one fourth of pesticide users had significant inhibition of cholinesterase activity (consistent with poisoning due to organophosphates or carbamates); in Thailand and Indonesia, fewer than 1% had significant inhibition of cholinesterase activity.

Causes

See Table 1.

Pathophysiology

It is beyond the scope of this book to describe the pathophysiology of all the major pesticides used at work. However, as an example, the pathophysiology of organophosphates is summarized below.

Organophosphates cause acute toxic effects by inhibiting the enzyme acetylcholinesterase, leading to the accumulation of the neurotransmitter acetylcholine. Symptoms resulting from mild poisoning include fatigue, headache, dizziness, blurred vision, excessive sweating, nausea, vomiting, diarrhea, and abdominal cramps. Symptoms of moderate to severe acute poisoning include weakness, fasciculations, shortness of breath, loss of balance, small pupils, and unconsciousness. (The pathophysiology of delayed neuropathy due to organophosphate exposure results from effects on a different tissue esterase.)

Potential chronic adverse effects of pesticide exposure include cancer as well as reproductive and developmental, immunological, and neurological and behavioral effects.[2] Many pesticides have been registered for use in the U.S. with inadequate or incomplete data on chronic health effects.

More than 50 pesticides have been shown to be carcinogenic in experimental animals. Increased rates of lung cancer have been demonstrated consistently in studies of commercial pesticide applicators, where chlorinated hydrocarbon insecticides have been the most frequently used pesticides and application is in enclosed spaces. Results of case-control studies have revealed excess cases of lymphoma, leukemia, and soft-tissue sarcoma in foresters and farmers exposed to organochlorine insecticides such as chlordane, Lindane (hexachlorobenzene), and DDT. Of these results, the association is strongest between non-Hodgkin's lymphoma and insecticides as well as phenoxyacetic acid herbicides.

Some pesticides can cause reproductive effects in animals and humans. The nematocide dibromochloropropane (DBCP), which was banned from use in the continental U.S. in the 1970s but is still sold to and used by a number of developing countries, can reduce sperm count and sperm motility and may also cause genetic changes in germ cell lines. There is a higher occurrence of female births and spontaneous abortions following fathers' exposure to DBCP. Chlordecone (Kepone), an insecticide and fungicide not manufactured in the U.S. since 1975, can cause decreased sperm count and motility among production workers. Ethylene dibromide adversely affects spermatogenesis in some animals, but effects in humans have not been conclusively demonstrated. Similarly, carbaryl has adversely affected gametogenesis and fertility in rodents, but human studies again are not conclusive. Several animal studies also show suppression of the hypothalamic-pituitary-ovarian axis.

Pesticides have also caused adverse developmental effects in animals. Of approximately 200 pesticides tested, almost half cause birth defects. However, significant developmental toxicity in humans has yet to be conclusively demonstrated.

Immunologic effects of pesticide exposure are not well documented. Occupational exposure to carbamates, captans, chloronitrobenzenes, and organophosphorus compounds induces contact hypersensitivity responses or asthmalike reactions. Certain populations may be at greater risk for immunologic effects; these populations include those heavily exposed at work or elsewhere, atopic or asthmatic individuals, genetically predisposed individuals, patients on drug therapy, people otherwise immunologically compromised, the very young, and the very old.

A wide variety of neurological effects have been demonstrated or suggested as a result of pesticide exposure. Effects may arise from acute or chronic exposure and can persist for many years. Exposure to organochlorines, pyrethroids, organophosphorus esters, and carbamate insecticides can cause acute neurological effects. Acute exposure can initiate long-term neurological problems not recognizable at the time of acute toxicity. Exposure to organochlorine and organophosphate insecticides can lead to chronic deterioration in neurological function. The most common chronic symptoms are fatigue, lethargy, and partial transient paralysis and/or weakness in the peripheral muscles of the hands and feet. Some organophosphates can cause delayed neuropathy of distal muscles. A peripheral neuropathy, with symptoms of tingling and/or numbness, can be caused by some organochlorine insecticides. Central nervous system effects, with mental confusion, irritability, recent memory loss, agitation, anxiety, difficulty in performing "thinking" tasks, and linguistic disorders, have been demonstrated with exposure to organochlorines, organophosphorus esters, and carbamates.

Prevention

A wide variety of measures can prevent pesticide poisoning. These measures apply the prevention strategy outlined in chapter 1, including anticipation, surveillance, analysis, and control.

Aspects of prevention covered below are (a) premarket testing; (b) selection and use of pesticides; (c) surveillance and biological monitoring; (d) treatment, including first aid and emergency treatment; (e) government regulations pertaining to manufacture, formulation, packaging, storage and transport, sale and use, disposal, and community right to know; (f) the international code of conduct; and (g) education and training. Many workers at risk, especially in economically underdeveloped areas, lack access to first aid and medical care, as well as biological monitoring and medical surveillance.

Premarket testing. Pesticides are intentional poisons whose purpose is to eliminate pests—insects, rodents, other animals, microbials, fungi, and weeds. It is thus reasonable to anticipate adverse health effects in humans and essential to evaluate the acute and chronic toxicity of pesticides for humans and animals before marketing them. Evaluation of toxicity requires in vitro and in vivo assays for toxicity, mutagenicity, carcinogenicity, and teratogenicity. If such assays yield positive results, restrictions on use should be considered.

Selection and use of pesticides. An increasing number of alternatives to synthetic pesticides for pest control, such as the use of natural predators and the broad approach of integrated pest management, should be considered. If pesticides are to be used, careful attention should be given to selecting the appropriate pesticide and choosing the least toxic alternative whenever possible—an application of the principle of positive engineering control of hazards (*see* part 1).

Surveillance and biological monitoring. It has been suggested that biological monitoring programs for agricultural workers exposed to organophosphate pesticides include the following components:

> "1. *Identification of high-risk populations requiring surveillance.* Mixers, loaders, applicators, flaggers, and others who directly handle pesticides should be provided with a full surveillance program. . . . When a review of local field labor patterns by crop and by periods of pesticides application identifies farm workers at significant risk of residue exposures, studies comparing the cholinesterase activity of these workers with their preseason levels (or with other

workers in untreated fields) will allow assessment of actual exposures and of the need for periodic surveillance. The Rural Health and NIOSH units in Region IX of the Public Health Service have produced a county-based model for the development of crop and pesticide profiles in the identification of high-risk field exposures. In addition, reports of individual or crew poisonings should trigger investigations of application or field work activities which have not been previously identified as high risk.

"2. *Baseline cholinesterase determination*. Optimally, both red cell and plasma cholinesterase activity should be determined as a baseline for all high-risk workers. . . . When only one test is performed, red cell cholinesterase activity determinations are preferable. In routine surveillance of field workers' exposure or the surveillance of applicators in developing nations, where resources are extremely limited, the whole-blood field methods may be employed.

"3. *Periodic surveillance*. At least one subsequent cholinesterase activity measurement should be required for all high-risk workers each season, to be drawn at the peak of the application period. Additional tests should be required at the discretion of the supervising physician or responsible agencies. These data should be collected by or referred to the appropriate public health agency for analysis of the prevalence of cholinesterase activity inhibition, and to assess the adequacy of worker protection programs.

"4. *Criteria for removal from work*. Criteria for the testing of workers and for their eventual removal from work should be established. If red cell cholinesterase measurements are required, as recommended above, a decline of greater than 15% from the worker's baseline value may be regarded as significant

> ($p < .05$), and might be used as the criterion for retesting of cholinesterase activity. Hayes' suggested criteria for removal from work—a 15% decline from the red cell baseline and a 20% decline from the plasma baseline—may be difficult to implement in field situations, but they do reflect the association between moderate cholinesterase inhibition and disabling symptoms more accurately than the current [California Food & Drug Administration, CFDA] criteria of 40% and 50% inhibition, respectively. . . .

"Workers identified as cholinesterase-inhibited on routine surveillance should not be returned to work involving potential reexposure to organophosphates or carbamates until their cholinesterase activity has returned to the previously established baseline. If a baseline is not available, however, decisions regarding return to work are complicated by the fact that the patterns of recovery for cholinesterase activity are not well understood. . . .

"The use of sequential postexposure cholinesterase determinations to confirm cholinesterase inhibition appears to offer a feasible and useful alternative to reliance on the laboratory normal range in the absence of baseline values and when the clinical presentation does not require the use of atropine, which would confirm the diagnosis more definitively. Plasma cholinesterase should increase 15% to 20% between the initial test at the time of exposure and retesting two to three days later if a significant organophosphate-induced cholinesterase inhibition has occurred. Regeneration of plasma activity is more likely to be seen during this short period because of its more rapid rate of recovery. Further increases on subsequent determinations would confirm the diagnosis.

"Affected workers should be kept from work involving exposure to these chemicals until their red cell cholinesterase has regenerated. Erythrocyte, rather than plasma, values are recommended as the end-point because the former better reflect physiological effects on the nervous system. In determining when regeneration has been completed, consideration should be given to the fact that the red cell cholinesterase activity of a healthy person may normally vary by 10% upon retesting. If a value increases by more than 10% over a value drawn 10 days previously (using the conservative estimate of 1% per day for red cell cholinesterase recovery), the baseline may not yet have been reached. In most cases, agricultural workers are forced to return to work for economic reasons long before their red cell cholinesterase can be demonstrated to have completely regenerated and very often before their symptoms have completely resolved. [Editors' note: There should be wage retention during the period in which a worker is removed from work.]

"In addition to monitoring worker exposure in routine surveillance programs, cholinesterase determinations are also extensively used to establish reentry intervals (i.e., the time between application of pesticides and reentry to the field), to assess changes in application of field work practices, to evaluate personal protective equipment and field application equipment, to establish exposure limits, and to monitor compliance with crop treatment and reentry schedules."[3]

Treatment of Pesticide Poisoning. Workers, managers, and health care personnel should be trained to recognize the signs and symptoms of pesticide poisoning and to render first aid to affected individuals until these individuals can be treated by health care providers. Physicians and other health care personnel should be trained in the appropriate treatment of patients with pesticide poisoning. Health care personnel should also be trained in the critical role of facilitating measures to prevent pesticide poisoning.

Continuing with the example, workers acutely poisoned with organophosphates should not return to work until signs and symptoms have completely ceased and their cholinesterase levels have returned to at least 80% of normal (or baseline) values.

First Aid and Emergency Treatment. General measures for treating serious acute pesticide poisoning include (a) observation; (b) adequate aeration of the lungs; (c) bathing and shampooing the exposed individual to remove pesticides on the skin and hair; and (d) for some pesticides, such as the organophosphates, inducing vomiting (with ipecac), if there has been ingestion, followed by administration of activated charcoal.

Specific measures vary with the type of pesticide involved. For some types of pesticide poisoning, there are no specific measures. For organophosphate poisoning, treatment measures include administration of the specific antidotes atropine and pralidoxime (2-PAM). (These medications should never be given prophylactically.)

Government Regulations. Pesticide regulation is unusually complicated in that responsibility for worker and public health protection, as well as for environmental protection, is divided between the Environmental Protection Agency (EPA) and the Occupational Safety and Health Administration (OSHA). Pesticide laws generally cover all aspects of pesticide manufacture and formulation, transport, storage, sale, use, and disposal.

Under the Federal Insecticide, Fungicide, and Rodenticide Act (FIFRA), EPA is responsible for registering new pesticides to ensure that, when used according to label directions, they will not present unreasonable risks to human health or to the environment. However, this protection is based on a cost/benefit estimate that often underestimates both health risks to workers or the general public and environmental damage. EPA is also responsible for setting use conditions and ensuring the protection of all workers who mix, load, or apply agricultural or structural pesticides or who harvest crops treated with pesticides. The current standards are only minimal (*see* 40 CFR, Part 170: Worker Protection Standards for Agricultural Pesticides). However, more comprehensive agricultural worker safety regulations were expected to be published by EPA in 1991.

FIFRA allows states to adopt pesticide programs that are "at least as effective" or more protective than the federal law. Thus, some states may have more extensive regulations on worker protection, pesticide registration and use, and reporting of pesticide poisoning. (For example, California has more extensive worker protection standards, an independent pesticide registration process, and a mandatory requirement for physicians to report all cases of known and suspected pesticide poisonings. Texas also has adopted more extensive worker protection standards, including a comprehensive Farmworker Right to Know [Agricultural Hazard Communication] Law. State laws also vary widely on coverage of farm workers under workers' compensation and unemployment insurance.) There is no complete or accurate data base of all pesticide illnesses in the U.S.; states vary widely on their illness-reporting requirements. Also, there is no data base on pesticide use, hampering efforts to identify and quantify occupational exposure.

Under the federal Food, Drug, and Cosmetic Act (FDCA), the EPA sets tolerances, or maximum legal limits, for pesticide residues on raw food commodities and feed grains marketed in the U.S. OSHA, on the other hand, sets and enforces standards, including permissible exposure limits (PELs) for workers who manufacture, formulate, store, and transport pesticides. OSHA also enforces field sanitation regulations, which provide for drinking water and toilet facilities for field-workers.

Pesticide regulation also occurs at the state level. State agencies, primarily agriculture departments, are also involved in regulating the use, storage, and disposal of pesticides designed for agricultural and structural purposes. Other agencies, including state and local health departments, structural pest control boards, environmental protection departments, and air and water quality departments, may also have standards dealing with public health and environmental protection.

Regulations should cover at least the following:

Manufacture, Formulation, and Packaging: OSHA regulations generally cover workplaces and workers who manufacture, formulate, and package pesticides. Under the Hazard Communication (Right to Know) Standard, material safety data sheets (MSDSs) should be developed on each pesticide, including chemical and physical data, adverse acute and chronic health effects, and health and safety preventive measures. These MSDSs should be part of training programs for all workers who manufacture, formulate, warehouse, package, and transport pesticides.

Many specific health and safety provisions of OSHA that cover other types of chemical workplaces also apply to these pesticide facilities. For example, pesticides should be formulated in facilities with adequate ventilation. OSHA may specifically require that workers formulate and pack pesticides under ventilated hoods or in closed systems. Care should be taken to avoid spills, and spill containment and cleanup procedures should be meticulously followed when spills do occur. OSHA has specific requirements governing the use of personal protective equipment, such as respirators (after appropriate fitting and training), aprons, and eye protection, and the use of gloves and other protective clothing during pesticide manufacture, formulation, and packaging. Cotton is an effective protective material only where workers will not come into contact with liquid formulations. Impermeable gloves and other types of protective clothing may be required, depending on the particular pesticide.

Storage and Transport: Tanks and drums used to store pesticides should be made of appropriate materials, and precautions should be taken to contain leaks, if they occur. After they are used, tanks, drums, and all other pesticide containers should be disposed of and used for no other purpose. Pesticide storage areas should be marked, locked, and made inaccessible to the public. Electrical and other safety hazards should be eliminated from areas where pesticides are stored. Transport vehicles should be appropriately labeled. Work practices to avoid spills or other accidents should be carefully followed during loading, transport, and unloading. Vehicles used to transport pesticides should be properly maintained and never used to transport people, food, or water. Procedures for dealing with spills or other accidents during transport should be developed in advance, and vehicle operators should be trained in these procedures.

Sale and Use: Under FIFRA, pesticide manufacturers are required to prepare labels that prescribe use conditions and identify human health and environmental hazards. Users should carefully read and follow use instructions and safety precautions on pesticide labels, including those concerning protective clothing and equipment. However, although these labels provide the only information most agricultural workers receive on health hazards, they often lack complete information about chronic health effects and are generally printed only in English. Many farm workers never even see pesticide containers or their labels. Moreover, although workers covered by OSHA are entitled to MSDSs on pesticides, agricultural workers covered by the EPA (or state agriculture departments) have not yet been given the right to know about the hazards of pesticides.

Sale of pesticides in unlabeled or inadequately labeled containers is prohibited by federal and state laws. The mixing, loading, and applying of pesticides pose particularly serious health risks. People who are mixing pesticides must follow label directions carefully. In many cases, labels will specifically require that closed systems be used and specific protective clothing and equipment be worn. Because dermal absorption is likely and is the most hazardous route of entry for pesticides, bare hands should never be used to mix pesticides. Appropriate protective clothing and equipment should be used as indicated on the label. Equipment in good repair should be used to apply pesticides.

In general, appropriate personal protective equipment and clothing should be provided and worn, as indicated on the pesticide label. However, for many agricultural workers, personal protective equipment may be very uncomfortable in extremely hot or humid weather and may pose even more serious health risks to workers. Yet current standards may require workers to reenter fields before the legal reentry period, providing that they wear protective clothing.

Children are a significant part of the agricultural work force. They are particularly vulnerable to the hazards of pesticides and should never be permitted to mix, load, or apply them. Children are also at risk for exposure to pesticide residues on treated crops. Economic reality in the U.S., however, often leads to children working in the fields with their parents or being brought to fields because of lack of adequate child care facilities. Different state laws address this aspect of child labor.

EPA worker protection standards prohibit exposing workers directly or indirectly (through drift) to pesticides. Generally, field-workers should not enter treated fields without protective clothing until it is safe to do so. However, EPA has established safe reentry times for only 17 of the approximately 40 000 pesticides registered for use in the U.S. (*see* Table 2 at end of entry). For pesticides without a prescribed reentry interval, it is necessary to wait until the dust has settled or the spray has dried. Some states have set more protective and extensive reentry standards.

After using pesticides and before eating, drinking, or smoking, workers should carefully wash with soap and water. The employer should provide washing facilities and should allow workers the time to use them, although this is often violated in practice. Contaminated clothing should not be laundered with other family laundry. Containers used for mixing or storing pesticides should never be used later to store or transport food or water.

Disposal. Care should be taken to dispose of unused pesticides and containers in a manner that will protect people and the environment. EPA or state agencies may have more specific provisions on recycling drums and disposal procedures.

Community Right to Know. Title III of SARA (U.S. Superfund Amendments and Reauthorization Act) provides for the community's right to know about use, production, spills, and other releases of pesticides as it does with other toxic substances (*see* Environmental Health Law in part 3).

International Code of Conduct. In 1985, the Food and Agricultural Organization of the United Nations adopted the

International Code of Conduct on the Distribution and Use of Pesticides. The primary purposes of the code include assisting countries that have not yet developed their own infrastructures and recommending standards of conduct for all pesticide manufacturers and exporters to minimize adverse human health and environmental effects. The code covers such areas as pesticide safety testing, worker and public health and safety protection, advertising, and technical assistance to developing countries on sound pest management practices, including alternatives to pesticides. These standards, however, are only discretionary and are not enforceable by any international agency.

Education and Training. It is critical that all individuals potentially exposed to pesticides be educated about the hazards and trained in measures to minimize exposure. Training programs should be mandated in government regulations. Training should reach all individuals potentially exposed but should focus first on those who are most heavily exposed.

Training should be done regularly, with periodic refresher programs. Training programs should be modified to take into account that many agricultural workers speak a different language and may not be able to read well or at all in any language. Education should also extend to government decision makers and to members of the general public, because many people can be exposed to pesticides in consumer products, in their homes, and in other ways.

BSL, EW

References

1. Mehler L, Edmiston S, Richmond D, O'Malley M, Krieger RI. *Summary of Illnesses and Injuries Reported by California Physicians as Potentially Related to Pesticides*. Sacramento, Calif: California Dept. of Food & Agriculture, Div. of Pest Management, Environmental Protection & Worker Safety (1220 N St., Sacramento, CA 94271-0001).

2. The remainder of this section has been adopted from Baker SR, Wilkinson CF, eds. *The Effect of Pesticides on Human Health. Advances in Modern Experimental Toxicology.* Vol. 18. Princeton, NJ: Princeton Scientific Publications; 1990.

3. Coye MJ, Love JA, Maddy KT. Biological monitoring of agricultural workers exposed to pesticides: I. Cholinesterase activity determinations. *J Occup Med.* 1986;28:619-627.

Further Reading

California Department of Health Services, Hazard Evaluation Service. *Pesticides: Health Aspects of Exposure and Issues Surrounding Their Use.* Berkeley, Calif: Dept of Health Services (2151 Berkeley Way, Berkeley, CA 94704-1011); 1988.

Committee on Government Operations. *Report on Exports Banned by US Regulatory Agencies.* 38th Report. Washington, DC: US Government Printing Office; 1978. House Report No. 95-1686.

Consolidated List of Products Whose Consumption and/or Sale Have Been Banned, Withdrawn, Severely Restricted or Not Approved by Governments. 2nd ed. New York, NY: United Nations; 1984:93-194.

Control of Pesticides Residues in Food: A Directory of National Authorities and International Organizations. Uppsala, Sweden: Swedish National Food Administration; 1983.

Craddick B, ed. *Federal and State Employment Standards and U.S. Farm Labor.* Austin, TX: Motivation Education & Training (Research Office, 55 North IH 35, Rm. 117, Austin, TX 78702; available in English and Spanish); 1988.

Food and Agricultural Organization. *FAO Guidelines for the Regulation and Control of Pesticides, Iincluding a "Model Scheme" for the Establishment of National Organizations.* Rome, Italy: Food and Agricultural Organization; 1985.

International Group of National Associations of Manufacturers of Agrochemical Products. *Guidelines for the Safe and Effective Use of Pesticides*. Brussels, Belgium: International Group of National Associations of Manufacturers of Agrochemical Products; 1983.

International Organization of Consumers Unions. *The Pesticide Handbook: Profiles for Action*. 2nd ed. Penang, Malaysia: International Organization of Consumers Unions; 1986.

Morgan DP. *Recognition and Management of Pesticide Poisonings*. 4th ed. Washington, DC: Environmental Protection Agency; 1989. EPA 540/9-88-001.

Office of Technology Assessment. *Neurotoxicity: Identifying and Controlling Poisons of the Nervous System*. Washington, DC: US Government Printing Office; 1990. OTA-BA-436.

Texas Center for Policy Studies and the National Center for Policy Alternatives. *The Pesticides Crisis: A Blueprint for States*. Washington, DC: National Center for Policy Alternatives (2000 Florida Ave., NW, Washington, DC 20009); 1988.

United Nations Industrial Development Organization. *Formulation of Pesticides in Developing Countries*. New York, NY: United Nations; 1983.

Weir D, Schapiro M. *Circle of Poison: Pesticides and People in a Hungry World*. San Francisco, Calif: Institute for Food and Development Policy; 1981.

Wilk V. *The Occupational Health of Migrant and Seasonal Farmworkers in the U.S.* 2nd ed. Kansas City, Mo: Farmworker Justice Fund; 1986. (Distributed by National Rural Health Care Association, 301 E. Armour Blvd., Ste. 420, Kansas City, MO 64111.)

World Health Organization, Environmental Health Criteria. Geneva, Switzerland: World Health Organization. #9, DDT and its derivatives, 1979; #29, 2,4-D, 1984; #34, Chlordane, 1984; #38, Heptachlor, 1984; #39, Paraquat and diquat, 1984; #40, Endosulfan, 1984; #44, Mirex, 1984; #45, Camphechlor, 1984; #63, Organophosphorus insecticides, 1986; #64, Carbamate pesticides, 1986.

World Health Organization. *Recommended Health-based Limits in Occupational Exposure to Pesticides.* Geneva, Switzerland: World Health Organization; 1982. WHO Technical Report Series 677.

World Health Organization. *Safe Use of Pesticides.* Geneva, Switzerland: World Health Organization; 1985. WHO Technical Report Series 720.

World Health Organization and United Nations Environment Programme. Public Health Impact of Pesticides Used in Agriculture. Geneva, Switzerland: World Health Organization, and Nairobi, Kenya: United Nations Environment Programme; 1989.

Table 1.— Major Types of Pesticides

Type	Examples	Acute Signs and Symptoms	Laboratory Confirmation
Organophosphates	Parathion Malathion	Headache, dizziness, blurred vision, sweating, nausea and vomiting, weakness, tremor, incoordination, abdominal cramps, skin rash	Reduction in plasma and red blood cell cholinesterase activity levels

Type	Examples	Acute Signs and Symptoms	Laboratory Confirmation
Carbamates	Methomyl Carbaryl Thiram	Nausea and vomiting, abdominal pain, diarrhea, sweating, salivation, blurred vision, flulike symptoms, and, with some carbamates, skin rash	Reduction in plasma and red blood cell cholinesterase activity levels
Pentachlorophenol (PCP)		Skin, eye, nose, and throat irritation; sweating, weakness, nausea, and headache	Blood, urine, and adipose tissue levels of PCP
Organochlorines	DDT Heptachlor	Headache, weakness, apprehension, excitability, disorientation, paresthesias, seizures, and skin rash	Pesticide or metabolite levels in blood or urine
Nitrophenols	Dinitrophenol Dinoseb	Headache, nausea, flu-like symptoms, sweating, thirst, yellow staining of skin	Pesticide levels in serum and urine
Dipyridyls	Paraquat Diquat	Eye, skin, and upper respiratory tract irritation (nausea, vomiting, diarrhea, and abdominal pain after ingestion)	Dipyridyl levels in blood and urine
Chlorphenoxy herbicides	2,4-D 2,4,5-T	Skin irritation and rash (mouth, skin, and gastro-intestinal tract irritation after ingestion)	Levels in blood and urine
Phthalimide derivatives	Captan Captafol	Skin and respiratory tract irritation	

Table 2.—EPA Pesticide Reentry Intervals*

Chemical (Trade Name)	Waiting Period (Days)	Main Crops Used On
Azinphos-methyl (Guthion)	1	Fruits, nuts, melons, ornamental shrubs, shade trees
Carbofuran (Furadan)	1†	Grains, tobacco, peanuts, sugarcane, potatoes, grapes, sunflowers
Carbophenothion (Trithion)	2	Fruits, nuts, cotton (also cattle)
Chlorpyrifos	1‡	Corn, fruits, nuts, vegetables, cotton
Demeton (Systox)	2	Most fruits, nuts and vegetables
Dicrotophos (Bidrin)	2	Cotton, coffee, soybeans
Endrin	2	Cotton and orchard crops
EPN	1	Corn, rice, cotton, grapes
Ethion	1	Cotton, most vegetables
Fosetyl (Aliette)	7§	Ornamental shrubs, pineapple, hops
Mevinphos (Phosdrin)	2-4‖	Vegetables, fruits, field crops
Monocrotophos (Azodrin)	2	Cotton, tobacco, sugarcane, peanuts, potatoes
Oxydemeton-methyl (Metasystox-R)	2	Flowers, ornamental shrubs, some vegetables
Parathion-ethyl (Parathion)	2	Wide range of uses, including vegetables of all kinds, especially corn and potatoes, and tobacco
Parathion-methyl (Methyl parathion)	2	Same as above, and cotton
Phosalone (Zolone)	1	Apples, cherries, almonds, grapes, artichokes, other fruits and nuts
Propargite (Omite)	7¶	Fruits, nuts, ornamentals, cotton, corn, grapes

*EPA *Code of Federal Regulations Title 40, Part 170 - Worker Protection Standards for Agricultural Pesticides.*
Table adapted from chart which appears in "Pesticides and You: A Guide for Farmworkers, Small Farmers and Rural Communities." Published by Rural America, 1302 18th Street, N.W., Washington, D.C. 20036 (phone: 202-659-2800), 1980; a table entitled "List of Pesticides Having Reentry Intervals" prepared by James D. Adams, Office of Pesticide Programs, U.S. Environmental Protection Agency, October 31, 1985 (see Appendix II); and the *1986 Farm Chemicals Handbook*.
†14 days on seed and sweet corn.
‡4 days on citrus, grapes, and peaches.
§Hops only.
‖Proposed but not yet implemented.
¶Grapes only.

Pigment Disorders

ICD-9 709.0

Identification

Pigment disorders can be identified by lightening or darkening of the skin. Hypopigmented or hyperpigmented macules often occur in areas of previous dermatitis or trauma to the skin or following exposure to certain chemicals.

Occurrence

Postinflammatory hyperpigmentation and hypopigmentation are the most common pigment disorders. Any dermatitis or trauma to the skin, such as thermal or chemical burns, may lead to an increase or decrease in pigmentation in that area. Heavily pigmented individuals have postinflammatory pigment abnormalities most notably, and these abnormalities are slower to resolve. (*See* Contact Dermatitis, Irritant, for data on incidence rates for occupational skin diseases.)

Causes

Monobenzylether of hydroquinone, an antioxidant used in rubber manufacturing, was the first chemical implicated in inducing work-related loss of pigment. Certain phenolic compounds used as antioxidants or germicidal disinfectants also produce pigment loss. Hypopigmented macules that are not in sites of direct contact may also occur after exposure to these chemicals. Hyperpigmentation can also be caused by exposure to aerosols of metallic silver, mercury compounds, and arsenic.

Pathophysiology

Chemically induced leukoderma occurs because the chemical either blocks the synthesis of melanin (the pigment responsible for skin color) or, in rare instances, causes direct injury to melanocytes (cells responsible for melanin synthesis). In postinflammatory hypopigmentation, the transfer of melanosomes (pigment-containing granules) from melanocytes to keratinocytes is blocked. Postinflammatory hyperpigmentation results from an increase in melanin synthesis and the release of large pigment granules into the dermis.

Prevention

Chemical leukoderma can be prevented by implementing engineering controls and by using protective clothing and gloves to prevent skin contact with hydroquinones and phenolic compounds. Monobenzylether of hydroquinone is rarely used now in rubber gloves. Postinflammatory hyperpigmentation and hypopigmentation are best prevented by the prompt recognition and treatment of inflammatory skin diseases.

Other Issues

Currently, phenolic germicides, present in detergents, may be the most common cause of occupational leukoderma. Workers at risk are those who clean and sanitize floors and equipment (eg, in hospitals). Workers at risk for developing leukoderma due to the hydroquinones are in the cosmetics industry, where bleaching creams are made. Heavily pigmented workers (eg, blacks, Asians, and Hispanics) are at greatest risks for developing postinflammatory hyperpigmentation or hypopigmentation.

MB, KA

Further Reading

Adams RM. *Occupational Skin Disease.* Philadelphia, Pa: WB Saunders; 1990.

Maibach HI, ed. *Occupational and Industrial Dermatology.* 2nd ed. Chicago, Ill: Year Book Medical Publishers; 1988.

Nordlund JJ. Vitiligo. In: Thiers BH, Dodson RD, eds. *Pathogenesis of Skin Disease.* New York, NY: Churchill Livingston; 1986;99-127.

Pleural Diseases, Asbestos-Related

Identification

Three kinds of pleural disease result from exposure to asbestos fibers: (a) pleural plaques/discrete pleural thickening, (b) diffuse pleural thickening, and (c) benign exudative pleural effusion.

Pleural plaques/discrete pleural thickening. The presence of bilateral pleural thickening is a strong marker of previous asbestos exposure. The individual is usually asymptomatic unless other respiratory abnormalities are present or the plaques are extensive. Many pleural plaques are not detectable on routine radiograph. On chest x-ray, most plaques occur in the middle of the diaphragm, in the posterolateral chest wall between the seventh and ninth ribs, or in the lateral chest wall between the sixth and ninth ribs. Plaques can calcify and can be seen along the diaphragm and lateral wall margins. They can also calcify along the mediastinum and left heart border. Plaques that are not seen on routine chest x-ray can be detected on computed tomography (CT), which is also helpful in distinguishing between pleural and parenchymal location of nodules. Pleural plaques due to asbestos exposure often appear alone, in the absence of underlying interstitial disease. The differential diagnosis of discrete pleural plaques is limited; these plaques may be seen with mesothelioma (malignant or benign fibrous mesothelioma), metastatic cancer, lymphoma, or myeloma.

Asbestos exposure frequently, though not always, causes bilateral plaques; hence, if the plaque formation is unilateral, other etiologies should be considered. The most frequent non-asbestos etiology of plaques is previous trauma. Calcified pleural plaques can occur with trauma or infection, either of which results most often in unilateral discrete pleural thickening. They also can occur with a rib fracture, x-ray therapy, scleroderma, and chronic mineral oil aspiration.

Diffuse pleural thickening. The clinical importance of distinguishing diffuse pleural thickening from discrete pleural plaque formation is not well established. Diffuse pleural thickening results from visceral pleural thickening, as opposed to the parietal pleural plaques noted above. The individual with mild diffuse pleural thickening is usually asymptomatic. If the pleural thickening is extensive, the person can experience dyspnea on exertion, chest tightness, and difficulty taking a deep breath. On chest x-ray, diffuse thickening generally involves the costophrenic angle and has ill-defined borders on all x-ray views. It is also much less likely to calcify than circumscribed plaques. Pulmonary function tests can yield normal results or can demonstrate a restricted lung capacity. The diffusion capacity is normal when corrected for the actual lung volume. Diffuse pleural disease in the absence of parenchymal asbestosis can cause significant pulmonary impairment and has reportedly caused respiratory failure.

Diffuse thickening is less specific for asbestos exposure than discrete pleural plaques. Other conditions that can result in diffuse thickening include infection (tuberculosis or other bacterial infections), connective tissue diseases (especially scleroderma, rheumatoid arthritis, and systemic lupus erythematosus), and, rarely, sarcoidosis, uremia, or drug reactions.

Benign exudative pleural effusion. A benign exudative effusion can occur suddenly. It is the most common asbestos-related abnormality within the first 20 years after initial exposure to asbestos. The volume of effusions is usually small but may be several liters. In general, effusions resolve within 1 year without residual abnormalities, but they may leave blunted costophrenic angles or diffuse visceral pleural thickening. Fluid examinations are always sterile exudates that may be serous or serosanguinous. Pulmonary function will be abnormal if the pleural effusion is large enough to create areas of atelectasis and shunting, or if there is concomitant parenchymal disease. The diagnosis is made by establishing exposure to asbestos, performing thoracentesis, ruling out other causes of effusion (usually requiring a pleural biopsy), and not finding evidence of a malignancy for 3 years. The main differential diagnoses are malignant mesothelioma and lung cancer.

Occurrence

Some occupational groups with 40 or more years of exposure to asbestos have a prevalence of pleural plaque formation as high as 80%. The presence of plaques depends on the concentration and duration of exposure. The mean latency period is about 20 years from time of first exposure to radiographic evidence of a plaque. There is a relationship between pleural thickening and cumulative asbestos dose, independent of the time elapsed since first exposure, but this relationship is less well established than that for parenchymal fibrosis. Plaques tend to calcify with time, although calcification is not always readily apparent on the plain chest x-ray. The prevalence of diffuse pleural thickening in occupationally exposed groups has not been well described.

In contrast to pleural plaques and diffuse pleural thickening, asbestos-induced benign exudative effusions often occur early, many within 10 years from first exposure to asbestos. The prevalence of this disorder is not well known; in one study, it was dose related.

All occupational groups exposed to asbestos (directly, in handling it, or indirectly, by being close to workers handling it) are at risk of developing asbestos-related pleural diseases. In the general population, surveys have revealed x-ray evidence for pleural plaques ranging up to 3% among those surveyed. In the

U.S., the vast majority of pleural plaques occur in people who are occupationally exposed, directly or indirectly, to asbestos.

Recent studies have demonstrated reduced pulmonary function in workers with circumscribed plaques in comparison with workers similarly exposed but without plaques. The relationship between development of plaques and subsequent risk for development of parenchymal fibrosis or lung cancer has not been well quantified. Other asbestos-related diseases can occur in the absence of pleural diseases.

Causes

Asbestos-related pleural diseases are caused by inhalation of asbestos fibers of any type (*see* Asbestos-Related Diseases).

Pathophysiology

The exact pathogenesis of pleural disease due to asbestos exposure is not known. Pleural plaques are discretely elevated, gray-white areas on the parietal pleura (rarely occurring on the visceral pleura). Microscopically, they are composed of relatively acellular strands of collagen interposed between normal tissue with intact lamellae and a covering layer of mesothelial cells. Asbestos bodies are not visible, but fibers can be demonstrated by electron microscopy and microchemical analysis. The mechanism by which pleural plaques undergo calcification is not clear.

The pathology of diffuse asbestos-related pleural thickening results from the thickening and fibrosis of the visceral pleura, often fusing with the parietal pleura. The pathogenesis is unclear. Diffuse pleural thickening appears to be a common sequela of old asbestos-induced pleural effusion, with costophrenic angle blunting and diffuse visceral fibrosis. It is not known how often diffuse visceral pleural fibrosis occurs after an asbestos-induced pleural effusion.

Prevention

Strategies necessary to prevent asbestos-related pleural diseases are the same as those needed to prevent asbestos-related parenchymal disease (*see* Asbestos-Related Diseases).

DCC

Further Reading

Becklake MR. Asbestos-related diseases of the lungs and pleura: current clinical issues. *Am Rev Respir Dis.* 1982;126:187-194.

Rosenstock L, Hudson LD. The pleural manifestations of asbestos exposure in occupational medicine. *State of the Art Reviews in Occupational Medicine.* 1987;2:383-409.

Schwartz DA, Galvin JR, Dayton CS, Stanford W, et al. Determinants of restrictive lung function in asbestos-induced pleural fibrosis. *J Appl Physiol.* 1990;68(5):1932-1937.

Pneumoconiosis—*See* Asbestosis, Cobalt-Induced Interstitial Lung Disease, Coal Workers' Pneumoconiosis, and Silicosis

Porphyrias—*See* Hepatic Porphyrias

Pregnancy Outcome, Adverse

Identification

Adverse pregnancy outcome includes spontaneous abortion, stillbirth, congenital defects, prematurity, and low birth weight. The World Health Organization defines spontaneous abortion as any nondeliberate interruption of an intrauterine pregnancy before the 28th week of gestation, timed from the date of the last menstrual period, in which the fetus is dead when expelled. After 28 weeks, if the fetus is dead at birth, it is considered a stillbirth. The term *congenital defects* refers to abnormalities of appearance or function that are present at birth. Prematurity is a birth before 37 weeks of gestation. The term *low birth weight* is used if a baby weighs less than 2500 grams at birth, regardless of gestational age.

Occurrence

It is estimated that 15% to 20% of recognized pregnancies end in spontaneous abortion. (A recent study monitoring urinary human chorionic gonadotropin [hCG] to detect early pregnancies found that 35% of all pregnancies abort.) Most spontaneous abortions occur in the second or third month of gestation.

No surveillance system for spontaneous abortion exists in the United States. In Finland, all birth outcomes are registered, including early fetal loss of a recognized pregnancy, and this registry can be linked to registries of occupation to look for increased risk of miscarriage. Much literature in this field comes from use of these registries.

About 4% to 5% of all newborns have a detectable birth defect, 7% are of low birth weight, 5% are premature, and another group, less well defined, has developmental or functional problems in childhood. Two surveillance systems exist in the U.S. to detect changes in the frequency of birth defects. One, in metropolitan Atlanta, collects detailed information but is geographically limited; the other is a national program that monitors birth defects from cases in hospitals in selected areas. These surveillance systems have not detected significant changes in the occurrence of major birth defects. It is possible that both systems would fail to detect birth defects secondary to occupational exposures; neither is able to track birth defects by industry, so an increase in one geographic area or industry might be lost in the aggregate data. Because the etiology of most birth defects is not known, it is also possible that an increase in environmental causes would be masked by a decrease in infectious causes, such as congenital rubella.

Causes

Table 1 lists the possible nonoccupational causes of spontaneous abortion, but for many of these, there is no firm evidence nor is the degree of risk understood. Congenital defects are caused by genetic factors, environmental factors, or both. Most congenital defects of known cause are genetic, and although many environmental agents have been implicated as teratogens, relatively few have been associated with a specific defect.

Table 1.—Possible Nonoccupational Causes of Spontaneous Abortion

Strong Evidence	Weaker Evidence
Parental age	Gravidity
Previous spontaneous abortion	Oral contraceptives
Maternal smoking	Spermicidal agents
Maternal alcohol consumption	Prior induced abortion
Conception with intrauterine device (IUD) in place	Stress
Ionizing radiation of gonads	Nutrition
Maternal medication/drug use	Hormonal factors
Maternal illness	
Uterine abnormalities, other gynecologic conditions	

Table 2 lists the nonoccupational causes of congenital defects. Table 3 lists the known and suspected occupational causes of adverse pregnancy outcome; among these are agents that have been well studied and to date have not been clearly linked with adverse reproductive outcome. Table 3 also includes a scale for measuring the certainty of the causal association between the occupational exposure and the reproductive effect; details on individual substances can be found in Barlow and Sullivan[1] and in a report prepared by the Office of Technology Assessment (OTA).[2] The scale does not incorporate details of dose; an association is listed as a strong one even if the data exist only at high doses in animals. In applying the details of these tables to an individual exposure situation, dose-response must be considered in more detail. Table 4 lists industries that have been associated with adverse reproductive outcome, without identifying a specific agent.

Pathophysiology

After fusion of the sperm and ovum, the embryo undergoes growth and development. Weeks 1 and 2 are the period of rapid division of the zygote, implantation, and formation of

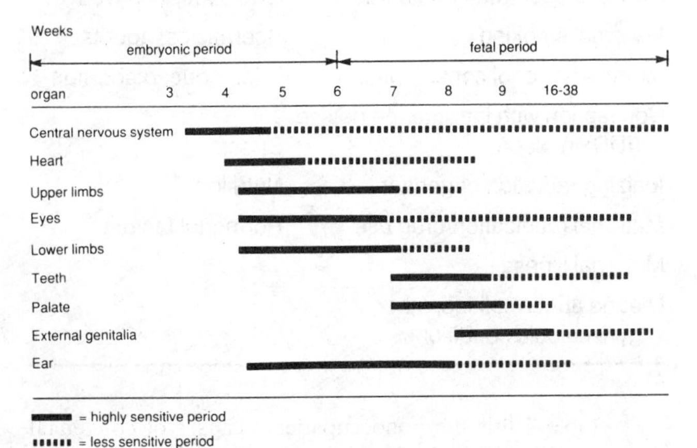

Fig. 1.—Periods of embryonic and fetal development that are highly sensitive to environmentally induced damage.

Table 2.—Nonoccupational Causes of Congenital Defects

A. Inherited disorders

B. Drugs and chemicals (exposure during pregnancy)*
- Alcohol
- Androgens
- Anticoagulants
- Antithyroid drugs
- Chemotherapeutic drugs
- Diethylstilbestrol (DES)
- Isotretinoin
- Lithium
- Phenytoin
- Ribavirin
- Streptomycin
- Tetracycline
- Thalidomide
- Trimethadione
- Valproic acid

C. Infections during pregnancy*
- Cytomegalovirus (CMV)
- Rubella
- Syphilis
- Toxoplasmosis
- Varicella

D. Ionizing radiation exposure during pregnancy*

*Exposure could occur in the occupational setting. For the drugs listed, the data are available from high-dose animal experiments or therapeutic use in humans; a threshold has not been defined but may exist for some adverse reproductive outcomes.

Table 3.—Occupational Agents Causing Adverse Pregnancy Outcome

Agent	Outcome Seen in Human Studies	Outcome Seen in Animal Studies	Strength of Association*
Anesthetic gases	Spontaneous abortion	Birth defects	1,3,5,7
Antineoplastic drugs	Spontaneous abortion	Birth defects, fetal loss	2,3
Arsenic	Spontaneous abortion, low birth weight	Birth defects, fetal loss	2,3
Benzo(a)pyrene	None	Birth defects	1
Cadmium	Low birth weight, skeletal disease	Fetal loss, birth defects	2,3
Carbon disulfide	Menstrual disorders, spontaneous abortion		
Carbon monoxide (CO)	None specific to CO	Birth defects, low birth weight	2
Chlordecone	None	Fetal loss	2,5
Chloroform	Eclampsia	Fetal loss	1
Chloroprene	None	Birth defects	2,3,5
Ethylene glycol ethers	None	Fetal loss	2
Ethylene oxide	Spontaneous abortion	Fetal loss	1,3
Formanides	None	Fetal loss, birth defects	2
Inorganic mercury	Spontaneous abortion	Fetal loss	1,3
Lead	Spontaneous abortion, prematurity, neurological dysfunction in child	Birth defects, fetal loss	2,4
Organic mercury	Cerebral palsy, CNS malformation	Birth defects, fetal loss	2,4
Ozone	None	Fetal loss	1
Physical stress	Prematurity	None	4

Agent	Outcome Seen in Human Studies	Outcome Seen in Animal Studies	Strength of Association*
Polybrominated biphenyls (PBBs)	None	Fetal loss	2
Polychlorinated biphenyls (PCBs)	Low birth weight	Low birth weight	2,4
Radiation	CNS defects, skeletal anomalies	Fetal loss, birth defects	2,4
Selenium	Spontaneous abortion	Low birth weight	2,7
Tellerium	None	Birth defects	2
2,4-D	Skeletal defects	Birth defects	1,3,8
Video display terminals	Spontaneous abortion	Birth defects	1,8
Vinyl chloride	CNS defects	None	3,6
Xylene		Fetal loss	

* 1 = limited positive animal data; 2 = strong positive animal data
3 = limited positive human data; 4 = strong positive human data
5 = limited negative animal data; 6 = strong negative animal data
7 = limited negative human data; 8= strong negative human data

the bilaminar embryo. Exposure to toxins in this period does not usually result in specific defects. Figure 1 illustrates the periods of sensitivity for major organ systems, during which an environmental exposure could cause damage to the embryo or fetus.

Because of these critical periods, the timing of an exposure largely determines the effect of a toxin. Delivering an insult to an organism just before or during the early stages in the development of an organ can cause a defect, whereas the same organ would not be susceptible to injury later in fetal development. A variety of agents given at the same developmental stage could cause the same defect, and the same agent given at different periods could cause different defects.

Table 4.—Industries with a Reported Increased Risk of Adverse Reproductive Outcome in Exposed Women, without Linkage to Specific Exposures

Industry	Reported Outcome
Rubber industry	Spontaneous abortion
Leather industry	Spontaneous abortion
Chemical industry	Spontaneous abortion
Electronics industry (in solderers)	Spontaneous abortion
Metal work	Spontaneous abortion
Laboratory work	Spontaneous abortion, birth defects
Construction	Birth defects
Transportation	Birth defects
Communications	Birth defects
Agriculture and horticulture	Birth defects
Jobs with mixed solvent exposures	Birth defects, spontaneous abortion
Textiles	Spontaneous abortion

In the first 3 weeks, the most probable effect of significant exposure is severe damage and death of the embryo; organogenesis has not yet begun. The period from 4 to 9 weeks is the time when classic birth defects might be induced. After 9 weeks, organogenesis is basically complete, although the central nervous system continues to develop through the prenatal period. Exposures after 9 weeks may cause postnatal functional abnormalities. Carcinogens could potentially exert an effect at any stage in development.

The effect of exposure also varies with the dose of a toxin, with the degree of abnormal development increasing as the dose increases. At lower exposures, damage may be repairable; high doses may lead to death of the embryo; and doses in between may cause a congenital defect.

In addition, a toxic agent may affect the embryo even if one of the parents is exposed before conception. An exposure can damage germ cells in the ovary or the sperm. Damage to the ova can persist for years, while damage to the sperm can be repaired by cell turnover if the germinal epithelium remains intact. Exposure prior to conception can affect the development of the fetus because the toxin persists in the maternal body. For example, PCBs are known to be stored in fat for a significant period of time, and lead is stored in bone. The storage of toxins exists in a steady state with the blood, so the fetus can be exposed to these body stores through the maternal circulation.

Spontaneous abortion in particular has multiple causes and several mechanisms. In some spontaneous abortions, the fetus shows a cessation of development or a focal malformation; in these cases, the defect apparently causes intrauterine death. Chromosome abnormalities incompatible with life have been found in approximately 35% of aborted fetuses studied. In at least 50% of cases, the fetus appears normal at detailed examination; one can hypothesize that the abortion results from a disturbance to the pregnancy without the induction of an anomaly.

Prevention

Primary prevention. Two basic approaches have been used to prevent exposure of a pregnant woman to an identified teratogen: reduction or elimination of her exposure to the toxin, or transfer of the woman to a different job without that exposure. Engineering control of exposure or product substitution are the preferred solutions (*see* part 1). Because there is a time lag between conception and the recognition of a pregnancy—a time when the embryo or fetus is especially vulnerable to the effects of exposure—significant exposure may occur if medical removal is the only preventive strategy.

The only guideline for exposure that specifically addresses prevention of a reproductive hazard is the Nuclear Regulatory Commission's standard for radiation exposure during pregnancy, which limits exposure to 0.5 rem for the duration of the pregnancy. Other standards may not be low enough to protect a developing fetus, if the fetus is more sensitive than the adult.

In the face of uncontrollable exposure, medical removal may be needed. A job transfer can be used effectively for prevention, given that pregnancy lasts for a limited time and the woman can return to her job after delivery. In addition, if a woman becomes pregnant in a high-risk job, a transfer can be accomplished more quickly than product substitution or the institution of appropriate engineering controls. However, the implementation of "fetal protection" policies has often resulted in discrimination against women in hiring and job placement, and decisions about job placement for pregnant or "potentially pregnant" workers have been made without adequate knowledge or counsel. This issue is discussed in more detail below.

After the child is born, occupational exposures of the parents can continue to affect the child's growth and development. An understanding of these as well as of prenatal exposures can afford opportunities for primary prevention. Hazards can occur from substances that are (a) brought home on the clothes of a parent, (b) used in the homes of self-employed people, or (c) excreted in breast milk. As a general rule, good work practices, which include a change of clothes and care not to contaminate street clothes, will control the first two possibilities. Breast-feeding is discussed in more detail below.

Secondary prevention. After a hazardous exposure has occurred, a congenital defect or developmental disorder could theoretically be prevented by terminating the pregnancy. Each time this question arises, the physician must analyze the circumstances and details of the exposure, providing appropriate consultation and advice in each case. The risk of an exposure must be evaluated against the background risk in any pregnancy. In most cases, an occupational exposure does not add significantly to the background risk.

Other Issues

Job transfers. Suggesting a job transfer may raise difficult social and economic issues. As discussed above, a man may recover testicular function if he is removed from exposure. Most workers' compensation statutes limit recovery of damages for infertility, because the presence of the disease does not result in disability for work. There is not yet much experience in using the workers' compensation statutes for medical removal as outlined above. Therefore, removal from exposure must be negotiated with the employer in each case. A clinician who suggests this step needs to be sensitive to the dilemma that can potentially be created, because job transfer may mean income loss. As in similar situations during pregnancy, issues of benefits and legal rights should be referred to a lawyer with expertise in workers' compensation and employment law. A recommendation to a woman to attempt a job transfer or to leave her job for the duration of her pregnancy must be discussed in great detail. The woman should be informed of the magnitude of the risk, insofar as it is known, and of her legal rights. Some employers have a policy regarding pregnancy, and many will follow the recommendations of a physician regarding job placement.

In the U.S., pregnant workers have limited legal rights. However, an employer is not required to transfer a pregnant worker to a safe job. The federal Pregnancy Discrimination Act (Pub L No. 95-555) does not require any employer to offer specific benefits, but it does require that employers provide coverage for pregnancy equal to their coverage of any other temporary medical disability. For example, if an employer guarantees any employee that a job will be available when the employee returns from sick leave, then no exception can be made for leave due to pregnancy.

The law also prohibits discrimination against a pregnant worker in ways unrelated to fringe benefits. An employer cannot refuse to hire or promote a woman because she is or may become pregnant. The law does not require that unpaid maternity leave be available, so a pregnant woman is not protected from loss of her job if she opts to leave work because of a potentially dangerous exposure. Some states have mandated additional benefits for pregnant women.

The U.S. Supreme Court has recently declared a "fetal protection" policy to be unconstitutional. A woman may not be denied employment because she may become pregnant in a job that may pose a risk to the fetus.

Breast-feeding. Although obstetricians and pediatricians encourage breast-feeding based on data showing that breast milk provides immunoglobin IgA, an amino acid profile, and fats essential for the developing infant, these advantages could be outweighed by the transmission of toxic chemicals. Chemicals make their way into breast milk by a passive transfer that depends on three major characteristics: polarity of the chemical, lipid solubility, and molecular weight. A compound with a low molecular weight that is polarized and lipid soluble can be transferred readily.

The infant's dose depends on the metabolic fate of the substance in the mother. If a substance is rapidly metabolized or excreted by the mother, the dose in breast milk will decrease rapidly when exposure stops. Solvents, although very fat soluble, are excreted through the lung, liver, and kidneys as well as through the breast milk, so the maternal body burden decreases rapidly after exposure has ceased. Excretion through breast milk can be a major route for toxins that the body has no other way of handling; this is the case for PCBs, PBBs, DDT, and related halogenated hydrocarbons. The baby's dose may be high even after the mother has been removed from exposure.

Only for a very few substances have the acute health effects of a given dose to an infant been defined; chronic effects of low-dose exposures are virtually unknown. This uncertainty makes it difficult to decide whether to recommend breast-feeding. Details of pharmacokinetics and specific recommendations can be found in Mattison.[3]

LW

References

1. Barlow SM, Sullivan FM. *Reproductive Hazards of Industrial Chemicals.* New York, NY: Academic Press; 1982.

2. Office of Technology Assessment. *Reproductive Health Hazards in the Workplace.* Washington, DC: U.S. Government Printing Office; 1985.

3. Mattison DR, ed. *Reproductive Toxicology: Progress in Clinical and Biological Research.* Vol 117. New York, NY: Alan R Liss; 1983.

Further Reading

Bracken MB, ed. *Perinatal Epidemiology.* New York, NY: Oxford University Press; 1984.

McDonald AJ. Work in pregnancy. *Br J Ind Med.* 1988;45:577-580.

Sager PR, Clarkson TW, Nordberg GF. Reproductive toxicity of metals. In: Friberg L, ed. *Handbook on the Toxicology of Metals.* New York, NY: Elsevier-North Holland Biomedical Press; 1988.

Steeno OP, Panskihila A. Occupational effects of male fertility and sterility. *Andrologia.* 1984;16:5-22, 93-101.

Stein ZA, Hatch MC, eds. Reproductive problems in the workplace. *State of the Art Reviews in Occupational Medicine.* 1986;1:3.

Welch LS. Decision-making about reproductive hazards. *Semin Occup Med.* 1986;1:97-106.

Radiation, Ionizing, Adverse Effects

ICD-9 990, E926.3, E926.5, E926.8, E926.9

Identification

Ionizing radiation consists of short-wavelength, high-frequency electromagnetic radiations (x-rays and gamma rays) as well as various particulate radiations (electrons, protons, neutrons, alpha particles, and other atomic particles). Both types of radiation are produced when atoms disintegrate, whether naturally or in such devices as nuclear reactors or cyclotrons. Ionizing radiation differs from nonionizing radiation in its capacity to disrupt atoms in the absorbing material, thereby giving rise to ion pairs and the breaking of chemical bonds.

Ionizing radiation can cause many types of adverse health effects, depending on the dose of radiation that is absorbed, the rate at which it is absorbed, the quality (linear energy transfer) of the radiation, and the conditions of exposure. Most of the adverse health effects of ionizing radiation, including various types of tissue injury (eg, erythema of the skin, cataract of the lens, impairment of fertility, depression of hemopoiesis), are produced only when the relevant threshold doses are exceeded. Certain other health effects, however, are assumed to have no threshold and, therefore, to increase in frequency with any increase in dose; included in this category are mutagenic and carcinogenic effects.

Below are four main types of adverse health effects (which are also covered under Pathophysiology):

1. Effects on genes and chromosomes: The frequency of gene mutations and chromosome aberrations increases with the dose of ionizing radiation; the changes are not pathognomonic of radiation ex-

posure, however, because similar effects can result from other causes.

2. Effects on tissues: Mitotic inhibition and other cytological abnormalities are detectable immediately after exposure to a large dose of ionizing radiation in the bone marrow, the lining of the gut, and other tissues with rapidly dividing cells. These effects are followed months or years later by fibrosis and atrophy. A dose large enough (eg, >2 Sieverts [Sv] absorbed in a single brief exposure) to kill a major percentage of stem cells in the marrow or intestinal mucosa will cause the acute radiation syndrome. The typical prodromal symptoms include anorexia, nausea, and vomiting during the first few hours after exposure, followed by a symptom-free interval until the main phase of the illness. The main phase of the intestinal form of the illness typically begins 2 to 3 days after exposure, with abdominal pain, fever, and increasingly severe diarrhea, dehydration, toxemia, and shock, leading to death within 7 to 14 days. In the hemopoietic form, the main phase typically begins in the second or third week after exposure, with a reduction in white blood cells and platelets; if damage to the bone marrow has been severe enough, death from septicemia or exsanguination may occur between the fourth and sixth week after irradiation.

3. Effects on growth and development of the embryo: Intensive exposure during organogenesis can cause various malformations, including disturbances in brain development and mental retardation.

4. Effects on the incidence of cancer: The incidence of various types of cancer is increased by exposure to ionizing radiation, depending on the dose and conditions of exposure. However, the induced cancers have no distinguishing features by which they can be individually identified as having been caused by irradia-

tion. Such cancers include many types, the more prominent being leukemia and cancers of the thyroid, lung, female breast, respiratory tract, and digestive tract. Also noteworthy historically are the skeletal cancers resulting from radium poisoning in early dial painters.

Occurrence

No U.S. population-based data on the occurrence of radiation effects exist at this time. In 1979, it was estimated that 62 000 U.S. workers were exposed to ionizing radiation in the nuclear energy fuel cycle; 36 000 were exposed at naval reactors; 500 000 were exposed in the healing arts; 100 000 were exposed in research; and about 7 million were exposed, at substantially lower average exposure levels, in manufacturing and other industrial sectors. The types of workers who may be exposed occupationally to ionizing radiation include atomic energy plant workers, dentists and dental assistants, electron microscopists, fire alarm makers, industrial fluoroscope operators and radiographers, inspectors using and workers located near sealed gamma ray sources, nuclear submarine workers, petroleum refinery workers, physicians and nurses (including radiologists), thorium ore producers, workers with thorium alloys, uranium miners, uranium mill workers, and x-ray technicians.

Causes

Apart from the effects that may be postulated to result from natural background radiation (including radon), the adverse health effects of ionizing radiation in the U.S. population today are attributable largely to excessive exposure to x-rays and/or radioactive isotopes. X-rays for diagnostic medical imaging represent the major source of exposure for most people in the United States. The most intensive occupational exposures occur, on average, among workers in nuclear power plants or in other parts of the nuclear fuel cycle.

Pathophysiology

The injuries caused by ionizing irradiation result from several types of effects:

1. Effects on genes and chromosomes: Although any molecule in the body can be altered by irradiation, DNA is considered to be the most critical molecular target, because damage to a single gene may, if unrepaired, alter or kill the affected cell. In the low-to-intermediate dose range, the frequency of mutations and chromosome aberrations appears to increase linearly with the dose of radiation; however, the dose required to double the frequency of mutations in human germ cells is estimated to be at least 1 Sv. It is not surprising, therefore, that heritable effects of radiation have yet to be demonstrated in humans.

2. Effects on tissues: *See* Identification above. In general, the susceptibility of cells to killing by irradiation increases with their rate of proliferation. In tissues such as the gastrointestinal mucosa and bone marrow, the killing of stem cells may interfere with the normal replacement of differentiated cells, leading to impaired function and radiation sickness (*see* description of acute radiation syndrome in Identification above).

3. Effects on growth and development of the embryo: Embryonal and fetal tissues are usually radiosensitive, in keeping with their proliferative character.

4. Effects on the incidence of cancer: The induced cancers do not appear until years or decades after exposure and, for the most part, have been detected only after relatively large doses (0.5 to 2.0 Sv). Because the dose-incidence relationship in the low-dose domain remains to be defined, the carcinogenic risks from low doses can be estimated only by interpolation or extrapolation from higher doses, based on assumptions about the dose-incidence relationship.

Prevention

The guiding principle in radiation protection is that the dose should be kept as low as is reasonably achievable. In addition, absolute limits have been set on the permissible doses to radiation workers and other members of the population. The dose equivalent (DE) limit that has been recommended to prevent the impairment of organ function has been set at 0.5 Sv (50 rem) per year for all organs other than the lens of the eye; for that organ, the recommended annual DE limit has been set at 0.15 Sv (15 rem). To restrict the risks of mutagenic and carcinogenic effects to levels that are acceptably low, additional DE limits have been recommended for the various organs of the body, as well as for the body as a whole; for the latter, the value has been set at 50 mSv (5 rem) per year.

To limit radiation exposure to desired levels, radiation sources, facilities where they are produced or used, and work procedures must be carefully designed. This can be accomplished through training and supervision of involved personnel, implementation of a well-conceived radiation protection program, and systematic oversight and monitoring of health physics. Also needed are careful provisions for dealing with radiation accidents, emergencies, and other contingencies; systematic recording and updating of each worker's exposures; thorough labeling of all radiation sources and exposure fields; appropriate monitoring of facilities and interlocks to guard against inadvertent irradiation; and various other precautionary measures.

General principles to be observed in every radiation protection program include the following:

1. A well-developed, well-rehearsed, and updated emergency preparedness plan to enable prompt and effective response in the event of a malfunction, spill, or other radiation accident; this should include appropriate provisions for isolation and decontamination of contaminated objects or people.

2. Appropriate use of shielding in facilities, equipment, and work clothing (such as aprons and gloves).
3. Appropriate selection, installation, maintenance, and operation of all equipment.
4. Minimization of exposure time.
5. Maximization of distance between personnel and sources of radiation (the intensity of exposure decreases inversely with the square of the distance from the source).
6. Appropriate training and supervision of workers to accomplish routine tasks with minimal exposure and to cope safely with unanticipated events.

AU

Further Reading

Cember H. *Introduction to Health Physics.* New York, NY: Pergamon Press; 1983.

Hall EJ. *Radiation and Life.* 2nd ed. New York, NY: Pergamon Press; 1984.

National Council on Radiation Protection and Measurements. *Recommendations on Limits for Exposure to Ionizing Radiation.* Bethesda, Md: National Council on Radiation Protection and Measurements; 1987. NCRP Report No. 91.

Shapiro J. *Radiation Protection: A Guide for Scientists and Physicians.* Cambridge, Mass: Harvard University Press; 1979.

Upton AC. Ionizing radiation. In: Levy BS, Wegman DH, eds. *Occupational Health: Recognizing and Preventing Work-related Disease.* Boston, Mass: Little, Brown; 1988:231-246.

Radiation, Nonionizing, Adverse Effects

ICD-9 E926.0, E926.1, E926.2, E926.4, E926.9

Nonionizing radiation is part of the electromagnetic spectrum but has lower energy than ionizing radiation. It includes many familiar types of transmissions: ultraviolet, visible light, infrared, and microwave/radio frequency. Laser is an amplified form of nonionizing radiation.

The risk of injury from radiation is proportional to the amount of energy absorbed. This, in turn, depends on the amount of energy generated by the source, an unobstructed path to a person, the distance from the source, and the elapsed time of exposure. Therefore, the risk of injury usually can be reduced if the amount of energy generated can be reduced, if the source can be shielded (or the person wears protective clothing or other equipment), if the distance from the source can be increased, or if the time of exposure can be reduced. The amount of energy in electromagnetic radiation varies with the wavelength. Thus, ultraviolet radiation (wavelength of 200 to 400 nanometers [nm]) has more energy than infrared (700 to 1000 nm) and microwave (0.1 to 1000 centimeters [cm]) radiation.

ULTRAVIOLET RADIATION (UVR)

Identification

Low-intensity ultraviolet radiation (UVR) from sunlight, with acute overexposure, is responsible for the common sunburn; with chronic overexposure, it is responsible for solar elastosis, solar keratosis, skin cancer, and cataracts. Bursts of sun exposure have been linked to an increased risk of cutaneous and intraocular melanoma. Exposure to high-intensity UVR can cause conjunctivitis and/or keratitis, a work-related condition known as "ground-glass eyeball" (or "welder's flash" or "flash burn"; *see* Eye Injury), as well as a sunburn effect. This is especially true of arc welding, which can affect welders or bystanders with just seconds of unprotected exposure. Symptoms follow a latent period of 6 to 12 hours and consist of a foreign-body sensation, photophobia, lacrimation (tearing), and blepharospasm (spasm of the eyelid) lasting up to 48 hours.

Occurrence

Low-intensity UVR is produced naturally by the sun and artificially by incandescent, fluorescent, and discharge forms of light sources. Locations at high altitude and close to the equator experience solar UVR exposure of higher intensity. Effects are also potentiated by photosensitizers, including certain drugs (eg, chlorpromazine, tolbutamide, and chlorpropamide) and plants containing furocoumarins and psoralens (eg, figs, lemon and lime rinds, celery, and parsnips). High-intensity UVR is commonly seen in industrial settings. It is emitted by welding arcs, plasma torches (used in heavy industrial cutting processes), electric arc furnaces, germicidal and black-light lamps, and some lasers. There are approximately 1 to 5 million welders in the U.S. potentially at risk for welder's flash.

Causes

Ultraviolet radiation injury is caused by excessive absorbed energy from electromagnetic radiation in the UVR region (200 to 400 nm). Risk of injury is proportional to time of exposure and to absorbed energy, expressed as microwatts per square centimeter.

Pathophysiology

The chronic skin effects of UVR are a result of basophilic degeneration of connective tissue and fragmentation of elastic tissue, with loss of collagen. The epidermis atrophies, and many abnormal cells are evident histologically in a disorderly pattern in areas of exposure, providing what are considered to be high-risk areas for skin cancer formation. The acute effects of UVR on the eye are primarily thermal. They increase with proximity of the eye to the UVR source and stronger arcs, generally requiring over 20 seconds of unprotected exposure at a distance of 7 feet to lead to welder's flash.

Prevention

Overexposure to low-intensity UVR should be avoided by wearing protective clothing and sunglasses. Benzophenone and anthranilate sunscreens absorb UVR and offer partial protection to bare skin. Opaque sunscreens, such as titanium dioxide, offer the most complete protection and may be essential if there is photosensitization. Prevention of high-intensity UVR damage requires the isolation of high-intensity ultraviolet sources and the use of goggles and other shields with proper filters. Exposure guidelines have been established by ACGIH but not by OSHA.

Other Issues

Individuals vary in their susceptibility to photosensitizers, depending on the tendency of their skin to concentrate photosensitizing agents. Some prescription medications are photosensitizers. Having light skin and working at high altitude (where solar radiation is more intense) are risk factors for sun-induced skin cancer.

Further Reading

Emmett EA, Buncer C, Suskind RB, et al. Skin and eye diseases among arc welders and those exposed to welding operations. *J Occup Med.* 1981;23:85.

Hu H. Effects of ultraviolet radiation. *Med Clin North Am.* 1990;74:509-514.

U.S. Department of Health, Education, and Welfare. *A Recommended Standard for Occupational Exposure to Ultraviolet Radiation.* Rockville, Md: National Institute for Occupational Safety & Health; 1977. HSM Publication 73-11009.

INFRARED RADIATION (IRR)

Identification

The primary effect of IRR on biologic tissues is thermal. The lens of the eye is particularly vulnerable to this effect, and cataracts can be produced by chronic IRR exposure at levels far below those that cause skin burns.

Occurrence

In industry, significant levels of IRR are produced directly by lamp sources and indirectly by sources of heat. Glassblowers and furnace workers are particularly at risk ("glass blower's cataract") after chronic IRR exposure that lasts for more than 10 years. Other occupations at risk include workers who handle molten metal (such as foundry workers, blacksmiths, and solderers), oven operators, workers who use baking and drying heat lamps, and movie projectionists.

Causes

Infrared radiation injury is caused by excessive absorbed energy from electromagnetic radiation in the IRR region (700 to 1000 nm). Risk of injury is proportional to time of exposure and to absorbed energy.

Pathophysiology

The lens of the eye has no heat sensors and a poor heat-dissipating mechanism; thus, it is particularly prone to the thermal effect of IRR. Typically, cataracts form at the rear surface of the lens after years of exposure.

Prevention

The IRR source should be properly shielded, and eyes should be protected with IRR filters. There are no exposure guidelines.

Further Reading

US Department of Health, Education, and Welfare. *Research Report: Determination of Ocular Threshold Levels for Infrared Radiation Cataractogenesis.* Rockville, MD: National Institute for Occupational Safety & Health; 1980. DHHS publication 80-121.

MICROWAVE/RADIO-FREQUENCY RADIATION (MW/RFR)

Identification

High-power microwave/radio-frequency radiation (MW/RFR) produces a thermal effect on biologic tissues. Upon exposure, this mechanism can be responsible for cataract formation, testicular degeneration, and, at extremely high intensities, death from hyperthermia. It can also interfere with medical electronic devices, especially cardiac pacemakers of the ventricular-synchronous type. Chronic exposure to low-power MW/RFR has been shown to cause alterations in electroencephalograms, behavior, and immune cell function in animals. It has also been linked in several controversial studies to headache, depression, cataracts, and cancer in humans. The significance of these associations is unknown and is the subject of investigation.

Occurrence

Microwave/radio-frequency radiation encompasses a wide range of wavelengths used in radar, television, radio, and other telecommunications systems. It is also used in various industrial operations, including heating, welding, and melting metals; processing wood and plastic (radio-frequency sealers); and creating high-temperature plasma.

Causes

Injury from MW/RFR is caused by excessive absorbed energy from electromagnetic radiation in the MW/RFR region (0.1 to 1000 cm, or 10 megahertz [MHz] to 300 gigahertz [GHz]). Risk of injury is proportional to time of exposure and to absorbed energy.

Pathophysiology

The thermal effect of MW/RFR generally requires relatively high power densities (above 100 watts per square meter).

Prevention

Appropriate shielding and engineering controls should be employed to reduce exposure at the MW/RFR source. OSHA has established PELs (29 CFR 1910.97).

Other Issues

Workers with pacemakers should proceed under supervision and with caution.

Further Reading

Foster KR, Guy AW. The microwave problem. *Sci Am.* 1986;225(3):32.

Roberts NJ Jr, Michaelson SM. Epidemiological studies of human exposures to radiofrequency radiation: a critical review. *Int Arch Occup Environ Health.* 1985;56:169.

Wilkening GM, Sutton CH. Health effects of nonionizing radiation. *Med Clin North Am.* 1990;74:489-507.

LASER RADIATION

Identification

Laser beams consist of an artificially generated type of radiation that concentrates a large amount of energy in a small cross-sectional area. This property poses a great risk to the eye, and skin burns also can occur. "Laser" is an acronym of Light Amplification by the Stimulated Emission of Radiation. Lasers employ light in a single wavelength that may range from ultraviolet to infrared.

Occurrence

The frequency of injuries from lasers is unknown. Lasers of various types are being used increasingly to provide accurate measurements for distance and leveling survey work in construction and to act as a heating agent in welding and as a cutting instrument in microelectronics and microsurgery. They are also used in communications and in the military.

Causes

Unlike other forms of radiant energy, the energy of lasers is not appreciably reduced by distance, and lasers are highly directional. Injury occurs when a person enters the beam path. Therefore, injury can be avoided by keeping the laser beam away from people.

Pathophysiology

The retina, lens, iris, and cornea are extremely vulnerable to injury from the thermal effects of laser radiation in the near-ultraviolet, near-infrared, infrared, and visible frequency ranges. Ocular damage can also occur indirectly from exposure to diffuse reflections from a high-power laser. Lasers producing nonvisible radiation can be particularly hazardous, given that exposure may not be readily apparent.

Prevention

Laser installations should be isolated, and laser beams must be terminated by a nonreflective, fireproof material. Goggles can help only if they afford protection specific to the wavelength of the laser being used. Proper worker education and regular eye exams are important. ACGIH has published guidelines for the safe use of lasers.

High-energy lasers not only create risk of injury from the laser beam but also can generate harmful fumes and gases from the laser source, either by interaction with air (eg, ozone) or by their intended use (eg, welding fumes).

HH

Further Reading

Sliney DH, Vorpahl KW, Winburn DC. Environmental health hazards from high powered infrared and laser devices. *Arch Environ Health*. 1975;30:174.

Renal Diseases, Chronic: Tubular and Interstitial

ICD-9 585
SHE/O

Most known occupational nephrotoxins affect the renal tubule or the supporting interstitium. The manifestations of toxicity vary with the agent, the chronicity, and the intensity of exposure and, in certain cases, the age of the affected person. For example, both lead and cadmium damage the renal tubules and interstitium (as opposed to the glomerulus), yet their clinical manifestations are markedly different. To complicate matters further, the renal tubular toxicity of these heavy metals is more distinctive—and thus more easily attributed to the toxin—than is the nonspecific progressive loss of kidney function due to interstitial nephritis, which may be clinically more important.

Identification

Chronic interstitial nephritis may be asymptomatic until renal failure is advanced. Its course differs from that of acute interstitial nephritis (such as due to methicillin), which presents with fever, edema, and proteinuria 7 to 14 days after reexposure to the drug. Chronic interstitial nephritis is marked by a declining creatinine clearance and normal serum creatinine until approximately 75% of renal function is lost. Reduced renal function from chronic interstitial nephritis may render the worker more vulnerable to other medical or environmental risk factors for renal disease. Two important occupational types are lead nephropathy and cadmium nephropathy:

1. Lead nephropathy in adults usually spares proximal tubular function (although children with lead poisoning develop aminoaciduria). Adult lead nephropathy presents with mild increases in serum creatinine (1.5 to 2.5 mg/dl), mild decreases in urine-concentrating ability, variable hypertension, and sometimes other stigmata of lead poisoning (lead colic, anemia, and peripheral neuropathy). No clinical test definitively proves that the nephropathy is due to lead; kidney biopsy is uninformative in this respect, and lead-inclusion bodies seen in animals are not present in humans. A presumptive diagnosis is based both on a careful occupational history documenting high exposure to lead and on the absence of other explanatory causes. The presence of hypertension in a lead-exposed worker with nephropathy supports the likelihood that lead has contributed to disease. Chelation (with calcium ethylene diamino-tetra acetic acid [EDTA]) is used to document renal lead burden, usually for research rather than clinical purposes.

2. Cadmium nephropathy is diagnosed presumptively by the presence of increased urinary beta-2-microglobulin or retinol-binding protein in the urine of workers with substantial cadmium exposure. The small proteins are not detected by routine urine dipstick methods unless proteinuria is severe; diagnosis usually requires a special radioimmune assay.

Other evidence of proximal tubular dysfunction may include aminoaciduria, reduced tubular reabsorption of phosphate, and/or glucosuria in the presence of a normal serum glucose. Urine cadmium increases above 10 μg/g of creatinine with the onset of renal disease. Increased urine cadmium is a marker of both overexposure and nephropathy. Following prolonged cadmium exposure, overall renal function declines, serum creatinine increases, and creatinine clearance decreases. Renal failure is uncommon, although the extent to which cadmium contributes to increased renal mortality or morbidity is unclear. Cadmium nephropathy seldom causes symptoms, unless the worker develops kidney stones. The lifetime prevalence of kidney stones is high (19% to 40%) in cadmium workers. The renal tubular and glomerular effects of cadmium are largely irreversible, even if occupational exposure to cadmium ceases.

Other substances found in the workplace that are toxic to the renal tubules include solvents, glycols, and metals (uranium, bismuth, chromium, and antimony). The evidence for their toxicity comes primarily from toxicological data, massive overexposure, and, in certain cases, pharmacological administration. It is unknown to what extent these substances contribute to occupational renal disease.

Occurrence

Chronic interstitial nephritis is rare but is more common in women than in men.

Causes

Nonoccupational causes include recurrent urinary infections, obstruction (congenital or acquired), abuse of analgesics (especially phenacetin, acetaminophen, and caffeine powders), and drug-induced immunotoxicity (methicillin, penicillin, ampicillin, rifampin, sulfonamides). Uncommon nonoccupational causes include systemic immune disorders (lupus erythematosus), renal transplant rejection, metabolic diseases (intrarenal crystallization from gout, hypokalemic nephropathy), primary glomerulonephritis (focal, membranous, Goodpasture's), infiltrative processes (amyloid, leukemia, lymphoma), sarcoidosis, and nephritis due to therapeutic radiation.

Regional epidemics of interstitial nephritis have occurred in Bulgaria, Yugoslavia, and Romania (Balkan nephritis) and in Japan (Itai Itai disease). Balkan nephropathy is of unknown etiology, yet it afflicts approximately 25 000 persons in Eastern Europe. Itai Itai (literally "ouch ouch") disease affects elderly women in Japan in areas where rice is contaminated with cadmium. The name reflects the bone pain from osteopathy and spontaneous fractures in malnourished, postmenopausal women with cadmium-induced nephropathy. Occupational groups at risk of cadmium nephropathy include workers in brazing, nickel-cadmium battery manufacturing, primary and secondary smelting, scrap metal recovery, electroplating, and pigment manufacturing. Populations exposed to lead include workers in lead battery plants, soldering, radiator repair, primary smelting, glass and ceramic manufacturing, and paint stripping. Nonoccupational exposure to lead occurs through paint flakes, leaded gasoline, and "moonshine" alcohol.

Pathophysiology

The mechanism by which chronic lead or cadmium exposure results in chronic interstitial disease is unknown, but it may involve either direct toxicity or immune processes. Phenacetin and related analgesics (acetaminophen) concentrate in the renal papilla, causing papillary necrosis.

Prevention

Preventable nonoccupational causes of chronic interstitial nephrosis include recurrent infection and analgesic nephropathy. Children and men with urinary tract infections should be evaluated for underlying urinary tract abnormalities. Women with recurrent kidney infections should be evaluated for diabetes, stones, and obstruction and should treat urinary tract infections promptly. Analgesic nephropathy can be prevented by discouraging daily use of high doses of analgesics, particularly mixtures of acetaminophen, aspirin, and caffeine. The OSHA lead standard has substantially reduced lead exposures in large industries, but small companies and secondary smelters still have potentially high exposures. Leaded paint remains an important environmental source of lead for urban children. Cadmium exposure is nephrotoxic even within the current OSHA PEL for cadmium dust of 200 $\mu g/m^3$ (100 $\mu g/m^3$ for fume). ACGIH has proposed reducing the cadmium PEL to g/m^3. Engineering controls (enclosure/ventilation), appropriately fitted respirators, and employee training, coupled with personnel and area monitoring, must be used to reduce exposure to the lowest level feasible. Exposures should not exceed 10 to 20 $\mu g/m^3$ to protect workers over a 40-year working lifetime.

For secondary prevention, urinary beta-2-microglobulin can be measured to determine whether cadmium exposure is adequately controlled or to identify workers with early tubular nephropathy. Medical removal of such workers may prevent progression of injury but will not correct the underlying problem of overexposure. No good early detector for lead nephropathy exists; serum creatinine, blood urea nitrogen, and blood pressure are all insensitive to early disease.

MT

Further Reading

Cotran RS, Rubin RH, Tolkoff-Rubin NE. Tubulointerstitial diseases. In: Brenner BM, Rector FC, eds. *The Kidney.* Vol 11. Philadelphia, Pa: W.B. Saunders; 1986:1143-1173.

Kjellstrom T. Renal effects. In: Friberg L, Elinder CS, Kjellstrom T, Nordberg GF, eds. *Cadmium and Health: A Toxicological and Epidemiological Appraisal.* Boca Raton, Fla: CRC Press; 1987;11:22-99.

Rosenstock L, Cullen MR. Diseases of the kidney and urinary tract. In: Rosenstock L, Cullen MR, eds. *Clinical Occupational Medicine*. Philadelphia, Pa: W.B. Saunders; 1986:80-90.

Renal Failure, Acute

ICD-9 584
SHE/O

Although acute renal failure rarely results from work, an occupational association, when it occurs, can be more easily recognized than when it occurs with chronic renal failure. This is because of the short time between the precipitating event and the onset of symptoms.

Identification

Acute renal failure is characterized by reduced or absent urine production (oliguria or anuria) and rapidly rising serum creatinine (increasing >0.5 mg/dl/day) beginning 1 to 6 days after an acute renal insult and lasting 1 to 3 weeks. Complete recovery is usual. Rarely does acute renal failure progress to end-stage renal disease. Some causes of acute renal failure are suggested by characteristic urinary signs: pigmented casts suggest hemoglobinuria and myoglobinuria; oxalate crystals seen under polarized light suggest ethylene glycol exposure.

Occurrence

Occupationally induced acute renal failure is rare. No accurate data on its occurrence are available.

Causes

Acute tubular necrosis occurs commonly from nonoccupational causes in hospitalized patients, usually due to hypotension (following surgical or medical shock) or drug toxicity (gentamicin, aminoglycoside antibiotics). Other nonoccupational causes are (a) obstruction of urinary outflow (due to prostatic hypertrophy, cellular debris from chemotherapeutic agents, or intrarenal crystals); (b) renal vein obstruction; (c) hypertonic radiographic contrast material; and (d) suicidal or accidental ingestion of drugs, pesticides (paraquat), solvents (chlorinated hydrocarbons or glycols), or heavy metals (mercury, arsenic, bismuth, antimony). Ethylene glycol (antifreeze) causes an estimated 40 to 60 deaths per year in people who drink it accidentally or as a substitute for ethanol. Diethylene glycol has caused epidemics of acute renal failure when mixed with sulfanilamide. The solvent dioxane, which is metabolized to diethylene glycol, has caused renal failure and death in workers following skin and inhalation exposure. Other occupational causes include high concentrations of hydrocarbon solvents (in enclosed spaces), potent nephrotoxins such as carbon tetrachloride or chloroform, and inadvertent intense exposure to toxic metals (uranium hexafluoride gas, mercury vapor from heated elemental mercury). Intrarenal obstruction from hemoglobin or myoglobin follows crush injuries or high-voltage electrocution. Hemoglobinuria occurs with arsine (AsH_3) poisoning. Myoglobinuria can be induced by severe unaccustomed exertion (eg, soldiers on a march). Renal failure is the dominant clinical complication of heat stroke in approximately 10% of survivors.

Pathophysiology

Acute renal failure involves a combination of ischemia, direct toxicity to the renal tubules, and obstruction.

Prevention

Ethylene glycol poisoning could be reduced if antifreeze compounds were made unpalatable, thus discouraging ingestion by substance abusers. Occupationally induced acute renal failure is largely restricted to subgroups at special risk: workers in enclosed spaces; arsine-exposed workers in the metal smelting, refining, or etching industries; and workers who handle uranium hexafluoride gas. A combination of engineering controls and strict work practices is necessary to ensure safety in these high-risk settings. Clusters of acute renal failure due to myoglobinuria and heat stroke can be prevented by ensuring that preplacement testing and physical conditioning programs are appropriately designed and conducted. High-risk groups, such as police, fire fighters, and military recruits and reservists, should review their training procedures and eliminate practices likely to cause life-threatening myoglobinuria or heat stroke. These practices include severe unaccustomed exercise, insufficient hydration, insufficient acclimatization, and exposure to heat.

MT

Further Reading

Cornish HH. Solvents and vapors. In: Doull J, Klaassen CD, Amdur MD, eds. *Toxicology.* New York, NY: Macmillan; 1980:468-496.

Fowler BA, Weissberg JB. Arsine poisoning. *N Engl J Med.* 1974;291:1171.

Lilis R, Landrigan PJ. Renal and urinary tract disorders. In: Levy BS, Wegman DH, eds. *Occupational Health: Recognizing and Preventing Work-related Disease.* 2nd ed. Boston, Mass: Little, Brown; 1988:465-476.

Schrier RW. Nephropathy associated with heat stroke and exercise. *Ann Intern Med.* 1967;67:356-376.

Raynaud's Syndrome—*See* Hand-Arm Vibration Syndrome

Respiratory Tract Cancer—*See* Laryngeal Cancer, Lung Cancer, Mesothelioma, and Nasal (or Sinonasal) Cancer

ICD-9 506.0-506.3
SHE/O

Respiratory Tract Irritation

Identification

Many agents cause primary respiratory tract irritation. There is no unique syndrome associated with these agents. The symptoms associated with the agents depend on locus of action. For agents acting in the upper respiratory tract, the symptoms can include sneezing, nasal discharge, hoarseness, cough, and phlegm (and possibly wheezing), as well as nonrespiratory effects, such as eye watering. For agents acting in the lower respiratory tract, the predominant symptoms are inspiratory cough, chest "tightness," headache, and progressive shortness of breath. Symptoms associated with the most extreme reactions of chemical pneumonitis and pulmonary edema include breathlessness progressing to respiratory failure. Severe overexposure can also result in life-threatening laryngeal edema. In general, mild irritant responses are reversible on removal from exposure, but recovery from severe reactions is prolonged by the need to recover from the acute inflammatory response. Chronic sequelae of edema and pneumonitis, such as lung fibrosis, bronchiolitis obliterans (resulting in chronic airways obstruction), and airways hyperreactivity (asthma) have been reported.

Occurrence

There are no data that indicate the prevalence or incidence of irritant reactions in the respiratory tract. Generally, any environment where such agents are used will, if uncontrolled, result in widespread reports of symptoms. Even in settings where controls are operating, accidents occur and overexposures result in symptoms. As a rule, every exposed person will react to irritants if there is a sufficient airborne concentration and length of exposure. Some people with asthma or other forms of airways hyperresponsiveness may react to lower concentrations of irritants.

Causes

Apart from their irritant reactions, many of these agents have disparate characteristics. Unfortunately, no common features separate respiratory irritants from agents that do not cause irritant reactions. Examples of agents that act primarily as irritants are chlorine and derivatives (hydrochloric acid, phosgene, chlorine dioxide), fluorine and derivatives (hydrofluoric acid, silicon tetrafluoride), bromine and iodine, sulfur dioxide and sulfuric acid, nitrogen oxides (nitrogen dioxide being the most hazardous), ozone, isocyanates, ammonia, acetic acid, acrolein, formaldehyde, and cadmium oxide.

Pathophysiology

The important features to distinguish irritants are their solubility in water and, for particulates, the particle size. Irritants that are highly water soluble dissolve readily in the mucosa of the upper airways; those that are relatively less soluble will dissolve in the lower airways and parenchyma. Agents that usually dissolve in the upper airways will, under conditions of severe overexposure, overwhelm the defenses and can penetrate to the lower airways as well. Larger particles—those greater than 10 microns in diameter—deposit in the upper airways; those between 2 and 10 microns deposit in the middle airways; and those less than 2 microns generally reach the air exchange units.

The irritant effects are generally attributed to the direct excitation of neural receptors in the mucous membranes of the respiratory tract. The associated neural responses include sensation of pain and reflex responses such as cough and reflex bronchoconstriction. Acute pulmonary edema results from a change in the permeability of the pulmonary vasculature and histamine release, which leads to bronchoconstriction, increased capillary pressure, and extravasation of fluids. Following such episodes, individuals may be left with persistent airways hyperresponsiveness. This condition has been labeled reactive airways dysfunction syndrome, or RADS.

Prevention

There are no unique approaches to the control of respiratory irritants. General control of gas and vapor exposure through proper maintenance and local exhaust ventilation is the first line of defense. Probably the most problematic situation is accidental overexposure. Although such accidents cannot be entirely prevented, proper education of the workforce regarding the risks and their avoidance is highly desirable. Furthermore, each worker should be carefully trained to avoid the risks associated with improper attempts to rescue a co-worker from an overexposure environment. Rescue attempts should be made only with proper (and therefore readily available) rescue equipment, including air-supplied respirators or self-contained breathing apparatus.

Surveillance is unlikely to provide early warning, but investigation and analysis of overexposure events may provide information on reasons that primary prevention systems are failing.

Other Issues

Exposure to relatively low levels of some irritants will acclimatize individuals to a higher odor threshold, thereby reducing awareness of potential overexposure.

DHW

Further Reading

Brooks SM, Weiss MA, Bernstein IL. Reactive airways dysfunction syndrome (RADS). *Chest.* 1985;88:376-384.

Gavrilescu N. Irritant gases and vapors. In: Parmeggiani L, ed. *Encyclopedia of Occupational Health and Safety.* 3rd ed., rev. Geneva, Switzerland: International Labour Office; 1983.

Greaves IA. Occupational pulmonary disease. In: McCunney RJ, ed. *Handbook of Occupational Medicine.* Boston, Mass: Little, Brown; 1987; chap 8.

Wegman DH, Christiani D: Respiratory disorders. In: Levy BS, Wegman DH, eds. *Occupational Health: Recognizing and Preventing Work Related Diseases.* Boston, Mass: Little, Brown; 1988; chap 21.

Sexual Dysfunction—*See* Infertility and Sexual Dysfunction

Sick Building Syndrome—*See* Building-Related Illness

Silicosis

ICD-9 502
SHE/O

Identification

The term silicosis refers to several forms of lung fibrosis that arise from inhaling crystalline forms of silica (silicon dioxide [SiO^2]): (a) nodular silicosis (also called "classic" or "pure" silicosis), with characteristic hyaline and nodular lung lesions that may aggregate into large fibrotic masses (conglomerate silicosis or progressive massive fibrosis [PMF]); (b) acute silicosis, a rapidly developing disorder having features of alveolar proteinosis and fibrosing alveolitis; (c) mixed dust fibrosis, showing fibrotic lesions—some of which may be typical "silicotic" nodules and others more irregular in shape—that arise from inhaling dusts of silica and other agents (eg, iron oxide, coal, welding fumes); and (d) diatomaceous earth pneumoconiosis, with fibrosing alveolitis and a prominent cellular reaction. Diagnosis is based on symptoms, exposure history, physical examination, chest x-ray, pulmonary function tests, and sometimes pathological examination of tissue.

Occurrence

Classic nodular silicosis remains the single most common pneumoconiosis worldwide. Estimates of current incidence are lacking, but before ventilation controls and other preventive measures were introduced in developed countries, prevalence rates exceeding 30% to 50% were not uncommon for workers in hard rock mining, foundries, and sandblasting operations. Although such rates may still occur in some developing countries, reduced exposures to respirable silica have greatly decreased the prevalence of classic silicosis. When it does occur, the disease is generally less severe and seldom causes PMF. The exception has been in silica flour mills that produce finely divided, almost pure silica powder; recent surveys in U.S. silica flour mills have shown a high prevalence of x-ray changes and rapidly progressive disease leading to PMF. Acute silicosis has similarly been observed among sandblasters and surface miners.

The number of cases occurring in the U.S. is unknown. About 250 individual workers' compensation claims for silicosis are awarded in 29 states each year. Silicosis is rarely reported by employers, as required by OSHA and MSHA. However, a survey in New Jersey based on hospital discharge summaries and physician reports discovered 60 cases per year for the period 1979 to 1987. Only one third of these cases had applied for compensation, and not one was reported by employers. Extrapolating to the U.S. as a whole, about 2360 cases per year are expected to occur in the next two decades.

Cases of silicosis in men outnumber those in women by 9 to 1. Most women with silicosis acquired it in the ceramics industry. Other industrial environments in which it occurs include mines (including ore processing), foundries, quarries, and silica flour mills.

As of September 1, 1988, 19 states and the District of Columbia have required silicosis to be reported. These states are Arkansas, Georgia, Indiana, Iowa, Kansas, Kentucky, Maine, Maryland, Michigan, Missouri, New Jersey, New Mexico, New

York, Oklahoma, South Carolina, Texas, Virginia, West Virginia, and Wisconsin.

Causes

Crystalline silica exists in three major forms: a-quartz (hexagonal crystals), tridymite (hexagonal), and cristobalite (cubic). Amorphous (noncrystalline) silica is relatively nontoxic and occurs mainly as diatomaceous earth and vitreous silica. Diatomaceous earth is composed of the skeletons of prehistoric marine organisms, and its commercial use depends on its being heated to high temperatures. When amorphous silica or a-quartz is heated, tridymite (>867°C) and cristobalite (>1470°C) are produced, and the crystalline forms persist on cooling. Regarding the biological activity of the various forms of silica, that of cristobalite is greater than that of tridymite, which is greater than that of a-quartz, which is much greater than that of amorphous silica; thus, conversion to more toxic forms by heating is an important health consideration. Collectively, these crystalline and amorphous forms of silica are termed "free silica" to distinguish them from silicates (asbestos, talc, and micas), which have different health risks.

Free silica is found chiefly as a-quartz in a wide variety of rocks, where it occurs in various percentages: sandstone, chert, flint (100% silica); granite (20% to 70%); and slate (40%). Digging, blasting, cutting, polishing, or otherwise generating respirable aerosols from these rocks presents health risks. Sandblasting and grinding operations that use silica-containing agents are similarly hazardous. Foundry workers, particularly those who grind fused sand from the surface of moldings, are also at significant risk.

Pathophysiology

Classic nodular silicosis in its "simple" form comprises discrete nodules about 2 to 6 mm in diameter, showing hyalinized fibrosis centrally surrounded by concentric whorls of fibrosis encapsulated by a zone of moderately cellular fibrosis. Microscopy with polarized light demonstrates birefringent material, chiefly crystalline silica, within these nodules. Growth of the nodules over time may lead to conglomeration and the formation of large fibrotic masses (PMF). Pleural thickening is common, particularly over the upper lobes and adjacent to areas of PMF. Pleural effusions are seldom a feature of silicosis and, when present, may suggest the development of tuberculosis. Silica also causes airway changes with mucous gland hypertrophy, goblet cell hyperplasia, and chronic inflammation in the walls of bronchi and bronchioles. Hilar lymphadenopathy and enlargement of mediastinal and supraclavicular nodes occur commonly after long periods of exposure, and histology shows the same silicotic nodules as found in the lungs; calcification within nodes ("eggshell" calcification) is also common in long-standing cases. Occasionally, enlarged nodes may cause compression of mediastinal structures.

Radiographic changes of simple silicosis are often accompanied by normal measurements of lung function or, at most, by mild impairment of airflow at low lung volumes. Cough and phlegm are common, however, among workers exposed for many years, and these symptoms reflect airways irritation. There may be accelerated decline of pulmonary function with simple silicosis. Lung function usually deteriorates more rapidly when conglomerate lesions develop. Because conglomeration often progesses, even after cessation of exposure, lung function usually declines at an accelerated rate and a severe obstructive-restrictive impairment can occur, leading eventually to chronic respiratory failure and cor pulmonale.

Acute silicosis is a rapidly developing disorder following inhalation of high concentrations of finely divided crystalline silica. Initial changes are the appearance of lipoproteinaceous material and inflammatory cells within alveolar spaces. Irregular and interstitial fibrosis usually follows, often within a few months of initial exposure. Lung function shows a restrictive defect that commonly progresses on cessation of exposure. Respiratory impairment is often severe.

When the silica content is high in mixed dust fibrosis, the resulting pneumoconiosis resembles nodular silicosis; a low silica content results in more irregular lesions that resemble tuberculosis. Mixed dust fibrosis appears to result from modification of the effects of small quantities of silica by accompanying nonfibrogenic dusts.

Diatomaceous earth pneumoconiosis is uncommon and resembles desquamative interstitial pneumonitis. The initial stages show a cellular form of interstitial pneumonitis, with subsequent peribronchial and perivascular fibrosis leading to diffuse interstitial fibrosis. Lung function is well preserved initially but shows restrictive impairment as the disease progresses.

All forms of silicosis are considered dose dependent; that is, the development of disease depends on the total dust burden in the lungs, its silica content, and the presence of other dusts. Uncommon complications of silicosis include scleroderma, rheumatoid arthritis, and glomerulonephritis.

Prevention

Recognition of crystalline silica as a hazard and imposition of a high standard of dust control or substitution with a less harmful agent are essential for reducing the risk of silicosis. The use of wet processes, enclosure of dusty operations, and ventilation control systems have proven effective measures in various work settings. The use of personal protective equipment (respirators) should be reserved for processes that are not amenable to adequate dust control or substitution methods.

The current OSHA permissible exposure limit (PEL) for crystalline silica as quartz is 0.1 mg/m^3 on a time-weighted average. The MSHA PEL for silica exposure in coal mining is determined by a formula that reduces the overall dust exposure limit based on the percentage of quartz if the dust is over 5% quartz; the formula is 10 mg per percentage of quartz. In other facilities, the MSHA PEL is determined by the formula: 10 mg per percentage of quartz + 2. With each of these formulas, the total silica exposure varies with the dust level and composition.

Other standards have been proposed, including by NIOSH, which has a 0.05 mg/m^3 recommended air level for a 10-hour time-weighted average exposure to respirable free silica, and one by ACGIH, which recommends 0.1 mg/m^3 for respirable quartz and 0.05 mg/m^3 for respirable tridymite and cristobalite.

Medical screening of workers exposed to a silica hazard should include routine baseline respiratory status checks (chest radiographs, lung function tests, and tuberculin testing) and radiographs and lung function tests at regular intervals (1 to 3 years). A chest x-ray is the most sensitive, specific, and readily available test for silicosis. There are no clear guidelines for removing workers from exposure if the workers show progression (increasing profusion and/or size) of nodular opacities or have an accelerated decline in lung function. Prudent medical practice would suggest that those with definite pneumoconiosis of at least moderate degree (ILO profusion > 1/1; opacities > p/q) are at greatest risk of disabling silicosis, tuberculosis, and perhaps lung cancer and therefore should have no further silica burden in their lungs.

The sudden appearance of a confluent opacity on the chest film always raises the possibility of active tuberculosis, and such a finding warrants immediate investigation and removal from the workplace.

Other Issues

Silicosis is the only pneumoconiosis that predisposes an individual to tuberculosis. Most commonly, tuberculosis occurs in association with moderate to severe silicosis. Earliest features may be a recent cough with little phlegm or an abrupt change in the chest radiograph. Sputum cultures will generally confirm the diagnosis, although the organism is often absent from sputum smears early in the disease. Treatment will thus often be initiated on clinical grounds before bacteriologic confirmation is made.

Necrobiotic nodules, lesions up to several centimeters in diameter, occur in association with rheumatoid arthritis or in the presence of circulating rheumatoid factor alone. Such nodules may be mistaken for tuberculosis, may cavitate, and may be sites of recurrent or chronic infection.

Carcinoma of the lung has been associated recently with silicosis in humans, and silica is an animal carcinogen. Although epidemiological data linking silica with lung cancer are limited and many earlier studies were negative, investigators during the last 10 years have strengthened the view that silica is a human carcinogen; it is unclear whether a fibrotic reaction to silica is necessary for the development of lung cancer.

A case definition for surveillance of silicosis has been adopted.[1] Silicosis is one of the conditions reported by some states under the SENSOR program.

IG

Reference

1. Silicosis: cluster in sandblasters—Texas, and occupational surveillance for silicosis. *MMWR.* 1990;39(25):433-437.

Further Reading

Forastiere F, Lagorio S, Michelozzi P, et al. Silica, silicosis and lung cancer among ceramic workers: a case-referent study. *Am J Ind Med.* 1986;10:363-370.

International Agency for Research on Cancer. *Evaluation of the carcinogenic Risk to Humans. Silica and Some Silicates.* Lyon, France: International Agency for Research on Cancer; 1987;42:39-144.

Landrigan PJ. Silicosis. In: Rosenstock L, ed. Occupational pulmonary disease. *State of the Art Reviews in Occupational Medicine.* 1987;2(2):319-326.

Parkes WR. Disease due to free silica. In: *Occupational Lung Disorders*. 2nd ed. Boston, Mass: Butterworth; 1982:134-172.

Silicosis and Silicate Disease Committee. Diseases associated with exposure to silica and nonfibrous silicate minerals. *Arch Pathol Lab Med*. 1988;112:673-720.

Valiante DJ, Rosenman KD. Does silicosis still occur? *JAMA*. 1989;262:3003-3007.

Sinonasal Cancer—*See* Nasal (or Sinonasal) Cancer

Skin Cancer

ICD-9 172,173

Identification

Cutaneous neoplasms may be benign, premalignant, or malignant. Malignant cutaneous tumors include basal and squamous cell carcinomas and malignant melanomas. Basal cell carcinomas occur most commonly on sun-exposed areas (eg, face, neck, and hands) and typically appear as translucent papules with pearly borders, telangiectasia, and central ulceration. Squamous cell carcinomas may be recognized by their rough, irregular, scaly surface; indistinct borders; and, frequently, ulceration. Malignant melanoma is the only common cutaneous neoplasm that often metastasizes. Melanomas appear as pigmented macules, papules, or nodules that have irregular (often notched) borders, irregular contours, and variegations in color (including black, blue, grey, or red).

Occurrence

There are more than 600 000 new cases of basal and squamous cell carcinoma and 32 000 new cases of melanoma in the U.S. each year. There are 6500 deaths due to melanoma and 2000 deaths due to other skin cancers each year.

Causes

Workers who are exposed to ionizing radiation, ultraviolet (UV) light, polycyclic hydrocarbons, arsenic, and tar are at highest risk for developing basal and squamous cell carcinomas of the skin. These skin cancers occur more often in latitudes where there is more intense sunlight.

Pathophysiology

Cutaneous neoplasms are induced by agents that can directly damage the DNA of epidermal cells, inhibit DNA repair, and/or possibly decrease immune surveillance of the epidermis. The mechanisms whereby these effects lead to unregulated growth (ie, neoplasia) are being determined by techniques of molecular biology.

Prevention

Outdoor workers in sunny climates should always be provided with hats, topical sunscreens, and, if practical, long-sleeved shirts. Use of arsenical insecticides should be discontinued. Exposure to other chemical carcinogens can be reduced by using protective clothing, gloves, adequate ventilation, and closed systems when carcinogenic agents are being handled.

Other Issues

Exposure to ultraviolent radiation (in sunlight) is currently the leading cause of occupational skin cancer. Workers exposed to UV light may develop actinic keratoses, which are premalignant lesions, or basal and squamous cell carcinomas. Oncogenic polycyclic hydrocarbons, which are found in coal tar and petroleum products, may cause skin precancers and cancers in workers who come in contact with tar, pitch, soot, crude paraffin, or asphalt. Workers exposed to beryllium or silica may be at risk for developing foreign-body granulomas or granulomatous inflammatory nodules.

MB, KA

Further Reading

Adams RM. *Occupational Skin Disease.* Philadelphia, Pa: WB Saunders; 1983.

Skin Infections

Identification

Diagnoses of skin infections are made by clinical appearance, Gram stain, potassium hydroxide preparation, Tzanck smear, and culture. Lesions vary and are determined by etiologic agents. Bacterial infections, most commonly caused by streptococci or staphylococci, may appear as superficial crusted plaques and papules (impetigo); perifollicular papules and pustules (folliculitis); or deep, tender, erythematous nodules or plaques with or without lymphangitic streaks (carbuncles and cellulitis). Fungal infections commonly appear as annular plaques with central clearing and an erythematous, raised, scaly border. Viral infections vary tremendously in clinical appearance; for example, *Herpes simplex* infections most commonly present as grouped vesicles on an erythematous base.

Occurrence

No reliable population-based data for the U.S. are known to exist.

Causes

Pyodermas, induced by streptococci or staphylococci, are the most common bacterial skin infections. These infections may occur as a result of trauma or as a complication of other occupational dermatoses. Fungal infection with dermatophytes (ringworm) or *Candida albicans* are often found in a local environment of moisture, warmth, and maceration. They therefore occur frequently in body folds and in warm climates. *Herpes simplex* is the most frequently identified work-related viral skin infection encountered in the U.S.

Pathophysiology

Bacteria and viruses gain access through disruptions in the normal epidermal barrier caused by trauma or underlying skin diseases (eg, atopic dermatitis and irritant or allergic contact dermatitis). Yeast and fungi can proliferate within the epidermis in susceptible individuals, especially in areas that are moist and warm. Infectious agents are acquired from contact with infected customers, animals, carcasses, hides, or materials found in the workplace. After gaining access, organisms are able to proliferate intracellularly or extracellularly. Clinical manifestations are caused by multiple factors, including direct cellular injury or necrosis, elaboration of toxins, release of inflammatory mediators, and induction of an immune response.

Prevention

Preventive measures depend on the etiologic agents and include identifying workers at risk, using properly selected protective clothing, washing hands, identifying and avoiding sources of infectious agents in the workplace, and reducing environmental factors associated with cutaneous infections (excess heat, high humidity, and frequent exposure to water).

Other Issues

Fungal and yeast infections often occur in workers whose local environment is humid or hot or whose work involves frequent exposure to water (eg, factory workers, dishwashers, and canning industry workers). Health care workers—especially those who are exposed to oral secretions, such as dentists, dental technicians, nurses, respiratory therapists, and anesthetists—are at high risk for herpetic infections of the hands. Day-care workers are also at increased risk. Sheep handlers are at high risk for orf (contagious ecthyma). Barbers and cosmeticians who contact customers suffering from contagious skin diseases are particularly at risk for bacterial and fungal infections.

MB, KA

Further Reading

Adams RM. *Occupational Skin Disease.* Philadelphia, Pa: WB Saunders; 1990.

Arndt KA, Bigby M. Skin disorders. In: Levy BS, Wegman DH, eds. *Occupational Health: Recognizing and Preventing Work-Related Disease.* 2nd ed. Boston, Mass: Little, Brown; 1988.

Skin Injuries

(*See also* Burn Injury; Cold-Related Disorders; Radiation, Ionizing, Adverse Effects; Radiation, Nonionizing, Adverse Effects) **ICD-9 910-919**

Identification

Occupational skin injuries result from acute trauma or a brief exposure to an injurious chemical. The clinical appearance is variable, depending on the cause.

Occurrence

Occupational skin injuries occur in 1.4% to 2.2% of full-time workers annually. They are more frequent in construction, manufacturing, and agriculture. Lacerations and punctures are the most common skin injuries and are easily identified. Thermal and chemical burns, abrasions, cold injuries, and acute radiation injuries also occur.

Causes

There are many causes of occupational skin injuries, including sharp, rough-surfaced, or hard objects that workers either hit or are hit by; caustics, acids, or other corrosive chemicals; hot objects, liquids, gases, or flames; electricity; and ionizing and nonionizing radiation.

Pathophysiology

Acute trauma or exposure to strong irritants or to physical agents causes direct injury to cells of the epidermis and, occasionally, the underlying tissues. Both the type and extent of damage vary considerably according to etiology.

Prevention

See Part 1: Hazard Control.

Other Issues

Hydrofluoric acid is capable of producing rapidly progressive and deeply destructive burns. Hydrofluoric acid burns cause severe pain and may be accompanied by systemic effects (eg, hypocalcemia). Initial management of these burns includes removing all clothing contaminated with hydrofluoric acid and washing affected areas copiously with cold water. Further therapy is directed at removing remaining fluoride ions to prevent further tissue destruction and systemic absorption. Acceptable measures include benzethonium chloride or benzalkonium chloride soaks, calcium gluconate gel application, and calcium gluconate injection.

MB, KA

Further Reading

Centers for Disease Control. Major work-related diseases and injuries. *MMWR.* 1986;35:61-63.

Smell and Taste Disorders ICD-9 781.1

Identification

Complaints of affected individuals include an abnormal, reduced, or absent sense of taste or smell. Thresholds and ability to identify various tastes can be tested. Formal testing may be performed with commercially available standardized tests, which usually involve placing increasing numbers of drops of specific substances on the tongue or exposing the subject to small bottles containing distinctive smells.

Occurrence

No data on the frequency of taste or smell disorders in the general population are available. In some occupational groups, individuals exposed to irritants are up to 15 times more likely to have some smell disorders than unexposed individuals.

Causes

Neurological disorders affecting the brain or the cranial nerves may impair the sense of taste, as may factors affecting cell turnover. Drugs with sulfhydryl groups, including acetylcholinesterase inhibitors, penicillamine, and antineoplastic and antirheumatic drugs, are common causes.

A variety of occupational exposures, including irritants (eg, cement dust, sulfuric acid, formaldehyde, acrylates, chlorine), solvents (eg, trichloroethylene, benzene, ethyl acetate, carbon disulfide, paint solvents), and metals (eg, lead, cadmium, chromium, nickel), impair the sense of smell. In addition, factors that influence cell turnover, such as ionizing radiation, drugs, or infections, may also impair the olfactory sense. Trauma may shear olfactory nerve bundles at the base of the skull.

Pathophysiology

Taste buds on the tongue, the soft palate, the epiglottis, and the larynx are innervated by the seventh, ninth, and tenth cranial nerves. The information is projected through the medulla, pons, thalamus, and hypothalamus to the cortex. Cells in taste buds generally have a lifespan of 10 days. Renewal may be affected by nutritional and hormonal states, radiation therapy, and increasing age.

Olfactory receptors, which contain cells with a turnover time of 30 days, are located from the superior turbinate to the nasal septum. Information on odors is transmitted through the first cranial nerve, through the cribriform plate in the ethmoid bone, into the olfactory bulb and the limbic system. Information is simultaneously transmitted through the fifth cranial nerve. Neurological disorders of either of these cranial nerves or of brain structures may affect smell.

Disturbances may arise locally, from atrophy as a result of physical or chemical injury to the receptors, from damage to neural structures, from systemic influences (eg, malnutrition, metabolic disturbance, ionizing radiation, or drugs), or from modification of the receptors by local environmental change (eg, saliva alteration, drugs, or metabolic agents).

Prevention

These disorders are prevented by reducing exposure to causative substances. Early diagnosis and removal from exposure are important.

Other Issues

Aging (beyond age 50) is associated with increasing impairment of taste and smell. Cigarette smoking may increase or decrease the probability of taste disorders. Impaired sense of smell may be an early indicator of central nervous system impairment from exposure to organic solvents.

MH

Further Reading

Feldman JI, Wright N, Leopold DA. The initial evaluation of dysosmia. *Am J Otolaryngol.* 1986;7:431-444.

Ryan C, Morrow L, Hodgson MJ. Cacosmia and neurobehavioral dysfunction associated with occupational exposure to mixtures of organic solvents. *Am J Psych.* 1988;145:1442-1455.

Sandmark B, Broms I, Lofgren L, Ohlson CG. Olfactory function in painters exposed to organic solvents. *Scand J Work Environ Health.* 1989;15:60-63.

Schiffman SS. Taste and smell in disease. *N Engl J Med.* 1983;308:1275-1279, 1337-1343.

Solvents, Organic, Adverse Effects

ICD-9 981, 982

Identification

Organic solvents are volatile compounds or mixtures that are used for extracting, dissolving, or suspending materials such as fats, waxes, and resins that are not soluble in water.[1] In addition to the illnesses associated with solvent exposure, workers may suffer from the neurotoxic effects of solvent exposure. The major categories of organic solvents and their associated health effects are shown in Table 1. Diagnosis of solvent-related health effects and determination of their association with organic solvent exposure can be difficult; critical aspects of identification are signs and symptoms suggestive of or consistent with these adverse health effects, an appropriate occupational and nonoccupational exposure history, and laboratory tests indicating (a) dysfunction associated with these compounds, and/or (b) the presence of these substances or their metabolites in blood, urine, and exhaled air.

Table 1.—Major Categories of Organic Solvents and Selected Examples of Specific Adverse Health Effects

Category	Examples	Selected Specific Adverse Health Effects
Hydrocarbons		
Aliphatic	n-bexane	Peripheral neuropathy
Aromatic	Benzene	Aplastic anemia, leukemia
Halogenated hydrocarbons	Carbon tetrachloride	Liver and kidney damage, cardiac sensitization
	Methylene chloride	Carboxyhemoglobin formation (same effects as carbon monoxide)
	Freons	Arrhythmias
Alcohols	Methyl alcohol	Optic atrophy, metabolic acidosis, respiratory depression
Glycol derivatives	Ethylene glycol monomethyl ether (2-methoxyethanol)	Toxic encephalopathy, macrocytic anemia
Ketones	Methyl-n-butyl ketone (MBK)	Peripheral neuropathy
Miscellaneous	Carbon disulfide	Psychosis, suicide, peripheral neuropathy, parkinsonian-like syndrome, coronary heart disease

Occurrence

The magnitude of adverse health effects due to organic solvents is not known. Approximately 10 million workers in the U.S. are potentially exposed to organic solvents at work. At high risk for adverse health effects are workers who are chronically exposed to high concentrations of airborne organic solvents, including workers in the plastics, printing, graphics, metal, and dry-cleaning industries, and workers who manufacture or use adhesives, lacquers, and paints.

Causes

See Identification above and Table 1.

Pathophysiology

Virtually all organic solvents depress the central nervous system with acute exposure, causing symptoms ranging—with degree of exposure—from lethargy to unconsciousness and death. The chronic effects of organic solvents on the central nervous system are the subject of intense research interest. According to NIOSH, studies of solvent-exposed workers have shown chronic changes in peripheral nerve function as well as neurobehavioral effects, including "disorders characterized by reversible subjective symptoms (fatigability, irritability, and memory impairment), sustained changes in personality or mood (emotional instability and diminished impulse control and motivation), and impaired intellectual function (decreased concentration ability, memory, and learning ability)."[1] Organic solvents also affect the skin, causing contact dermatitis and other disorders. Most organic solvents, except for aliphatic hydrocarbons, irritate the eyes, nose, and throat. There is evidence that exposure to organic solvents can cause spontaneous abortion and congenital malformations. A number of organic solvents are suspected carcinogens, based on their carcinogenicity in animals.

Prevention

Efforts should focus on primary prevention by reducing or eliminating exposure. Exposed workers should be educated about the solvents with which they work, the hazards these solvents can cause, and ways to prevent adverse health effects. Air monitoring for solvents, by personal and area measurements, should be performed periodically in all workplaces where there is solvent exposure. The most effective primary prevention measures are engineering measures. These include (a) effective design, installation, operation, and maintenance of closed systems and, where they cannot be used, exhaust ventilation systems; (b) isolation of hazardous processes or procedures, which can often be an effective approach, if the process lends itself to this type of measure; and (c) substitution of highly toxic or carcinogenic solvents with safer solvents, such as toluene instead of carbon tetrachloride, and use of alternatives—even soap and water in some circumstances. Gloves, aprons, and other personal protective equipment should be used to prevent direct skin contact. Chemical safety goggles and face shields should be used if there is a potential for splashing. Use of respirators should not be routine and should be reserved only for short-term maintenance situations, emergencies, implementation of engineering measures, and similar situations. Workers who may be or are being exposed should have preplacement and periodic examinations that focus on those organ systems most likely to be affected by organic solvents.

OSHA has promulgated PELs and NIOSH has established RELs for many organic solvents (*see* part 3).

BSL

Reference

1. National Institute for Occupational Safety and Health. *Organic Solvent Neurotoxicity.* Washington, DC: U.S. Government Printing Office; 1987. Current Intelligence Bulletin #48.

Further Reading

World Health Organization. *Recommended Health-Based Limits in Occupational Exposure to Selected Organic Solvents.* Geneva, Switzerland: World Health Organization; 1981. Technical Report Series 664.

Stomach Cancer

ICD-9 151

Identification

Early gastric cancer occurs without symptoms. Subsequently, weight loss (which occurs in 95% of patients), anorexia (25%), early satiety, and both dyspepsia and nonspecific pain (70%) are signs of a poor prognosis (less than a 10% 5-year survival rate). Nonspecific laboratory findings include anemia, the presence of fecal occult blood, and achlorhydria. X-ray diagnosis is commonly used but is less sensitive and clearly less specific than gastroscopy with biopsy. Exfoliative cytology should be included if no clear lesions are seen. A staging system, which was developed in the context of mass screening programs in Japan, reduced mortality effectively.

Occurrence

Approximately 25 000 cases occurred in the U.S. in 1986. The death rates in the U.S. for men and women were approximately 8.2 and 3.9 per 100 000 population, respectively, as compared with 63.1 and 30.3 in Japan and 0.9 and 0.7 in Syria. The disease is generally considered to have an environmental etiology because of the wide variability of incidence rates, which are consistently twice as high in men as in women. Incidence rates in the U.S. have declined dramatically over the past 50 years.

Excess incidence rates of stomach cancer in North America have been reported in some regions where wood and paper pulp industries are located. Laborers in the construction industry in California had a significantly high proportionate mortality ratio (PMR) from stomach cancer in one study.

Causes

Exposure to asbestos, wood dust, metals (nickel and chromium), and probably nitrosamines is associated with increased risk of stomach cancer. In addition, rubber and petroleum refinery workers and coal, arsenic, and silica miners may be at increased risk, as may be jewelers, glassblowers and grinders, carpenters, woodworkers, loggers, plywood mill workers, and workers exposed to solvents, dyes, inks, cutting oils, and ethylene oxide through involvement in the chemical industry, the printing industry, or other workplaces using these materials. Race, diet, low socioeconomic status, nitrosamine exposure, and atrophic gastritis have all been associated with increased incidence and mortality from stomach cancer. Pernicious anemia and blood group A greatly increase the risk of gastric cancer.

Pathophysiology

See An Overview of Occupational Cancer in part 3.

Prevention

Primary prevention is by prevention of exposure to carcinogenic agents, such as asbestos, nickel, chromium, nitrosamines, and tars. Secondary prevention includes early detection; however, no attempts at early detection in the U.S. have been reported. Because of the low incidence rates of stomach cancer, even among occupationally exposed people, screening has not been attempted. In Japan, where baseline incidence is much greater, 5-year survival increased from under 5% to over 30% after introduction of mass screening programs.

HF

Further Reading

Gamble JF, Ames RG. The role of the lung in stomach carcinogenesis: a revision of the Meyer hypothesis. *Med Hypotheses*. 1983;11:359-364.

Neugut AI, Wylie P. Occupational cancers of the gastrointestinal tract. In: Brandt-Rauf PW, ed. *Occupational Cancer and Carcinogenesis. State of the Art Reviews in Occupational Medicine.* 1987;2:109-135.

Sleisenger MH, Fordtran JS. *Gastrointestinal Disease.* 2nd ed. Philadelphia, Pa: W.B. Saunders; 1983.

Stress

ICD-9 308, 309

Identification

There is no consensus concerning a definition of occupational stress. Some researchers emphasize environmental factors, such as lack of control of work and workloads that require significant human adaptation. Others theorize about the role of individuals' personalities in coping with or succumbing to the work environment. And some examine the fit or lack of it between work and the individual. Stress has been considered a chronic condition by some who have examined the long-term exposure of workers to conditions that exceed the workers' ability to meet or master the demands of work. Other research has explored stress as an acute condition, focusing attention on short-term job stressors and traumas and on their related physiological and psychological outcomes.

Most conceptions of stress emphasize the damaging effect of persistent neurohormonal arousal. Although the stressor or external condition may vary (eg, lack of control over job demands, noise, major traumatic events), the widely observed involuntary and nonspecific response is similar in humans and other animals.

In this general adaptation syndrome, or "fight or flight" response, a host of reactions, such as increased catecholamine excretion, occur to mobilize the organism to respond. Stress occurs when this response is repeatedly aroused and frustrated, when an individual can neither fight nor flee. No diseases are identified as purely diseases of adaptation to stress, but stress contributes to or produces a host of maladaptive changes ranging from increased salivation and peptic ulcers to changes in blood pressure and heart rate.

Manifestations of stress can be classified as affective disorders (eg, anxiety, irritability), behavioral problems (eg, abuse of alcohol or drugs, difficulty sleeping), psychiatric disorders (eg, neuroses), and physical complaints (eg, headache, indigestion). The most widely studied effect is premature mortality from heart disease. There is increasing evidence that stress suppresses the immune system and can affect the outcome of other diseases, such as the common cold.

Stress is not widely accepted within the U.S. medical community as a cause of occupational disease and disability. (Stress is accepted in several other countries as a cause or a contributor to premature morbidity and mortality, as well as a factor in many injuries.) Severe workplace accidents, however, can be a cause of posttraumatic stress disorder (a subcategory of the major diagnostic axis of psychosocial stress in the *Diagnostic and Statistical Manual, Revised* [*DSM-III-R*][1] published by the American Psychiatric Association). For example, a worker witnesses a coworker being badly hurt or killed in an accident; the worker then experiences a strong reaction—an aversion to being at work, chronic anxiety, nightmares, and so forth—as a result of witnessing this event. Workers' compensation is then available to cover the cost of treatment. Most cases of compensation are of this nature, although there has been an increase in other sorts of stress-related cases, such as compensation for chronic harassment at work.

Related disorders found elsewhere in this book include traumatic injury, ataxia, headaches, memory loss, encephalopathy, burns, hypertension, collective stress disorder, noise-induced hearing loss, peptic ulcer, asthma, and low back pain.

Occurrence

Considering the lack of a consensus definition for stress, occurrence data are limited. However, many occupations have been identified as stressful using a variety of definitions, including rates of cardiovascular disease prevalence. Although a popular belief is that upper-level white-collar executive jobs are stressful, the most consistent finding in the scientific literature is that stress among lower-level and blue-collar factory workers results in many health problems—most commonly, cardiovascular disease such as hypertension and premature mortality from heart disease. Stressful jobs have been studied in the service, finance, and transportation industries, and include such occupations as secretaries, data entry workers, video display terminal operators, air traffic controllers, transportation workers (eg, bus drivers), waiters, police officers, low-level service workers, and social workers.

The occurrence or recognition of occupational stress is increasing. Workers' compensation claims for "work-related neuroses" more than doubled in California from 1980 to 1982. Nationwide claims for "mental stress" accounted for about 11% of all claims for occupational disease in the same period.

Causes

Causes may be *acute,* as in cases of job loss, adverse personnel decision, traumatic injury, or exposure to noxious chemicals, or *chronic,* as with job dissatisfaction or excessive workload. Combinations of pressure for high levels of productivity, machine-paced work, monotony, excessive and impersonal monitoring of performance, lack of control, boredom, sexual harassment, and noise levels are common stressors among blue-collar and lower-level workers.

Significant nonoccupational stressful life events, such as death or serious illness of a spouse or child, divorce, or legal or financial problems, contribute to an individual's overall condition. Work in isolation, such as late-night guard duty or monitoring panels alone at a power or chemical plant, can also be stressful. Minority workers may be relatively isolated at work because of language, culture, or race. Many women play dual roles as mother and worker and thus have a heavier total workload.

Workers' lack of control is widely held to be the major source of occupational stress. In jobs where there are high demands for constant productivity and sustained attentiveness, such as in piece-rate assembly-line work, computer-monitored word processing, and now "assembly-line" medical care, workers are unable to modify either the pace or the content of their activities. These conditions are particularly harmful if coupled with job hazards, physical demands (such as heavy lifting), or social isolation.

Examples of solutions to such hazards are to increase workers' control over pace or their ability to negotiate better shift schedules, and to increase the resources available to manage demands. Improved social relationships further increase workers' ability to manage stressful work environments, as well as providing human contact. When work is performed in a passive manner (eg, minding monitors in a facility where changes or accidents are rare), understimulation and the lack of control can be detrimental.

Thus, it is often the structure and organization of work, rather than any particular occupation, that produces stress. In evaluating causes, it is important to examine how work is organized and what the individual's response to work is, rather than merely examining the job title and assignment.

Pathophysiology

Investigations have focused on neurohormonal and cardiovascular function, immune responses, and behavior. Research has included laboratory studies of physiological responses, field studies of particular occupations, and epidemiological investigations. Catecholamine excretion in urine is the most commonly measured biochemical marker. Adrenaline, cortisol, glucose, prolactin, growth hormone, and lipids have also been investigated.

Stress has been found to elevate blood pressure, increase heart rate, and promote cardiac ischemia.

Prevention

Increases in workers' control and social support and decreases in demands will likely reduce stress-related illness. Although exercise, medication, therapy, meditation, changes in diet, and the like may improve the outcome and may be necessary in individual cases, structural changes are necessary when the work force in a given industry or occupation is affected. The restructuring of work can be a daunting task, but even the most stressful jobs can be changed. Modest but well-conceived changes can have significant benefits, not only in relieving stress but also in improving the function of an organization. In this regard, some stress control measures may not differ from progressive personnel and hazard control measures.

There are numerous examples of promising experiments in work organization that increase skills and control over work and also increase opportunities to form social networks.

Other Issues

A significant problem concerns the identification of occupational stress discussed above. To a greater degree than many other occupationally related diseases, identification of occupational stress requires a thorough analysis of the structure and content of work.

EH, JJ

Reference

1. Spitzer RL, ed. *Diagnostic and Statistical Manual of Mental Disorders*. 3rd ed., rev. Washington, DC: American Psychiatric Association, 1987.

Further Reading

Baker DB. The study of stress at work. *Ann Rev Public Health.* 1985;6:367-381.

Barnett RC, Biener L, Baruch G, eds. *Gender and Stress.* New York, NY: Free Press; 1987.

Cohen S, Tyrrell DAJ, Smith AP. Psychological stress and susceptibility to the common cold. *N Engl J Med.* 1991;325:606-612.

Division of Biomedical and Behavioral Science, National Institute for Occupational Safety and Health. Leading work-related diseases and injuries: psychological disorders. *MMWR.* 1986;35:613-614, 619-621.

Helzer JE, Robins LN, McEvoy L. Post-traumatic stress disorders in the general population. *N Engl J Med.* 1987;317:1630-1634.

Johnson JV, Hall EM. Job strain, work place social support, and cardiovascular disease: a cross-sectional study of a random sample of the Swedish working population. *Am J Public Health.* 1988;78:1336-1342.

Johnson JV, Johansson G, eds. *The Psychosocial Work Environment: Work Organization, Democratization, and Health.* Amityville, NY: Baywood Publishing; 1991.

Karasek R, Theorell T. *Healthy Work: Stress, Productivity, and the Reconstruction of Working Life.* New York, NY: Basic Books; 1990.

Schnall PL, Pieper C, Schwartz JE, et al. The relationship between "job strain," workplace diastolic blood pressure, and left ventricular mass index. *JAMA.* 1990;263:1929-1935.

Schottenfeld RS, Cullen MR. Recognition of occupation-induced posttraumatic stress disorders. *J Occup Med.* 1986;28:365-369.

Tendinitis, Tenosynovitis ICD-9 726, 727

Identification

Tendinitis is an inflammation of a tendon, the ropelike structure that attaches each end of a muscle to bone. Tenosynovitis is a tendon inflammation that also involves the synovium, the tendon's protective and lubricating sheath. Many forms of tendinitis have names that indicate the specific location of the inflammation, such as lateral or medial epicondylitis ("tennis elbow" and "golfer's elbow"), bicipital tenosynovitis (above the biceps muscle), and rotator cuff tendinitis (specifically affecting shoulder tendons). De Quervain's disease is a tendinitis that affects the thumb extensor or abductor tendons—those on the back side of the hand at the base of the thumb (the "anatomical snuffbox").

These conditions are characterized by localized pain, especially during active motions, often with swelling. Redness may also be present. Symptoms may progress to burning, aching, or swelling while the body part is at rest. Pain may also spread along the involved muscle. The onset of symptoms is usually gradual; sometimes, however, acute onset accompanies intensive work at unaccustomed tasks or motions, such as after transfer to a new job, return to work following a vacation, or increase in the work pace.

On physical examination, pain at the site of inflammation can be elicited by putting pressure on the tendon or its point of attachment to the bone. Usually, symptoms are precipitated or exacerbated by resisted motion of the muscle, with possibly fine crepitus (a grating sensation) but no pain on passive motion. Resisted motion of the antagonist muscle should not produce pain. In de Quervain's disease, sharp pain is produced by Finkelstein's test, in which the subject holds the thumb inside the fingers of the same hand, and the examiner stabilizes the forearm and bends the wrist, moving the hand toward the little finger. In rotator cuff tendinitis, there may be a painful arc within the range of motion—that is, pain between 70 and 90 degrees of shoulder abduction (sideways elevation of arm) but not below or above. There may be objective loss of strength in the muscle, resulting from pain that interferes with active or resisted motions in normal use. But physical examination methods are not well standardized, and the prevalence of findings may vary greatly among examiners.

Several related tendon disorders affect the upper body in occupational settings. If a shoulder tendon inflammation progresses to involve the bursa—a fluid-filled pocket within the shoulder or knee joint that cushions the joint structures—it is termed "bursitis." Tension neck syndrome is a condition with pain at the back of the neck, head, and shoulders thought to be produced by chronic spasm of neck and upper back muscles.

"Trigger thumb" or "trigger finger" occurs when the inflammation in a finger tendon causes the development of an enlarged nodule on the tendon or a narrowing of the tendon sheath. This results in the finger locking into position when bent because the nodule interferes with normal motion; it is often necessary to use the other hand to straighten out the finger. A ganglionic cyst is a fluid-filled swelling or lump that develops at the site of an inflammation. It is not always painful, but, like trigger finger, it may interfere with the smooth motion of the tendon. The fluid may be withdrawn from the ganglion with a needle, both to assist in diagnosis and to attempt to reduce or reverse it.

The differential diagnosis includes ruling out local infection, rheumatoid arthritis, history of bone fracture in the same area, and compression of local nerves (*see* Peripheral Nerve Entrapment Syndromes).

Occurrence

There is a lack of accurate data on the incidence and prevalence of tendon disorders in the general population. Many work-related cases have been described in the medical literature, but few of these reports include sufficient statistical analysis or control of potential confounding factors. In a study of automobile assembly workers in the 1940s, about seven cases of hand and wrist tendinitis per 100 workers per year came to medical attention. (De Quervain's disease accounted for 48% of these cases.)[1] This rate was probably an underestimate, because many workers do not report these conditions, at least not in the early stages. In occupational groups exposed to ergonomic stressors, the prevalence rate has been estimated to range from 4% to 56%. This rate varies according to the ergonomic features of the work, the diagnostic criteria used, and body location(s).

Work-related tendon disorders are more common at some body locations than at others, depending on ergonomic features of the work. The wrist and shoulder/neck regions tend to be affected more often than the middle of the arm; for example, in assembly-line food packing, the prevalences by body location in one study were wrist/hand 56%, neck 41%, shoulder 12%, and elbow/forearm 7%.[2] The extensor tendons of the wrist were affected three times more often than the flexor tendons, although 10% of the food packers had disorders in both sets of tendons. Lateral epicondylitis (at the outside of the elbow) has been estimated to be seven times more common than medial epicondylitis (at the inside of the elbow).

In one study, hand/wrist and elbow/forearm tendon disorders were four times more common among assembly-line food packers (who, at work, have rapid and repetitive hand movement and static muscle loading) than among sales clerks.[2] Male welders in a shipyard, who worked frequently with the arm overhead, had a prevalence of shoulder tendinitis 15 times higher than a group of male office clerks.[3] The risk of hand and wrist tendinitis associated with highly forceful manual exertions has been estimated to be six times higher than that in low-force jobs, and the risk associated with highly repetitive exertions three times higher than that in less repetitive work.[4] Many manual-intensive jobs involve simultaneous exposure to more than one of the factors listed above. Such combined exposures appear to have a greater than additive effect on risk. For example, work that is both forceful and repetitive has been estimated to increase the risk of hand and wrist tendinitis by 29 times.[4]

Causes

In general, high-risk occupational activities include any rapid, repetitive, or forceful manual tasks in service or manufacturing, especially in assembly-line or packing jobs. Sustained contractions of any muscle in a fixed position ("static loading"), such as needed to hold a tool or parts or to maintain a body posture, may strain the muscle-tendon system. Working with the wrist bent, the arm raised at the shoulder, or with any other joint in a nonneutral posture puts the muscle(s) at a biomechanical disadvantage, requiring more forceful muscular contractions to perform the task and increasing the internal forces exerted by the tendon. Using a pinch grip to hold parts or tools similarly creates a biomechanical disadvantage for the working muscles.

The specific cause of any tendon inflammation depends on the particular muscle-tendon system and the motion(s) that it controls. For example, repetitive or prolonged work with the arms at or above shoulder height may cause rotator cuff tendinitis and eventually chronic bursitis. With downward rotation of the forearm, especially with repetitive or forceful wrist flexion, the tendon attachment at the outside of the elbow may become inflamed. Repeated or forceful bending of the wrist toward the little finger and repeated or forceful thumb motion in opposition to the fingers or to operate a pull-trigger or push-button tool have been associated with de Quervain's disease. Bending back the wrist, especially in combination with rapid or forceful finger extension, such as in typing on a keyboard placed too high, may injure the extensor tendons on the back of the hands.

The force demands of a job are most obviously determined by the weight lifted or held in the hands. However, numerous other factors may contribute, including mechanical resistance of parts or controls to being moved; the need to grip an object with a pinch rather than a power grasp (full fist); torque produced by powered hand tools; tool handles that are too large or too small; low friction (slipperiness) between hand and parts or tool handles; infrequently sharpened knives or scissors; and gloves that fit poorly or are too thick. Some of these factors can increase the force required to perform manual work by as much as four to five times.

Exposure to vibration while performing manual work is another occupational risk factor. Vibration interferes with reflexes that enable the nervous system and brain to control precisely the amount of force exerted by a muscle. The forces that are exerted are higher than necessary, causing additional strain on the tendons. Because the higher grip forces result in increased transmission of vibration to the hand, a vicious circle may ensue in which damage to both tendons and nerves is progressively more severe (*see* Hand-Arm Vibration Syndrome).

Mechanical stress concentrations on top of a tendon may interfere with its movement and lead to injury. This may be caused by scissors, other tools, or mechanical levers with sharp edges that press on the sides or backs of the fingers.

Piece-rate work ("piecework," for which employees are paid per unit produced) has also been implicated in the development of tendon disorders. This type of wage system induces workers to increase both their work pace and often the hand forces they exert, as well as to reduce their total rest time, which is required for the tendons to recover from mechanical strain. In addition, piece-rate work is usually associated with greatly reduced variety in the work performed by each person.

Nonoccupational causes of tendon disorders may include amateur sports and recreational and household activities, especially when high forces are generated, when the activities are performed repeatedly, or when there is not an adequate training or "break-in" period.

In addition to ergonomic factors, common nonoccupational associations with tendinitis and tenosynovitis include diabetes mellitus and rheumatoid arthritis. (The prevalence of these conditions in the general population should not be permitted to deflect attention from the workplace and easily preventable causes.) Less common predisposing conditions include gout, hypothyroidism, calcium pyrophosphate deposition, collagen vascular disease, and indolent infection with tuberculus bacilli, atypical mycobacteria, or fungi.

Little or no trend by age has been found; however, this is difficult to evaluate in cross-sectional studies, because affected workers may be more likely to leave (the "healthy worker effect"). In high-exposure occupations, the contribution of nonoccupational factors, such as age, gender, and systemic diseases, to overall risk appears to be relatively small compared with the effects of ergonomic stressors in the workplace.

Pathophysiology

Inflammation results from irritation of a tendon or tendon group at either end of a muscle, sometimes including the tendon sheaths and other surrounding tissues. In many cases, the critical factor appears to be inadequate recovery time between muscular contractions. Each time a muscle contracts, it causes a slight strain or stretching of the tendon; if the next contraction occurs before the tendon has fully recovered, the strain accumulates. Cell and tissue injury eventually result. The time required for recovery is a function of the force and duration of each exertion, the frequency of rest periods, and the total work time with exposure to repetitive or forceful exertions. With very rapid or extremely forceful exertions, even a single contraction may be sufficient to cause tissue injury to the muscle and tendon.

When the nature of the work involves prolonged static postures and muscle contraction rather than rapid movements, the nature of the injury may be more complicated. In addition to the chronic strain produced by prolonged contractions, static loading reduces blood supply to the muscle and tendon; this may be an additional cause of cell injury and death and of eventual tendon degeneration.

Prevention

Primary prevention involves reducing the exposure to adverse ergonomic risk factors, especially forceful manual exertions, awkward or nonneutral body positions during work, static muscle loads, highly repetitive motions, and a rapid and unvarying work pace. Whenever possible, engineering controls (appropriate ergonomic design, selection, and use of tools, tasks, and workstations) should be used instead of administrative controls (*see* chap 1).

Particular attention should be paid to permitting work in a comfortable body posture and to redesigning tools, equipment, and work methods. A keyboard, workbench, or other work surface may need to be raised or lowered to permit the wrist to be kept straight and the arm by the side of the body. Tools should be selected and installed appropriately for the particular application. The forces transmitted by the tendons may be reduced by addressing all aspects of the job that contribute to its strength requirements (*see* Causes above). The need to use pinching grasps should be eliminated by designing tools to hold small parts, providing handholds or cutouts in tote bins and cartons, and determining the optimum handle diameter range for the hand sizes of the particular working population. Sources of external mechanical compression of soft tissues should be eliminated by measures such as padding the edges of a desk or bench where the wrists or elbows rest and using spring-loaded devices, rather than the backs of the fingers, to open tools such as scissors and tongs.

Piece-rate wages and machine-paced work should be avoided whenever possible, because of the pressure on workers to maintain a constantly high speed, even when in pain. If analysis of motion patterns demonstrates that there is a potential for greater variety in manual work by combining or alternating among various tasks, administrative controls, such as job rotation or work enrichment, may reduce repetitiveness and the duration of exposure to forceful exertions and awkward postures.

When vibrating hand tools are used or there is other exposure to segmental vibration, mechanical isolation and damping should be used to reduce the amplitude of the vibration transmitted to the hand and arm. Installing a tool on an articulating arm or overhead suspension will also help to reduce vibration transmission. Tools should be selected with low-intensity vibration that is well above the natural resonant frequencies of the upper extremity (30 to 300 Hz) (*see* Hand-Arm Vibration Syndrome).

The most important secondary preventive measure is training health care providers in the appropriate interview and clinical examination procedures necessary to identify work-related tendon disorders. Workers who report symptoms should receive immediate attention, as these conditions may appear deceptively minor in the early stages. Administrative obstacles in the workplace that prevent or deter workers from reporting symptoms to their physicians or to the plant occupational medical service should be identified and eliminated.

Once reported, cases should be treated conservatively, with total rest for the inflamed tendon(s), and jobs should be analyzed for ergonomic features that may be modified. Use of a splint or elastic bandage on the job should be considered only if it does not interfere with work or require the worker to exert more force or stress on another joint to perform the task. Return to work should be gradual to allow for the reconditioning of the muscle-tendon group. Follow-up is important to ensure that job modifications have been correctly selected to avoid continuing stress to the wrist and that symptoms and signs have not progressed.

Concerning tertiary measures, surgery is generally a last resort; it may be only temporarily effective if the worker is returned to an ergonomically stressful job that has not been modified. Possible loss of muscle strength and build-up of scar tissue following surgery make it imperative that job assignments be selected carefully to avoid recurrence.

LP

References

1. Thompson AR, Plewes LW, Shaw EG. Peritendinitis crepitans and simple tenosynovitis: A clinical study of 544 cases in industry. *Br J Ind Med.* 1951;8:150-159.

2. Luopajarvi T, Kuorinka I, Virolainen M, Holmberg M. Prevalence of tenosynovitis and other injuries of the upper extremities in repetitive work. *Scand J Work Environ Health.* 1979;5(suppl 3):48-55.

3. Herberts P, Kadefors R, Andersson G, Petersen I. Shoulder pain in industry: an epidemiological study on welders. *Acta Orthop Scand.* 1981;299-306.

4. Armstrong TJ, Fine LJ, Goldstein SA, Lifshitz YR, Silverstein BA. Ergonomics considerations in hand and wrist tendinitis. *J Hand Surg.* 1987;12(pt A):830-837.

Further Reading

Anyone for teno? *Br J Ind Med.* 1987;44:793-794. Editorial.

Hagberg M. Occupational musculoskeletal stress and disorders of the neck and shoulder: a review of possible pathophysiology. *Int Arch Occup Environ Health.* 1984;53:269-278.

Thorson EP, Szabo RM. Tendinitis of the wrist and elbow. *State of the Art Reviews in Occup Med.* 1989;4:419-431.

Viikari-Juntura E. Neck and upper limb disorders among slaughterhouse workers. *Scand J Work Environ Health.* 1983;9:283-290.

Viikari-Juntura E. Tenosynovitis, peritendinitis and the tennis elbow syndrome. *Scand J Work Environ Health.* 1984;10:443-449.

Thyroid Cancer

ICD-9 193

Identification

Most thyroid cancer presents as a hard, asymptomatic nodule in the thyroid gland. Thyroid scan shows diminished function. There are four histological types: papillary, follicular, anaplastic, and medullary.

Occurrence

Papillary thyroid cancer is the most common form. It occurs with a bimodal distribution with one peak in the 30s and 40s and a second peak later in life.

Causes

Exposure to ionizing radiation is a well-documented cause of thyroid cancer both from gamma or x-radiation of the thyroid and from exposure to radioactive isotopes of iodine. An elevated risk may exist from work at nuclear power plants or other nuclear facilities or from releases from such plants. A high incidence of thyroid neoplasms and hypothyroidism occurred among inhabitants of the Marshall Islands following aboveground detonation of an atomic bomb in 1954.

Pathophysiology

Ionizing radiation is a well-known carcinogen. Iodine is selectively absorbed by the thyroid so that exposure to radioactive iodine can result in high localized exposure of the thyroid to ionizing radiation.

Prevention

In the event of acute radiation exposure, oral intake of iodine blocks the uptake of radioactive iodine from inhalation. After radiation exposure, periodic examination of the thyroid for nodules is imperative.

LW

Further Reading

Mazzaferri EL. Thyroid cancer. In: Becker KL, ed. *Principles and Practice of Endocrinology.* Philadelphia, Pa: J.B. Lippincott; 1990:319-330.

Welch LS. Environmental toxins and endocrine function. In: Becker KL, ed., *Principles and Practice of Endocrinology.* Philadelphia, Pa: J.B. Lippincott; 1990:1686-1691.

Thyroid Disorders (Hypothyroidism)

ICD-9 240-246

Identification

Hypothyroidism has a varied presentation, depending on its severity. It can range from an asymptomatic decrease in serum thyroxine to a life-threatening condition of myxedema. Usual symptoms include lethargy, cold intolerance, constipation, weight gain, skin changes, and slowed mental functioning. A wide range of chemicals affect thyroid function in people and animals; most of them cause goiter (a nonspecific enlargement of the thyroid) and/or hypothyroidism.

Occurrence

Clinically apparent hypothyroidism was present in 2% of the U.S. population in one survey. The group screened had a mean age of 57, and the disease was 10 times more common in women than in men.

Causes

Lead has been reported to affect the pituitary-thyroid axis. Lead exposure in a battery plant resulted in decreased serum thyrotropin levels and low serum thyroxine, with flat or delayed response to thyrotropin-releasing hormone stimulation, suggesting anterior pituitary involvement. Cessation of exposure was followed by improvement over months to years. The affected people had nonspecific symptoms such as fatigue, muscle pains, and impotence, which may have been independently caused by lead exposure.

Thiouracil, thiocyanates, and thiourea are goiter-producing agents. Ethylene thiourea is used as an accelerator in the manufacture of rubber and as a chemical intermediate in the manufacture of pesticides, fungicides, dyes, and chemicals. It is in the same chemical class as propylthiouracil, which is used to treat hypothyroidism. Thiocyanate occurs naturally in plants in the cabbage family and causes goiter in animals. Consumption of milk from cows that feed on these plants has been associated with goiter in humans.

Halogenated aromatic hydrocarbons can also affect thyroid function in humans and animals. Polybrominated biphenyls (PBBs) caused primary hypothyroidism in 11% of men working in a PBB production facility. The pesticide hexachlorobenzene (HCB) caused goiter in humans. In the 1950s, a hexachlorobenzene epidemic occurred in Turkey: 40%

of those exposed—accounting for 60% of the women and 27% of the men—had thyroid enlargement.

Pathophysiology

Thiocyanates and thiouracil interfere with organic iodination in the thyroid, an essential step in thyroid hormone synthesis. The mechanism of action of lead on the thyroid is not known. Animal experiments suggest that the halogenated aromatic hydrocarbons impair the release of thyroid hormone.

Prevention

Occupational endocrine disorders can be prevented using standard industrial hygiene control of exposure: Substitute less toxic materials, use either positive engineering to eliminate release into the work environment or environmental controls such as local exhaust ventilation, or, if these are not feasible, use personal protective equipment. On-site changing rooms, washrooms, and separate eating facilities also help reduce exposure to toxins. Exposure and biological monitoring are useful in controlling episodes.

LW

Further Reading

Shapiro L, Surks M. Hypothyroidism. In: Becker KL, ed. *Principles and Practice of Endocrinology*. Philadelphia, Pa: J.B. Lippincott; 1990:363-369.

Welch LS. Environmental toxins and endocrine function. In: Becker KL, ed. *Principles and Practice of Endocrinology*. Philadelphia, Pa: J.B. Lippincott; 1990:1686-1691.

Tight Building Syndrome—*See* Building-Related Illness

Toxic Encephalopathy—*See* Encephalopathy, Toxic

Toxic Hepatitis—*See* Hepatitis, Toxic

Trauma—*See* Cumulative Trauma Disorders; Injuries, Fatal; and Injuries, Nonfatal

Tremor

ICD-9 332.1, 781.0, 985.0

Identification

A tremor is a rhythmic alteration in movement that is consistent in pattern, amplitude, and frequency and is due to reciprocal contraction of a muscle group and its antagonist. Tremors are objectively classified by location, speed, amplitude, relationship to rest and movement, etiology, and underlying pathological change. A physiological tremor, normally low in amplitude and invisible, occurs in all individuals whether asleep or awake. An abnormal tremor has a lower frequency of movement than a physiological tremor and occurs only when the person is awake. Abnormal tremors can be further categorized into two types: (a) Intention or action tremors (those that appear only with deliberate movement, as in multiple sclerosis) are coarse and irregular and are associated with cerebellum dysfunction and thus ataxia. (b) Static or nonintention tremors (those present at rest and associated with dysfunction of the basal ganglia and extrapyramidal pathways, such as in Parkinson's disease) are more rhythmic in nature and are often localized in one or both hands or, less frequently, in feet, jaw, lips, or tongue. Electromyography, psychomotor tests of eye-hand coordination, tapping performance tests, and neurological tests of cerebellar functioning are used to assess tremor objectively. The frequency and amplitude of tremor may be measured by an accelerometer connected to a tape recorder through a precision sound-level meter.

Occurrence

No reliable occurrence data exist.

Causes

Exposure to certain neurotoxins in the workplace can cause tremor (*see* Table 1 at end of entry). The progression of the motor signs of tremor has commonly been observed in workers exposed to inorganic mercury; with increasing dose, tremors of the finger and hands are followed by tremors of the eyelids and face, and potential involvement of the head and trunk. Testing with an accelerometer has revealed subclinical psychomotor and neuromuscular changes in mercury-exposed workers with normal neurological exams. In chronic organic mercury exposure, progressive weakness of the extremities is the most prominent sign of neuropathy; tremor also occurs but is not as common and usually occurs only after some delay following exposure. Exposure to zinc fumes causes metal fume fever (also known as brass worker's ague or smelter's chills), the symptoms of which may progress to a coarse intention tremor of the upper extremities. Carbon monoxide exposure or carbon disulfide intoxication may result in tremor resembling that of late extrapyramidal syndrome. Similarly, tremors are often evident in moderate to severe manganese poisoning. Chronic trichloroethylene exposure above the threshold limit value (TLV) of 100 ppm has produced tremor. Chronic exposure to methyl bromide, a colorless gas used as a fumigant, refrigerant, fire extinguisher, and insecticide, is known to lead to tremor; the effects are characterized by a mixture of cerebellar symptoms and peripheral neuropathy, whereas acute exposure causes delirium as well as visual and speech disturbances. Paraquat (1,1-dimethyl-4,4-bipyridylim dichloride) is a herbicide that has been observed to produce hand tremor 4 to 5 days after exposure as well as renal damage and mental disturbance. The organochlorine insecticide DDT is known to be neurotoxic after an acute exposure, causing tremors and seizures. Chlordecone (Kepone) exposure has been shown to be most associated with an intention tremor, although a static tremor occurs in people severely affected by it.

Nonoccupational causes of tremor, spasms, and convulsive movements include hyperthyroidism, parkinsonism, tardive dyskinesia, Tourette's syndrome, epilepsy, Wilson's disease, and dystonia. Tremor may also be due to alcoholic withdrawal, nicotine, caffeine, amphetamines, and barbituates. Hepatic and other encephalopathies may cause asterixis, a hand and finger tremor. Lithium, which is frequently given in the treatment of psychiatric disorders, may cause tremor.

Pathophysiology

Intention tremor is a manifestation of cerebellar disease, although the cerebellar afferent pathway was relatively preserved in persons who presented with tremor following organic mercury poisoning after ingesting contaminated fish from Minimata Bay, Japan. Degeneration of the basal ganglia and extrapyramidal pathway is seen in patients with nonintention tremor. A specific pathology for benign or essential tremors has not been described.

Prevention

Exposure to neurotoxins should be minimized. Effective industrial hygiene monitoring of respiratory, gastrointestinal, and dermal routes of chemical entry is discussed elsewhere in this book. Biological monitoring of workers for specific chemical exposures may be used as a guide to check on individual exposure levels, but careful attention must be paid to the interpretation of individual results. For more information, the American Conference of Governmental Industrial Hygienists (ACGIH) has issued biological exposure indices (BEIs) for a number of chemicals.

Removal from exposure and/or chelation of mercury may reverse the symptoms of tremor if performed early enough. N-acetyl-penicillamine is the preferred chelating agent for elemental mercury exposure. L-dopa therapy has been somewhat successful in treating parkinsonian-like tremor from manganese poisoning. Treatment of carbon disulfide-intoxicated persons with amantadine, diazepam, and L-dopa was not successful in lessening the tremor in a clinical case report. The toxicity of chlordecone has been observed to be proportional to the extent that the agent is retained in certain tissues; treatment of chlordecone-induced effects with cholestyramine in a 5-month trial significantly reduced the half-life of the pesticide in blood and fat.

Other Issues

Some nonworkplace factors may confound the identification of a work-related cause of tremor. Familial tremors, medium to coarse in amplitude and absent at rest, tend to progress in severity as a person ages and are often accentuated by physical stress, stimulants, and excitement. They tend to lessen upon ingestion of alcohol and sedatives, but they will worsen when the effect of these agents wears off.

RF, SP

Further Reading

Adams RD, Victor M, eds. *Principles of Neurology.* 2nd ed. New York, NY: McGraw-Hill; 1981.

Bradley WG, et al., eds. *Neurology in Clinical Practice.* Boston, Mass: Butterworth-Heinemann; 1990.

Chandra SV. Neurological consequences of manganese imbalance. In: Dreosti IE, Smith RM, eds. *Neurobiology of the Trace Elements.* Vol 2. Clifton, NJ: Humana Press; 1983.

Juntunen J. Alcohol, work and the nervous system. *Scand J Work Environ Health.* 1980;10:461-465.

Langolf GD, Chaffin DB, Henderson R, Whittle HP. Evaluation of workers exposed to elemental mercury using quantitative tests of tremor and neuromuscular functions. *Am Ind Hyg Assoc J.* 1978;39:976-984.

Peters HA, Levine RL, Matthews CG, Sauter SL, Rankin JH. Carbon disulfide-induced neuropsychiatric changes in grain storage workers. *Am J Ind Med.* 1982;3:373-391.

Roels H, Malchaire J, Van Wamkeke JP, Buchet JP, Lauwerys R. Development of a quantitative test for hand tremor measurement. *J Occup Med.* 1983;25:481-487.

Table 1.—Exposures Associated with Tremors

Neurotoxin	Major Uses or Sources of Exposure
Metals	
Manganese	Iron, steel industry
	Welding operations
	Metal-finishing operations
	Fertilizers
	Manufacturers of fireworks, matches
	Manufacturers using oxidation catalysts
	Manufacturers of dry cell batteries
Mercury	Scientific instruments
	Electrical equipment
	Amalgams
	Electroplating industry
	Photography
	Felt making
	Taxidermy
	Textiles
	Pigments

Neurotoxin	Major Uses or Sources of Exposure
Solvents	
Carbon disulfide	Manufacturers of viscose rayon
	Paints
	Preservatives
	Textiles
	Rubber cement
	Electroplating industry
Trichloroethylene	Degreasers
	Painting industry
	Paints
	Lacquers
	Varnishes
	Dry-cleaning industry
	Rubber solvents
	Adhesive in shoe and boot industry
	Process of extracting caffeine from coffee
Gases	
Methyl bromide	Insect fumigant
	Fire extinguisher
	Refrigerant
	Degreasing agent for wool
Carbon monoxide	Exhaust fumes of internal combustion engines, incomplete combustion
	Acetylene welding
Insecticides/Herbicides	
Chlorinated hydrocarbons (DDT, chlordecone)	Agricultural industry
Paraquat	Agricultural industry
	Field-workers

Trenchfoot—*See* Cold-Related Disorders

Tuberculosis
(and Silicotuberculosis)

ICD-9 011 (and 502)
SHE/O

Identification

Tuberculosis (TB) is an infectious disease that is usually transmitted from person to person via the airborne route, although transmission by the oral and parenteral routes occasionally occurs. The initial infection usually goes unnoticed, but the host develops delayed hypersensitivity to tuberculin within a few weeks, indicating that infection has occurred. Initial lesions usually heal, leaving no residual changes except pulmonary or tracheobronchial lymph node calcifications, which can often be seen on a chest x-ray. In fewer than 10% of patients, there is direct progression to pulmonary TB or—by lymphohematogenous dissemination of bacilli—to lymphatic, meningeal, or other extrapulmonary involvement.

Most adult cases arise from reactivation of a latent focus remaining from the initial infection. This may occur at any time later in life and at any body site seeded with organisms during the initial infection. If untreated, from one half to two thirds of an infected population will die within 2 years. However, early and appropriate chemotherapy nearly always results in a cure. Early symptoms include fatigue, fever, and weight loss; cough, chest pain, and hemoptysis (coughing up blood) generally appear later.

People who have been infected with tubercle bacilli usually react to a tuberculin skin test using 5 tuberculin units (5 TU) of purified protein derivative (PPD). The reaction may be suppressed in critically ill TB patients during certain acute infectious diseases, as well as in people who are immunosuppressed by disease or drugs. Reactions to tuberculin can also be caused by other mycobacteria and BCG vaccination. A diameter of ≥10 mm of induration is defined as a positive indicator of infection in most populations; however, among household contacts of infectious TB cases, people with human immunodeficiency virus (HIV) infection, and people with abnormal results on chest x-rays consistent with TB, a diameter of ≥ 5 mm of induration is considered a positive indicator.

A positive skin test only indicates mycobacterial infection, probably with *Mycobacterium tuberculosis*. A presumptive diagnosis of current disease is usually made by demonstration of acid-fast bacilli (AFB) in stained smears from sputum or other body fluids. A positive smear justifies therapy. The diagnosis is confirmed by culture. Without bacteriologic confirmation, current disease is presumptively diagnosed with a positive tuberculin reaction and clinical or x-ray evidence of an ongoing disease process compatible with TB.

Occurrence

Worldwide, about 8 million new cases and 3 million deaths occur annually, most in developing countries. Industrialized countries have shown downward trends in infection, morbidity, and mortality for many years. There has, however, been a resurgence of TB in the United States recently, which is apparently related to HIV infection. In people with latent tuberculous infection, HIV infection appears to increase substantially the risk of developing clinical TB.

In general, infection, morbidity, and mortality rates are higher in older people, in men, and in people at lower socioeconomic levels. In the United States, minority groups and foreign-born people from high-prevalence countries have higher rates of infection and disease than do non-Hispanic whites who were born in the U.S.

An increased incidence of TB in certain occupational groups has been observed since the 1880s in three types of occupations:

1. Occupations that attract people who are at high risk for TB. Because TB incidence is related to socioeconomic status—with the highest rates occurring among the poor, minority group members, and immigrants and refugees from countries with a high prevalence of TB—occupations in which these groups are disproportionately represented have higher TB rates than those of the general population. These occupations include unskilled laborers, food handlers, migrant farm workers, and lower-paid health care workers.

2. Occupations that increase susceptibility to the infecting organisms. The classic example of an occupational hazard that increases TB risk is exposure to silica dust. Occupational groups at risk for silicosis include miners, quarry workers, stonemasons, sandblasters, and foundry and pottery workers. There is a markedly increased TB incidence among people with x-ray evidence of silicosis. The major problem appears to be an increased risk of disease if infection occurs. Silica particles appear to exert their effect by impairing the ability of macrophages to inhibit the growth of, and kill, tubercle bacilli.

3. Occupations that may increase the likelihood of exposure to infection in environments conducive to transmission. Occupational groups that may be at increased risk of exposure are people working in correctional institutions, nursing homes, extended care facilities, acute care hospitals, mental hospitals, mycobacteriology laboratories, autopsy rooms, zoos, primate research centers, shelters for the homeless, drug addiction treatment centers, and high-prevalence countries. Social workers and others who work with

the poor may also be at higher risk. Tuberculosis in prisons appears to be an increasing problem and may be exacerbated in the future by the further increase in HIV infection among prison inmates. Nursing home residents have a 70% to 570% higher incidence of disease than demographically similar people living in the community. The risk to hospital and other institutional health care workers is probably lower than in the past; but in institutions that serve high-risk populations, especially public hospitals and clinics, the workplace risk may still be substantial, primarily because of exposure to patients with undetected TB. This is a particular problem when the clinical presentation is atypical, as is often the case in elderly patients and in patients with HIV infection. Laboratory workers are at higher risk for TB, presumably due to the aerosolization of tubercle bacilli with procedures such as pipetting, centrifugation, and mixing. Accidental injection of bacilli into the skin can occur when suspensions of organisms are drawn up into a syringe. Autopsies on humans and animals with TB can lead to accidental aerosolization or injection of tubercle bacilli. Workers in zoos and primate research centers are at risk for acquiring TB, especially from primates imported from high-prevalence areas. People who are stationed abroad in areas of high TB prevalence may be at high risk for TB, especially if they are health care providers or are in prolonged contact with members of the local population who have undiagnosed respiratory illness.

Causes

M tuberculosis and *M africanum,* primarily from humans, and *M bovis,* primarily from cattle, are the infectious agents. Transmission usually occurs by airborne droplet nuclei produced by people with infectious pulmonary (or laryngeal) TB. Infection and disease with *M bovis* in humans is rare but may occur where disease in cattle has not been controlled. Bovine TB usually results from ingestion of raw, unpasteurized milk or dairy products, but occasionally it results from airborne spread to farmers and animal handlers. The degree of communicability of a patient with pulmonary TB depends on the number of bacilli discharged and their aerosolization by coughing, sneezing, talking, or singing. Effective therapy reduces communicability promptly. Although TB ranks low among communicable diseases in infectiousness, the long exposure of some contacts, notably household contacts, leads to a 20% to 30% risk of becoming infected. Once infection occurs, there is a 2.5% to 5.0% chance of the infection progressing to disease within the next 2 years. Thereafter, the risk decreases, but persists for a lifetime. For infected infants, the lifetime risk of developing disease may be 10% or more. The risk of developing disease is highest in children under 5 years old, lowest in later childhood, and high again in adolescence and young adulthood.

Pathophysiology

If tubercle bacilli on aerosolized droplet nuclei reach the alveoli, the organisms are engulfed by macrophages. However, they resist killing by macrophages, remain viable, and continue to multiply. The bacilli are spread through the lymphatic channels to regional lymph nodes and through the bloodstream to more distant sites. Acquired immunity usually develops over a period of several weeks. The early immune response usually limits multiplication and spread of bacilli. The host generally remains asymptomatic, the lesions heal, and the immune response can be demonstrated by testing for hypersensitivity to tuberculin. In a few infected people, there may be direct or early progression to clinical illness. In some others, illness may develop after an interval of years or decades, when tubercle bacilli that have persisted in the body begin to replicate and produce disease.

Prevention

1. Primary prevention is accomplished by the following: (a) Maintaining a high index of suspicion for the disease, especially in high-risk populations. (b) Providing adequate ventilation and selected use of UV light in areas where there is a high risk of transmission (eg, mycobacteriology laboratories). (c) Giving BCG vaccination to people who are uninfected (tuberculin-negative). Because the risk of infection is very low in the United States, BCG is used only in rare circumstances when other control measures cannot be applied. In countries with a high risk of infection and minimal resources, BCG vaccine for infants and young children may be an important TB prevention measure. (d) Eliminating TB among dairy cattle by tuberculin testing and slaughter of reactors. (e) Pasteurizing or boiling milk. (f) Applying measures to prevent silicosis in industrial settings. (g) Improving social conditions that increase the risk of infection (eg, overcrowding) and of disease (eg, undernutrition).

2. Secondary prevention is accomplished by the following: (a) Performing sputum smear examination of people in high-incidence areas who have persistent cough lasting ≥3 weeks. This gives a high yield of infectious TB cases. (b) Performing tuberculin testing of all household members and close nonhousehold contacts of infectious cases. (c) Performing selective tuberculin testing on groups at high risk for infection and disease, such as immigrants from areas where TB is prevalent. Positive tuberculin reactors and symptomatic individuals (regardless of skin test reaction size) should receive chest x-rays and appropriate therapy.

3. Resources and facilities for early diagnosis and treatment of cases should be provided. Infectivity should be controlled by prompt, specific drug therapy, which

usually results in sputum conversion within a few weeks. Most patients can be treated on an outpatient basis. Hospitalized patients with sputum-positive pulmonary TB should be placed in respiratory isolation. Patients should be taught how to cover the mouth and nose when coughing or sneezing. Patients who are bacteriologically negative, who do not cough, or who are known to be on adequate chemotherapy need not be isolated. Preventive treatment with isoniazid (INH) is effective in preventing the progression of latent infection to clinical disease. Isoniazid is routinely advised for infected people, regardless of age, if one or more of the following is present: recent infection, close contact with a current case, TB that was never effectively treated with antituberculosis drugs, diabetes, silicosis, prolonged therapy with corticosteroids or other immunosuppressants, and diseases that cause immunosuppression, such as HIV infection.

Other Issues

1. Tuberculosis control among groups of low socioeconomic status, minorities, and people who are foreign born presents special problems, because these groups often suffer from more economic hardships, coexisting diseases, lack of shelter and transportation, chemical dependencies, population mobility, noncompliance with long-term therapy, language and cultural barriers, and fear of deportation.

   People in these groups who have a cough that persists for more than 3 weeks should have chest x-rays and/or sputum smears and cultures. Transportation to a clinic will often be necessary, or mobile x-ray units may need to be taken to the worksite or living area. Because of their mobility, people who are placed on treatment should be given records that they can carry with them to indicate their current treatment status. Whenever possible, directly observed intermittent therapy of 6 months' duration should be used. In cases in which a worker is moving and requires continued treatment or other follow-up, health care providers should route pertinent information to the city or state health department to apprise it of the worker's next destination and to ensure that information is available to other health care providers.

These groups can be screened for TB infection at the worksite when workers are hired or at other community sites. Screening of migrant workers for TB infection is best done in home base sites rather than at temporary work locations so that preventive therapy can be more easily completed by those found to be infected. Workers with significant tuberculin reactions should be placed on preventive therapy in accordance with American Thoracic Society/CDC guidelines, but only if it seems likely that they will be able to complete at least 6 months of treatment. Otherwise, preventive therapy provides little benefit. Supervised, twice-weekly preventive therapy should be used if self-administered therapy is not likely to be successful. Those with a significant tuberculin reaction who are not prescribed preventive therapy should be counseled about the significance of the skin test reaction and instructed to seek medical attention should they develop symptoms suggesting TB.

2. Prevention of exposure to hazardous levels of silica dust (and thus of silicotuberculosis) can be achieved by engineering controls, such as improved ventilation, process enclosure, wet mining techniques, and the use of substitutes for some industrial silica in manufacturing processes (*see* Silicosis). When these methods fail, additional worker protection may be provided by respiratory protection devices. Periodic tuberculin skin testing of workers exposed to silica dust is indi-

cated, and INH preventive therapy should be considered for those who have significant reactions. Workers who have signs and symptoms suggestive of current TB should be given chest x-rays and sputum cultures. Although modern treatment regimens are usually effective in curing TB, treatment failures and relapses are more frequent in these patients than in those without silicosis.

3. Preventing TB transmission in closed environments, such as in hospitals, shelters for the homeless, prisons, and nursing homes, requires that people who work in these settings be educated about the signs and symptoms of TB and that people with such signs and symptoms be promptly and appropriately evaluated for TB. Routine screening by tuberculin skin tests and chest x-rays of people admitted to nursing homes and prisons is generally recommended. Tuberculin skin testing at the time of employment should be required for all people who will work in these environments. The need for repeat skin testing should be determined based on the epidemiological situation—that is, the likelihood of exposure to infectious TB. Employees who are documented as having newly acquired infection should be offered preventive therapy. Laboratory and autopsy room workers must be educated about the risk of TB transmission via aerosolization and injection, and they should take appropriate precautions to minimize the risk (eg, working in well-ventilated areas, using UV light, and carrying out certain procedures only under a hood). Workers in zoos and primate centers should be certain that new animals are given tuberculin skin tests and kept in quarantine until they are declared free of TB. Workers in shelters for the homeless, drug addiction treatment centers, and high-prevalence countries should receive tuberculin skin tests annually if exposure to risk groups con-

tinues. Skin test converters should be considered for INH preventive therapy.

DS

Further Reading

American Thoracic Society/Centers for Disease Control. Diagnostic standards and classification of tuberculosis and other mycobacterial diseases *Am Rev Respir Dis.* 1981;123:343-358.

American Thoracic Society/Centers for Disease Control. Treatment of tuberculosis and tuberculosis infection in adults and children. *Am Rev Respir Dis*. 1986;134:355-363.

Centers for Disease Control. T*uberculosis and Migrant Farm Workers.* Washington, DC: Department of Health & Human Services; June 1985.

Patterson WB, Craven DE, Schwartz DA, Nardell EA, Kasmer J, Noble J. Occupational hazards to hospital personnel. *Ann Intern Med*. 1985;102:658-680.

Snider DE. The relationship between tuberculosis and silicosis. *Am Rev Respir Dis*. 1978;118:455-460.

Ultraviolet Radiation—*See* Radiation, Nonionizing, Adverse Effects

Vibration Syndrome or Vibration White Finger—*See* Hand-Arm Vibration Syndrome

White Lung—*See* Asbestosis

Part 3
Special Topics

Occupational History Form

An accurate occupational (and environmental) history is necessary for recognition, treatment, and prevention of occupational health problems. A standard occupational history form, such as the one that follows, can aid in the collection and analysis of relevant information.

OCCUPATIONAL HISTORY FORM

I. IDENTIFICATION

Name: ______________________ Soc. Sec. No.: ______________

Address: ______________________ Sex: M F

______________________ Birthdate: ______________

Telephone: Home __________ Work __________

II. OCCUPATIONAL PROFILE

Fill in the table below listing all jobs at which you have worked, including short-term, seasonal, and part-time employment, and including military service.

Start with your present job and go back to the first. Use additional paper if necessary.

Workplace (Employer name and city)	Dates worked		Average hours per week	Type of Industry	Describe your job duties	Known health hazards in workplace (Dusts, solvents, etc.)	Protective equipment used (gloves, masks, etc.)
	From	To					

Adapted from a form developed by the American Lung Association of San Diego and Imperial Counties, CA.

III. OCCUPATIONAL EXPOSURE INVENTORY

1. Please describe any health problems or injuries you have experienced connected with your present or past jobs:

2. Have any of your co-workers also experienced health problems or injuries connected with the same jobs? . . No Yes
   If yes, please describe:

3. Do you or have you ever smoked cigarettes, cigars, or pipes? . . . . No Yes
   If so, which and how many per day: ____________
   Are you still smoking? . . . . No Yes
   If no, when did you stop? ____________________

4. Do you smoke while on the job, as a general rule? . . . No Yes

5. Do you have any allergies or allergic conditions? . . . . . No Yes
   If so, please describe:

6. Have you ever worked with any substance which caused you to break out in a rash? . . . . No Yes
   If so, please describe your reaction and name the substance:

7. Have you ever been off work for more than a day because of an illness or injury related to work? . . . . No Yes
   If so, please describe:

8. Have you ever worked at a job which caused you trouble breathing, such as cough, shortness of wind, or wheezing? . . . . No Yes
   If so, please describe:

9. Have you ever changed jobs or work assignments because of any health problems or injuries? . . . . No Yes
   If so, please describe:

10. Do you frequently experience pain or discomfort in your lower back or have you been under a doctor's care for back problems? . . . . .No Yes
    If so, please describe:

11. Have you ever worked at a job or hobby in which you came into direct contact with any of the following substances by breathing, touching, or direct exposure?
    If so, please check the box beside the substance

☐ Acids	☐ Cold (severe)	☐ Perchloroethylene
☐ Alcohols (industrial)	☐ Cotton dust	☐ Pesticides
☐ Alkalis	☐ Dichlorobenzene	☐ Phenol
☐ Ammonia	☐ Ethylene dibromide	☐ Phosgene
☐ Arsenic	☐ Ethylene oxide	☐ Radiation
☐ Asbestos	☐ Fibrous glass	☐ Rock dust
☐ Benzene	☐ Formaldehyde	☐ Silica dust
☐ Beryllium	☐ Heat (severe)	☐ Solvents
☐ Cadmium	☐ Isocyanates	☐ Styrene
☐ Carbon tetrachloride	☐ Ketones	☐ Talc
☐ Chlorinated naphthalenes	☐ Lead	☐ Toluene
☐ Chloroform	☐ Manganese	☐ TDI or MDI
☐ Chloroprene	☐ Mercury	☐ Trichloroethylene
☐ Chromates	☐ Methylene chloride	☐ Trinitrotoluene
☐ Coal dust	☐ Nickel	☐ Vibration
	☐ Noise (loud)	☐ Vinyl chloride
	☐ PCBs	☐ Welding fumes
		☐ X-rays

    If you have answered “yes” to any of the above, please describe your exposure on a separate sheet of paper.

IV. ENVIRONMENTAL HISTORY

1. Have you ever changed your residence or home because of a health problem? . . . . No Yes
   If so, please describe:

2. Do you live next door to or very near an industrial plant? . . . . No Yes
   If so, please describe:

3. Do you have a hobby or craft which you do at home? . . . . No Yes
   If so, please describe:

4. Does your spouse or any other household member have contact with dusts or chemicals at work or during leisure activities? . . . . No Yes
   If so, please describe:

5. Do you use pesticides around your home or garden? . . . . No Yes
   If so, please describe:

6. Which of the following do you have in your home? (Please check those that apply.)

   ☐ Air conditioner ☐ Gas stove ☐ Fireplace
   ☐ Air purifier ☐ Electric stove ☐ Central heating
   ☐ Humidifier

Additional Questions for the Evaluation of Occupational Associations to the Present Illness or Injury [1]

Questions	Interpretation
Is your condition better or worse when you are off work for a few days or on vacation?	Identify patterns suggesting either improvement or exacerbation on withdrawal from exposure.
Is your condition better or worse when you return to work after a weekend or vacation?	Identify patterns suggesting return of the condition on reexposure in the workplace.
Does your condition get worse or better after you have been back at work for several days or shifts?	Identify patterns suggesting either tolerance or cumulative effects with multiple exposure.
Describe your workplace. (Please draw a diagram and indicate your work station.)	Evaluate the proximity to exposure, protection available (ventilation or barriers), mobility within the workplace, and location of coworkers who may also be affected.
What ventilation systems are used in your work space? Do they seem to work?	Obtain a general impression of adequacy of ventilation by air movement and odors.
Does the protective equipment you are issued fit properly? Do you receive instructions in its proper use? Do you ever fix or make changes in the equipment to make it more comfortable?	Consider the possibility that protective equipment is not fully effective. In the case of respirators (masks), ask if they are "fit-tested" to comply with Occupational Safety and Health Administration (OSHA) regulations.
Where do you eat, smoke, and take your breaks when you are on the job?	Identify opportunities for food- and cigarette-borne intake, and evaluate the adequacy of rest stations (isolation from heat, noise, fumes).

Where are your (and your spouse's or partner's) work clothes laundered?	Identify possibility of passive exposure at home or of prolonged skin contact.
How often do you wash your hands at work, and how do you wash them?	Identify the potential for contamination of hands or contact with solvents or drying agents.
What is your spouse's or partner's occupation?	Identify the potential for passive exposure (an occupational history for the spouse or partner may be indicated).
Have any of your fellow workers experienced similar conditions?	Identifying others who may have been affected may lead to inquiries that clarify the individual patient's problem. Prevention-oriented interventions or requests for investigation by the state or federal enforcement agency may be required.
Do you recall a specific incident or accident that occurred on the job? Were others also affected?	Identify unusual or transient conditions that may have resulted in an exposure not reflected in the occupational history, such as leaks, fires, or uncontrolled exothermic chemical reactions.
Are animals (pets, livestock, birds, or pests such as mice) present in the vicinity of the workplace? Has there been a change in their health, appearance or behavior?	Animals (and especially animal wastes) may be a source of infectious or allergic hazards. Animals may also respond to toxic exposures that affect humans.

[1]Modified and reproduced, with permission, from Occupational and Environmental Health Committee of the American Lung Association: Taking the occupational history. *Ann Intern Med* 1983;**99**:641.

Occupational Sentinel Health Events*

An occupational sentinel health event (SHE-O) is a disease, disability, or untimely death which is occupationally related and whose occurrence may: (1) provide the impetus for epidemiologic or industrial hygiene studies; or (2) serve as a warning signal that materials substitution, engineering control, personal protection, or medical care may be required. Following survey of scientific literature, a list of 50 disease conditions linked to the workplace was presented in 1983; these were codable within the framework of the ICD-9. Three criteria were used for inclusion: documentation of associated agent(s), of involved industries, and of involved occupations. The updated list contains 63 diseases or conditions and is useful for the practicing physician to assist in occupational disease recognition, for occupational morbidity and mortality surveillance, and as a periodically updated compendium of occupationally related diseases.

The broad categories of SHE-Os are represented. The first group includes those diseases or conditions which, by their inherent nature, are necessarily occupationally related. Such conditions include the pneumoconioses; it is unlikely that these diseases would occur in the absence of an occupational exposure to the inciting agent. The second set of conditions includes such diseases as lung cancer, leukemia, peripheral neuropathy, and ornithosis, which may or may not be occupationally related; further information on the industry and occupation of the case is required to establish a causal link between condition and occupation.

*Mullan RJ, Murthy LI. Occupational sentinel health events: an updated list for physician recognition and public health surveillance. *Am J Indus Med.* 1991;19:775-799.

Table 1.— Occupationally Related Unnecessary Disease, Disability, and Untimely Death

ICD-9	Condition	A[a]	B[b]	C[c]	Industry/ Process/Occupation	Agent
011	Pulmonary Tuberculosis (O)[n]	P[d]	P T[e]	P,T	Physicians, medical personnel, medical lab workers	*Mycobacterium tuberculosis*
011, 502	Silicotuberculosis	P	P,T	P,T	Quarrymen, sandblasters, silica processors, mining, metal foundries, ceramic industry	Silica + *Mycobacterium tuberculosis*
020	Plague (O)	P	—	—	Shepherds, farmers, ranchers, hunters, field geologists	*Yersinia pestis*
021	Tularemia (O)	P	—	P,T	Hunters, fur handlers, sheep industry workers, cooks, vets, ranchers, vet pathologists, lab workers, soldiers	*Francisella tularensis*, *Pasteurella tularensis*

ICD-9	Condition	A[a]	B[b]	C[c]	Industry/ Process/Occupation	Agent
022	Anthrax (O)	P	—	P,T	Shepherds, farmers, butchers, handlers of imported hides or fibers, vets, vet pathologists, weavers, farmers	*Bacillus anthracis*
023	Brucellosis (O)	P	P	P	Farmers, shepherds, veterinarians, lab workers, slaughterhouse workers, field offers	*Brucella abortus, suis*
031.1[f]	Fish-Fancier's Finger (O)	P	P	P	Aquarium worker/cleaner, breeder/owner,	*Mycobacterium marinum*
					Longshoremen	*Mycobacterium marinum*
054.6	Herpetic Whitlow (O)	P	P	P	Surgical residents, student nurses, nurses, dental assistant, physicians, orthopedic scrub nurse, psychiatric nurse	Herpes simplex virus
037	Tetanus (O)	P	P	P	Farmers, ranchers	*Clostridium tetani*

042[g]	Human Immunodeficiency Virus Infection (O)	P	P	P	Health-care workers	Human immunodeficiency virus
056	Rubella (O)	P	P	P	Medical personnel, intensive care personnel	Rubella virus
070.0 .1	Hepatitis A (O)	P	P	P	Day care center staff, orphanage staff, mental retardation institution staff, medical personnel	Hepatitis A virus
070.2 .3	Hepatitis B (O)	P	P	P	Nurses and aides, anesthesiologists, orphanage and mental institution staff, med lab personnel, general dentist, oral surgeons, physicians	Hepatitis B virus
070.4	Non-A, Non-B Hepatitis (O)	P	P	P	As above for Hepatitis A and B	Unknown
071	Rabies (O)	P	—	P	Veterinarians, animal and game wardens, lab researchers, farmers, ranchers, trappers	Rabies virus

ICD-9	Condition	A[a]	B[b]	C[c]	Industry/ Process/Occupation	Agent
073	Ornithosis (O)	P	—	P,T	Psittacine bird breeders, pet shop staff, poultry producers, vets, zoo employees, duck processing and rearing	*Chlamydia psittaci*
082.0	Rocky Mountain Spotted Fever (O)	P	P	P,T	Laboratory technicians, tick breeder, virologist, microbiologist, physician	*Rickettsia rickettsil*
100.8	Leptospirosis (O)	P	P	P,T	Farmer/laborer	*Leptospira*
115	Histoplasmosis (O)	P	P	P,T	Bridge maintenance workers	*Histoplasma capsulatam*
117.1	Sporotrichosis (O)	P	P	P	Nurserymen, foresters, florist, equipment operators	*Sporothrix schenkii*
147	Malignant Neoplasm of Nasopharynx (O)	P	P	P	Carpenter, cabinet maker, sawmill worker, lumberjack, electrician, fitter	Chlorophenols
155 M[h,i]	Hemangiosarcoma of the Liver	P	P	P	Vinyl chloride polymerization industry Vintners	Vinyl chloride monomer Arsenical pesticides

158, 163	Mesothelioma (Malignant neoplasm of Peritoneum and Pleura)	P	—	P	Asbestos industries and utilizers	Asbestos
160.0	Malignant Neoplasm of Nasal Cavities (O)	P	P,T	P,T	Woodworkers, cabinet and furniture makers	Hardwood dusts
					Boot and shoe industry	Unknown
					Radium chemists and processors, dial painters	Radium
					Chromium producers, processors, users	Chromates
					Nickel smelting and refining	Nickel
					Sawmill worker, carpenter	Chlorophenols
161	Malignant Neoplasm of Larynx (O)	P	P,T	P,T	Asbestos industries and utilizers	Asbestos
162	Malignant Neoplasm of Trachea, Bronchus, and Lung (O)	P	P	P	Asbestos industry and utilizers	Asbestos
					Topside coke oven workers	Coke oven emission
					Uranium and fluorspar miners	Radon daughters

ICD-9	Condition	A[a]	B[b]	C[c]	Industry/ Process/Occupation	Agent
					Chromium producers, processors, users	Chromates
					Nickel smelters, processors, users	Nickel
					Smelters	Arsenic, arsenic trioxide
					Mustard gas formulators	Mustard gas
					Ion exchange resin makers, chemists	Bis(chloromethyl) ether, chloro-methyl methyl ether
					Iron ore (underground) miners	Radon daughters
					Plan protection workers/ agronomists	Pesticides, herbicides, fungicides, insecticides
					Welders	Unknown
					Copper smelter and roaster workers	Inorganic arsenic, sulfur dioxide, copper, lead, sulfuric acid, arsenic trioxide
					Welders, gas cutters	Asbestos, hexavalent chromium

					Foundry—floor molders and casters	Polyaromatic hydrocarbons
					Dichromate production—floor molders/casters	Unknown
					Chromate production	Chromium dust
					Chromate pigment production workers	Lead chromate, zinc chromate
					Pigment production	Zinc chromate dust
					Steel industry—furnace/foundry workers	Unknown
					Rubber reclaim operations	Unknown
170	Malignant Neoplasm of Bone (O)	P		P	Radium chemists and processors, dial painters	Radium
187.7	Malignant Neoplasm of Scrotum	P	—	P,T	Automatic lathe operators, metalworkers	Mineral/cutting oils.
					Coke oven workers, petroleum refiners, tar distillers	Soots/tars/tar distillates
					Tool setters, fitters, cotton spinners, chimney sweeps, machine operators	Mineral oil, pitch, tar

ICD-9	Condition	A[a]	B[b]	C[c]	Industry/ Process/Occupation	Agent
188	Malignant Neoplasm of Bladder (O)	P	—	P	Rubber and dye workers	Benzidine, alpha- and betanaphthylamine, magenta, auramine, 4-aminobiphenyl, 4-nitrophenyl
189	Malignant Neoplasm of Kidney, Other, and Unspecified Urinary Organs (O)	P	P	P	Coke oven workers	Coke oven emissions
204.0	Lymphoid Leukemia, Acute (O)	P	—	P	Rubber industry	Unknown
					Radiologists	Ionizing radiation
205.0	Myeloid Leukemia, Acute (O)	P	—	P	Occupations with exposure to benzene	Benzene
					Radiologists	Ionizing radiation
207.0	Erythroleukemia (O)	P	—	P	Occupations with exposure to benzene	Benzene
283.1	Hemolytic Anemia, Nonautoimmune (O)	P	—	P	Whitewashing and leather industry	Copper sulfate

					Electrolytic processes, arsenical ore smelting	Arsine
					Plastics industry	Trimellitic anhydride
					Dye, celluloid, resin industry	Naphthalene
284.8	Aplastic Anemia (O)	P	—	P	Explosives manufacture	Trinitrotoluene
					Occupations with exposure to benzene	Benzene
					Radiologists, radium chemists and dial painters	Ionizing radiation
288.0	Agranulocytosis or Neutropenia (O)	P	—	P	Occupations with exposure to benzene	Benzene
					Explosives and pesticide industries	Phosphorus
					Pesticides, pigments, pharmaceuticals	Inorganic arsenic
289.7	Methemoglobinemia (O)	P	—	P,T	Explosives and dye industries	Aromatic amino and nitro compounds (e.g., aniline, trinitrotoluene, nitroglycerin)
					Rubber workers	Aniline, o-toluidine, nitrobenzene

ICD-9	Condition	A[a]	B[b]	C[c]	Industry/ Process/Occupation	Agent
323.7	Toxic Encephalitis (O)	P	P	P	Battery, smelter, and foundry workers	Lead
					Electrolytic chlorine production, battery makers, fungicide formulators	Inorganic and organic mercury
332.1	Parkinson's Disease (Secondary) (O)	P	P	—	Manganese processing, battery makers, welders	Manganese
					Internal combustion engine industries	Carbon monoxide
334.3	Cerebellar Ataxia (O)	P	P	—	Chemical industry using toluene	Toluene
					Electrolytic chlorine production, battery makers, fungicide formulators	Organic mercury
354M[j]	Carpal Tunnel Syndrome (O)	P	P	—	Meat packers, deboners	Cumulative trauma
354.0	Mononeuritis of upper limb and	P	P	—	Dental technicians	Methyl methacrylate monomer
.2	mononeuritis				Poultry processing—turkey	Cumulative trauma
.3	multiplex (O)				Meatpackers, deboners	Cumulative trauma

357.7	Inflammatory and Toxic Neuropathy (O)	P	P,T	P,T	Pesticide industry, pigments, pharmaceuticals formulators	Arsenic/arsenic compounds
					Furniture refinishers, degreasing operations	Hexane
					Plastic-coated-fabric workers	Methyl n-butyl ketone
					Explosives industry	Trinitrotoluene
					Rayon manufacturing	Carbon disulfide
					Plastics, hydraulics, coke industries	Tri-o-cresyl phosphate
					Battery, smelter, and foundry workers	Inorganic lead
					Dentists, chloralkali workers	Inorganic mercury
					Chloralkali plants, fungicide makers, battery makers	Organic mercury
					Plastics industry, paper manufacturing.	Acrylamide
					Ethylene oxide sterilizer operator	Ethylene oxide

ICD-9	Condition	A[a]	B[b]	C[c]	Industry/ Process/Occupation	Agent
366.4	Cataract (O)	P	P,T	—	Microwave and radar technicians	Microwaves
					Explosive industries, trinitrotoluene workers	Trinitrotoluene
					Radiologists	Ionizing radiation
					Blacksmiths, glass blowers, bakers	Infrared radiation
					Moth repellant formulators, fumigators	Naphthalene
					Explosives, dye, herbicide and pesticide industries	Dinitrophenol, dinitro-o-cresol
					Ethylene oxide sterilizer operator, microbiology supervisors, inspectors	Ethylene oxide
388.1	Noise Effects on Inner Ear (O)	P	P	—	Occupations with exposure to excessive noise	Excessive noise
443.0	Raynaud's Phenomenon (Secondary) (O)	P	—	—	Lumberjacks, chain sawyers, grinders, chippers, rock drillers, stone cutters, jackhammer	Whole body or segmental vibration

					operator, riveter	
					Vinyl chloride polymerization industry	Vinyl chloride
493.0, 507.8	Extrinsic Asthma (O)	P	P,T	P,T	Jewelry, alloy and catalyst makers	Platinum
					Polyurethane, adhesive, paint workers	Isocyanates
					Alloy, catalyst, refinery workers	Chromium, cobalt
					Solderers	Aluminum soldering flux.
					Plastic, dye, insecticide makers	Phthalic anhydride
					Foam workers, latex makers, biologists	Formaldehyde
					Printing industry	Gum arabic.
					Nickel platers	Nickel sulfate.
					Bakers	Flour
					Plastics industry, organic chemicals manufacture	Trimellitic anhydride
					Woodworkers, furniture makers	Red cedar (plicatic acid) and other wood dusts
					Detergent formulators	*Bacillus*-derived exoenzymes
					Crab processing workers	Unknown
					Hospital and geriatric department nurses	Psyllium dust

ICD-9	Condition	A[a]	B[b]	C[c]	Industry/ Process/Occupation	Agent
					Laxative manufacture and packing	Psyllium dust
					Prawn processing workers	Unknown
					Snow crab processing workers	Unknown
495.4	Maltworkers' Lung	P	P	—	Maltworkers	*Aspergillus clavatus*
495.5	Mushroom-workers' Lung	P	P	—	Mushroom farm/spawning, shed, farmer	Pasteurized compost
495.8	Grain-handlers' Lung	P	P	—	Grain handlers	*Erwinia herbicola (Enterobacter agglomerans)*
	Sequoiosis	P	P	—	Red cedar mill workers, woodworkers, sawmill, joinery	Redwood sawdust *Thuja plicata*
495.9	Unspecified Allergic Alveolitis	P	P	—	Cinnamon processing workers	Cinnamon dust, cinnamaldehyde
					Distillery, vegetable compost plant worker	*Aspergillus fumigatus*
					Sawmill worker	Unknown

					Paper manufacture/wood room	*Alternaria*, wood dust
					Snow crab processing worker	Unknown
500	Coalworkers' Pneumoconiosis	P	P	P	Coal miners	Coal dust
501	Asbestosis	P	P	P	Asbestos industries and utilizers	Asbestos
502M[k]	Silicosis	P	P	P	Quarryman, sandblasters, silica processors, mining, metal, and ceramic industries	Silica
					Cryolite refining	Cryolite (Na_3AlF_6), quartz dust
	Talcosis	P	P	P	Talc processors, soapstone mining/milling, polishing, cosmetics industry	Talc
503M[l]	Chronic Beryllium Disease of the Lung	P	P	P	Beryllium alloy workers, ceramic and cathode ray tube makers, nuclear reactor workers	Beryllium

ICD-9	Condition	A[a]	B[b]	C[c]	Industry/ Process/Occupation	Agent
504	Byssinosis	P	P	P	Cotton industry workers	Cotton, flax, hemp, and cotton-synthetic dusts
506.0 506.1	Acute Bronchitis, Pneumonitis, and Pulmonary Edema Due to Fumes and Vapors (O)	P,T	P,T	P,T	Refrigeration, fertilizer, oil refining industries	Ammonia
					Alkali and bleach industries	Chlorine
					Silo fillers, arc welders, nitric acid industry	Nitrogen oxides
					Paper and refrigeration industries, oil refining	Sulfur dioxide
					Cadmium smelters, processors	Cadmium
					Plastics industry	Trimellitic anhydride
					Boilermakers	Vanadium pentoxide
					Organic chemical manufacture	Trimellitic anhydride
570, 570.3	Toxic Hepatitis (O)	P	P	P	Solvent utilizers, dry cleaners, plastics industry	Carbon tetrachloride, chloroform, tetrachloroethane, trichloroethylene, tetrachloroethylene

					Explosives and dye industries	Phosphorus, trinitrotoluene
					Fire and waterproofing additive formulators	Chloronaphthalenes
					Plastics formulators	Methylenedianiline
					Fumigators, gasoline and fire extinguisher formulators	Ethylene dibromide
					Disinfectant, fumigant, synthetic resin formulators	Cresol
584, 585	Acute or Chronic Renal Failure (O)	P	P,T	P,T	Battery makers, plumbers, solderers	Inorganic lead
					Electrolytic processes, arsenical ore smelting	Arsine
					Battery makers, jewelers, dentists	Inorganic mercury
					Fluorocarbon formulators, fire extinguisher makers	Carbon tetrachloride
					Antifreeze manufacture	Ethylene glycol
					Chromate pigment production workers	Inorganic lead
606	Infertility, Male (O)	P	P	—	Kepone formulators	Kepone
					DBCP producers, formulators, and applicators	Dibromochloropropane (DBCP)

ICD-9	Condition	A[a]	B[b]	C[c]	Industry/ Process/Occupation	Agent
692	Contact and Allergic Dermatitis (O)	P,T	P,T	—	Leather tanning, poultry dressing plants, fish packing, adhesives and sealants industry, boat building and repair	Irritants (eg, cutting oils, phenol, solvents, acids, alkalis, detergents); Allergens (eg, nickel, chromates, formaldehyde, dyes, rubber products)
733.9 M[m]	Skeletal Fluorosis (O)	P	P	—	Cryolite workers (grinding room)	Cryolite (Na_3AlF_6)
					Cryolite refining workers	Cryolite (Na_4AlF_6)

[a]A = Unnecessary disease
[b]B = Unnecessary disability
[c]C = Unnecessary untimely death
[d]P = Can be controlled by prevention.
[e]T = Can be controlled by treatment.
[f]Original ICD rubric = Cutaneous Diseases Due to Other Mycobacteria
[g]From the *International Classification of Diseases,* 9th Revision, Clinical Modification (ICD[9[CM)
[h]M = Modified ICD rubric.
[i]Original ICD rubric = Malignant Neoplasm of Liver and Intrahepatic Bile Ducts
[j]Original ICD rubric = Mononeuritis of Upper Limb and Mononeuritis Multiplex
[k]Original ICD rubric = Pneumoconiosis Due to Other Silica or Silicates
[l]Original ICD rubric = Pneumoconiosis Due to Other Inorganic Dust
[m]Original ICD rubric = Other Disorders of Bone and Cartilage

[n](O) denotes that further industry or occupation information is needed to establish the relationship of disease to occupation (ie, the disease condition is not inherently occupational).

External causes of injury and poisoning (occupational), including accidents, are classified in the ICD-9 under the E codes.

Guide to Diagnosis of Respiratory Disease

Based on frequency of occurrence, health consequences for affected individuals, and the potential for prevention, occupational lung diseases are a high priority for attention. NIOSH places them first on its list of the 10 leading work-related diseases and injuries. Over two thirds of the 600 most commonly used chemicals and aerosols with regulated exposure limits have adverse pulmonary effects.

Although the toxins with potential pulmonary effects are diverse and their effects at the molecular and cellular levels are complex, the commonly measured responses of the lungs to insult are few. Questionnaires; chest radiographs; lung function tests, including measures of forced airflow from the lungs, lung volumes, and blood-gas determinations; and physical examinations are frequently used in recognizing and quantifying pulmonary disease. Their use in recognizing nonmalignant occupational lung diseases is discussed below.

Questionnaires

People suffering from lung diseases may cough, wheeze, or experience shortness of breath whether or not they have significant abnormalities detectable by laboratory testing. Thus, asking groups of people questions about respiratory symptoms in a systematic fashion and then analyzing the results has traditionally been one of the most powerful tools of occupational lung disease investigation. Comparing questionnaire responses of workers exposed to suspected hazards with those of workers not exposed can confirm that a problem exists. Questioning groups of workers with different levels of duration of exposure to suspected hazards can confirm that problems relate to the intensity or duration of exposure and can thus provide information about levels of exposure that may *not* be harmful.

Systematic questioning is aided by standardized questionnaires that have been used and studied in a variety of settings over many years. Questionnaires published by a committee organized by the American Thoracic Society (ATS),[1] as well as those developed by the British Medical Research Council[2] and NIOSH[3] have been used extensively in occupational lung disease investigations. Each presents a series of questions concerning respiratory symptoms and tobacco use. The ATS and NIOSH questionnaires also inquire into some hazardous workplace exposures. Each questionnaire has also been modified by investigators at different times to meet the needs of particular studies. If data are collected in a standardized fashion, the practice of using identical questions for different groups can magnify the value of the information derived from any one group by permitting comparison with other groups. Recording and reporting such data in the public health and scientific literature is encouraged.

Questionnaires can provide significant information quickly and inexpensively. However, the quality of the information depends on the ability of the questionnaire to detect abnormality (sensitivity), to confirm normality correctly (specificity), to elicit the same responses over time (consistency), and to avoid poor wording or poor construction, each of which can lead to underreporting or overreporting of abnormality (bias).

Chest Radiography (X-rays)

Chest radiographs are the standard method for recognizing the pneumoconioses. Some of the pulmonary responses to exposure to dusts such as silica, coal, and asbestos are visible on chest x-ray. These exposures cause dust deposition, inflammation, and scarring, which can result in characteristic patterns of abnormality on radiographs.

A standardized method of x-ray interpretation disseminated by the International Labour Organization (ILO) is often used to diagnose and classify dust diseases of the lungs.[4] By this method, opacities resulting from inflammation, dust deposition, or scarring are classified according to their shape, size, location, and profusion. Profusion is determined by comparing the worker's film with "standard" films distributed by the ILO. There are four major categories of increasing profusion of opacities: 0, 1, 2, and 3. Whichever standard film most closely matches that of the worker determines the major category of profusion. If the film is in a border area between two major categories, both categories are noted, with the category *most like* the film noted first. Thus, for example, a film that shows a higher concentration of opacities than the category 1 standard film but that is more like the 1/1 standard than like the 2/2 standard is classified as 1/2.

Some exposures such as coal and silica can result in the development of opacities greater than 1 cm in diameter. These opacities are categorized as A, B, or C, depending on their size. There are also conventions for classifying pleural abnormalities as well as for noting the appearance of x-ray changes suggestive of certain other diseases.

The ILO system was originally established to achieve consistency in film interpretation during health surveillance or epidemiological investigations. It has been used, also, for compensation determination and clinical evaluation. In the U.S., readers trained in this method of interpretation who pass a competency test administered by NIOSH are designated as "B readers."[5] Despite efforts to achieve standardization in the interpretation of chest x-rays through the use of the ILO system, however, there remains significant interreader and intrareader variability manifested as disagreement on the presence or absence of abnormality. Some of this variability can be reduced through attention to production of consistent, high-quality radiographs using appropriate techniques.

Radiographs may not show early changes resulting from exposure to dusts. For example, nearly 20% of asbestos workers with fibrotic changes on pathological examination do not have detectable radiographic abnormalities.[6] Physicians untrained or inexperienced in recognizing occupational lung disease may miss certain abnormalities, which are often quite subtle. Nevertheless, particularly at higher profusions, there is a reasonable correlation between lung pathology findings and radiographic interpretations in miners with coal workers' pneumoconiosis.

Chest radiographs are not an adequate tool for surveillance or diagnosis of all occupational lung diseases. Respiratory diseases such as occupational asthma or bronchitis are not generally detectable on chest x-ray; emphysema is accurately recognized only in advanced stages. Other methods of investigation are needed for these disorders, as well as for identifying functional changes associated with the development of the pneumoconioses.

Measures of Lung Function

The most widely available test of lung function records both the volume of air a subject exhales forcibly in 1 second following maximal inhalation and the total volume exhaled without a time limit. These values are compared with population averages for people of the same gender, height, and age, and they are reported both as absolute values and as a percent of predicted values. The test is spirometry; the tracing that relates exhaled volume to time is a spirogram; and the device used for measurement and recording is a spirometer. The American Thoracic Society has recommended guidelines for performance of spirometry and for the equipment that should be used in the industrial or diagnostic settings.[7] People with significant lung disease may have more difficulty than others in performing consistently on spirometry testing.[8]

The most common values reported are the forced expiratory volume in 1 second (FEV_1), the forced vital capacity

(FVC), and the ratio between these two values. People with diseases causing limitation of airflow out of the lungs have lowered FEV_1 and are said to demonstrate obstructive airways disease. Those with scarring of the lungs sufficient to diminish the lungs' ability to expand during inspiration have reduced FVC and are said to have restrictive lung disease. Often, people with advanced lung disease have reduced values for both the FEV_1 and FVC. The ratio of one to the other, compared with average values, can suggest whether obstruction or restriction predominates. Significant obstructive airways disease can result in an inability to empty the lungs fully, known as air trapping. This may give the false impression that a restrictive defect is present, because the FVC will be diminished Other tests may determine if the reduced FVC is due to air trapping or to true restriction.

There is no single pattern of abnormality in spirometry that will distinguish occupational from nonoccupational lung diseases. Spirometry alone cannot diagnose pneumoconioses or other occupational lung diseases, but it is useful in quantifying abnormality and defining a pattern of response in exposed workers. Spirometry is particularly useful in comparing groups of workers who are exposed to a suspected hazard with those who are not. In this instance, most individuals may fall within a normal range, but the average lung function of the exposed workers may differ significantly from the average of the unexposed.

Interpretation of spirometry results varies among institutions. Multiple "prediction equations" have been derived from examination of various populations, but most equations have been based on few subjects.

The results of population surveys among nonwhites have rarely been reported, and a nondiscriminatory way to interpret spirometry results obtained from nonwhites remains subject to varying opinions. On average, black Americans have lower vital capacities and FEV_1 than whites of the same gender, height, and age. However, this does not appear to be a fixed ratio at all ages and heights. Americans of Hispanic and Asian backgrounds also appear to have different population averages than whites. Application of normal values not derived from appropriate reference populations could result in significant misclassification. Clearly, prediction equations are needed that are based on either nonwhite race or some alternative characteristic (such as sitting height), which may be predictive of FEV_1 and FVC and be independent of racial background.

There is no clear or absolute separation between normality and abnormality in lung function testing. Population studies have indicated that many people who seem to have pulmonary disease will have spirometry values within the so-called normal range. Others without disease may fall in the abnormal range. How many fall on one side or the other depends on where the line separating normal from abnormal is drawn. People engaged in gainful employment are, on average, healthier than the average population. Thus, it is not unusual for working people to score above average (above 100% of predicted) on spirometry testing.

Some workers may experience abnormal symptoms, such as shortness of breath on exertion, or may recognize that their work capacity is diminishing while their spirometry remains within the normal range. Measurement of an individual's lung function over time and comparison of current values with those obtained previously might indicate a greater rate of loss than predicted by aging alone. Currently, there is no consensus on how to interpret such data obtained by comparison of an individual with his or her younger self rather than with a reference population, although this may hold promise if interpretive strategies are validated in the future. The normal variability in lung function testing over time for both healthy and unhealthy people complicates interpretation of longitudinal data. While aggregate data for groups of workers may show trends over time (diminishing function with exposure to pulmonary hazards), longitudinal changes in any individual must be interpreted with caution.

Spirometry with medication administration is used to detect abnormalities associated with asthma. A reduction in FEV_1 following administration of a pharmacological bronchoconstrictor (eg, methacholine or histamine) can demonstrate hyperresponsive airways. Improvement in FEV_1 following administration of a bronchodilator demonstrates a component of "reversibility" of airways abnormality and may indicate the presence of asthma. Occasionally, workers are exposed to specific workplace substances under controlled laboratory conditions to determine whether these substances precipitate asthmatic reactions. This diagnostic challenge testing is not a routine practice and should only be performed in specialized centers with personnel experienced in recognizing and treating complications that can arise from such procedures.

Peak flow meters, portable devices that estimate the rate of airflow from the lungs during forced exhalation, have also been useful in the recognition of asthmatic responses to workplace hazards. Peak flow measurement has been used at the worksite to demonstrate variation in flow rates during a work shift, a finding suggestive of occupational asthma.

Other tests of lung function in common use include arterial blood-gas testing (at rest and during exercise) and determination of the lung's diffusion capacity. Because the lung functions to pass oxygen from outside the body into the blood stream (where it is transported to body organs to support metabolic activities) and to eliminate waste gas (carbon dioxide), measuring oxygen and carbon dioxide in the arterial blood provides information about how effectively the respiratory system is performing. People with extremely disordered lung function may show abnormalities at rest. Others with milder abnormalities may only manifest those abnormalities during exercise, when the demand for oxygen goes up and the lungs may have more difficulty meeting those needs. Age, altitude, and nonpulmonary diseases can interfere with the interpretation of test results.

The diffusion capacity of the lung is an estimate of the quantity of oxygen the lung is able to pass from the air exchange units (alveoli) into the blood stream in a given period of time. Although oxygen is the gas of interest, the test generally measures the diffusion capacity for carbon monoxide, which is felt to reflect the diffusion capacity of oxygen and is technically feasible to measure. The test is much more technically difficult and less reproducible than spirometry, but it may demonstrate true abnormality in individuals with diffuse lung diseases where the airways are unobstructed and the abnormality is not so advanced as to cause identifiable restriction.

Tests of pulmonary function are widely but inconsistently used in determining disability. Different programs offering compensation or benefits to impaired workers use different procedures, tests, and reference populations. The particular rules and regulations governing any program of interest should be consulted for relevant information.

Physical Examination

The physical examination is an important component of individual health care services, but it tends to be insensitive and nonspecific in identifying and characterizing occupational lung diseases, particularly in their early stages. Health care providers are trained to examine the respiratory system by a process of visual observation (inspection), touching (palpation), tapping (percussion), and listening with a stethoscope (auscultation). Abnormalities detectable by inspection, palpation, and percussion tend to be either acute, such as an asthma attack or lung infection, or chronic and fairly advanced, such as the barrel chest of a patient with emphysema.

Auscultation can reveal abnormal lung sounds that are associated with lung dysfunction. For example, wheezing, present in people with an asthma attack, may indicate the partial obstruction, narrowing, or collapse of large airways. Persistent fine crackles, called rales, may result from fibrosis of the lung parenchyma. These sounds are heard at times in people with asbestosis. For many people with lung abnormalities, however, doctors are unable to discern any changes in breathing patterns or sounds on physical examination. Also, abnormalities noted by one observer at one time may be unrecognized by another observer or at other times.

Strategies for Testing

No single test or group of tests is adequate for the recognition of all occupational lung diseases. Different methods of examination are sensitive to different processes and provide different information about the lungs.

Appropriate selection of approach and tests is critical in the effective pursuit of prevention. Part 1 of this book includes a general discussion of screening, surveillance, and biological monitoring. The pulmonary tests described in this section can be useful in those activities, as well as in diagnosing the health problems of individual workers and in quantifying impairment.

Health surveillance for groups of workers exposed to known or suspected pulmonary hazards should include appropriate use of the tools described above. For example, questionnaires may be useful in screening workers exposed to hazards known to cause asthma. Workers reporting chest tightness, cough, wheezing, or shortness of breath in a pattern that suggests an occupational cause should have a diagnostic evaluation, and their work environments should be explored for a breakdown in proper hazard controls.

Test selection should be based on knowledge of the likely consequences of unintended exposure to excessive levels of the hazardous substance. For example, chest radiographs would be an inappropriate surveillance tool for detecting occupational asthma in isocyanate-exposed workers. In contrast, periodic x-rays might be useful for following foundry workers or sandblasters exposed to silica, because abnormalities on the film may precede significant symptoms or impairment.

Health surveillance should never substitute for environmental monitoring. Environmental controls should be implemented and maintained on the basis of current standards and scientific knowledge, even in the absence of disease demonstrable on tests such as those described here. If testing is performed as part of a carefully designed and analyzed study, it may be useful for investigating suspected pulmonary hazards for which standards either do not exist or may be insufficiently protective. It may also be useful for identifying individuals for whom current or past environmental controls have been inadequate in order to identify abnormalities at as early a stage as possible and to prevent the progression of disease.

GRW

References

1. Ferris BG. Epidemiology standardization project, II: recommended respiratory disease questionnaires for use with adults and children in epidemiologic studies. *Am Rev Respir Dis*. 1978;118(pt 2, no 6):7.

2. Medical Research Council. Standardized questionnaire on respiratory symptoms. *Br Med J*. 1960;2:1665. (For information, write Publications Group, Medical Research Council, 20 Park Crescent, London W1N 4AL, England.)

3. For information, contact the NIOSH Division of Respiratory Disease Studies, Epidemiology Branch, 944 Chestnut Ridge Road, Morgantown, WV 26505.

4. International Labour Organization. *1980 International Classification of Radiographs of the Pneumoconioses*. Geneva, Switzerland: International Labour Office Technical Publication, *See also* Morgan, RH. Radiology. In: Merchant JA, ed. *Occupational Respiratory Diseases*. Washington, DC: US Government Printing Office; 1986:137-153. DHHS (NIOSH) publication 86-102.

5. For information about this program, contact NIOSH/DRDS Receiving Center, 944 Chestnut Ridge Road, Morgantown, WV 26505.
6. Kipen HM, Lilis R, Suzuki Y, et al. Pulmonary fibrosis in asbestos insulation workers with lung cancer: a radiological and histopathological evaluation. *Br J Ind Med.* 1987;44:96-100.
7. American Thoracic Society. Standardization of spirometry: 1987 update. *Am Rev Rspir Dis.* 1987;136:1285-1298.
8. Eisen EA. Standardizing spirometry: problems and prospects. In: Rosenstock L, ed. *Occupational Pulmonary Disease. State of the Art Reviews in Occupational Medicine.* 1987;2:213-225.

An Overview of Occupational Musculoskeletal Disorders

Musculoskeletal disorders refer to the many problems that affect the muscles, tendons, ligaments, and joints of the musculoskeletal system, as well as the associated nerves that control and regulate muscular movement (the peripheral sensory and motor nerves). Work-related musculoskeletal disorders have been known to affect virtually every part of the body: back, hips, legs, knees, hands, wrists, elbows, arms, shoulders, and neck. Back and upper extremity disorders, the most frequent of these disorders, are covered in detail below.

The terminology used for these disorders is not well defined. Many scientists prefer to discuss the entire group of disorders using general terms, such as soft tissue disorders, cumulative trauma disorders, repetitive strain injuries, repetitive motion disorders, occupational cervicobrachial disorders, or overuse syndrome. This practice has the disadvantage of being unclear as to the specific diagnoses being included. It does, however, reflect the fact that several different clinical disorders often occur in workers with the same occupational ergonomic exposures, and that some of these workers experience severe fatigue or pain that may not conform to any specific recognized pattern. Furthermore, the criteria for most specific clinical diagnoses are not standardized and are rarely consistent from one researcher or physician to another; many hard-to-diagnose cases may be lumped together under a vague diagnosis such as arthritis.

Diseases in this group have several features in common, including a set of generic risk factors. Typically, both onset and duration are chronic or subchronic rather than acute. (One exception is tendinitis associated with a sudden change in motion patterns or work pace.) As with many chronic diseases, multiple factors usually contribute to the development of a case. Therefore, it is often useful, for both epidemiological and preventive purposes, to consider these problems as a group rather than individually. Nevertheless, diagnosis of an individual's disorder, if possible, is worthwhile, because it facilitates choice of the most appropriate treatment and job redesign measures.

Identification

The most common upper extremity disorders that have been linked to risk factors in the workplace are those affecting the muscle-tendon systems and those affecting the peripheral nerves. The most important of these disorders are: carpal tunnel syndrome, hand-arm vibration syndrome, tendinitis (including tendosynovitis and other related conditions), and peripheral nerve entrapment syndromes, such as thoracic outlet, pronator teres, cubital tunnel, and Guyon's canal syndromes. Other conditions that may occur due to occupational causes include bursitis, epicondylitis, trigger finger, Dupuytren's contracture, and degenerative joint disease. (The first three of these are mentioned in the entry on tendinitis.)

Precise diagnosis of these disorders may be difficult. Usually it requires a careful description of symptoms by the affected worker, as well as a physical examination and sometimes a laboratory test. The location of the pain and the type of signs and symptoms (eg, numbness, tingling, swelling, aching, burning pain) can usually help differentiate nerve compressions from other diseases, although tests of nerve impulse conduction or sensory nerve ability may also be used. Muscle-tendon disorders can often be distinguished from joint problems on physical examination by comparing discomfort and limitation of movement on active, passive, and resisted movements.[1,2] However, many individuals with work-related musculoskeletal disorders may have nonspecific symptom patterns, no signs, or negative findings on "objective" tests.

For back disorders, whether work related or not, it is often difficult to identify the particular tissue or structure that has been damaged. In many case series, the mechanism of disease cannot be determined for up to half of the cases. Thus, an individual's pain, limitation of movement, and disability (inability to perform daily tasks at home or at work) are often used as criteria to determine the existence and severity of a condition.

Epidemiology

Musculoskeletal disorders are generally underreported in the workplace; therefore, most available prevalence and incidence data probably underestimate their true magnitude. Difficulties and inconsistencies in diagnosis compound this problem. However, occupational and nonoccupational musculoskeletal disorders are widespread, and although they are not directly fatal, these disorders account for much disability, pain, and suffering. In the United States, more people are disabled from working as a result of musculoskeletal disorders than from any other group of diseases.[1]

Considering work-related cases alone, the U.S. Bureau of Labor Statistics (BLS) has estimated that the incidence of all "disorders associated with repeated trauma" in the private sector, excluding firms with fewer than 11 employees, is 10.0 per 10 000 full-time workers per year.[3] However, many industries and occupations have rates that are much higher. Especially hazardous industries include meat, poultry, and fish processing; leather tanning; heavy manufacturing, including manufacturing of furniture, aircraft, motor vehicles, appliances, and electrical equipment; and light manufacturing, including manufacturing of electronic products, shoes, textiles, apparel, and upholstery. Musculoskeletal disorders have also been found to be highly prevalent in clerical work, postal service, janitorial work, industrial inspection and packaging, various types of construction work, and numerous other industrial and service occupations. These jobs involve labor-intensive processes, although the work does not always require great muscular strength or fast, repetitive motions.

Back disorders often occur in occupations that require frequent heavy lifting, especially when combined with other stressors, such as prolonged sitting and exposure to whole-body vibration. At high risk are truck drivers, operators of cranes and other large vehicles, warehouse workers, nurses, nursing aides, and other patient-care workers.

The rates of musculoskeletal disorders reported to BLS have been increasing in recent years, although it is not clear to what extent this reflects a true increase in occurrence as compared with improved diagnosis and reporting of these disorders.

Prevention

Primary prevention should focus on engineering controls, especially the ergonomic design of workstations, equipment, tools, and work organization to fit the size, speed, and strength capabilities of the human body. The goal is to reduce or eliminate six generic risk factors for occupational musculoskeletal disorders: (a) rapid or repetitive movements, (b) forceful exertions, (c) mechanical force concentrations, (d) awkward or "nonneutral" postures, (e) vibration, and (f) local or environmental cold.[4] In high-risk jobs, often two or more of these factors are present and synergistically exert adverse effects. Control measures, to be effective, usually need to address all existing exposures.

1. Effects of rapid and repetitive work may be mitigated by slowing the overall work pace, providing longer and more frequent rest breaks, avoiding piece-rate wages in favor of hourly rates, and permitting operators to set their own work pace. Piece-rate wages and machine-paced work may be particular problems, because they motivate workers to maintain a constant

high speed, even when fatigued or in pain. Administrative controls, such as job rotation or work enlargement (adding more tasks to the job description), may reduce the duration of exposure or increase the variability of motions performed. However, careful ergonomic job analyses are required to determine whether there is a true potential for greater variety in motion patterns by combining or alternating among various tasks. Performing a larger number of tasks that are all ergonomically stressful is not an effective control strategy and should not be allowed to divert attention from ergonomic redesign efforts.

2. The force required to perform a job may be reduced by lighter loads or less frequent lifting, improved friction between hand and tool or part, and better quality control on parts. Lifting tasks should be free from any obstruction that forces a worker to lift over or around it or to jerk the load. Slip clutches reduce the torque produced by powered hand tools. Tools may be installed on overhead balancers or articulating arms mounted on the workbench; this decreases both the weight that must be supported and the torque transmitted to the arm. Parts, tools, and equipment should be designed so that awkward body postures or pinch grips do not have to be used, especially for repeated, prolonged, or forceful motions, because these postures result in a biomechanical disadvantage—a power grasp that increases the force necessary to perform a task. Regularly scheduled, frequent maintenance of tools and equipment, including sharpening of knives or scissors, avoids unnecessary mechanical resistance. Gloves (to protect the hand from abrasion, mechanical stress, cold, vibration, or chemicals) should be selected with caution to avoid increasing the force required to grip tools and parts; such gloves

should be available in a wide range of sizes to fit the entire working population.

3. Mechanical force or stress concentrations refer to highly concentrated forces applied by physical contact between any object in the workplace and the skin, with pressure on the underlying tissues, such as blood vessels and tendons. These concentrations may be reduced by rounding and padding edges and corners (such as on tool handles, table edges, and chair seats) that come into contact with the body.

4. Work postures may be kept close to anatomically neutral by avoiding work locations that are above shoulder height, farther forward than arm's length, or behind the midline of the body. Eliminating the need for twisting motions, and designing and selecting tools that permit work to be performed with the wrist straight and the rest of the body in a comfortable position, will also contribute to keeping work postures neutral. Providing adequate illumination and locating work close to eye height will eliminate the need for bending and twisting the neck. Prolonged static muscular constructions may be minimized by suspending tools overhead and providing seating wherever possible. Chairs should have lower-back support and adequate seat padding, and they should be designed and selected for compatibility with the specific task being performed. In general, equipment and furniture should be adjustable in height and other relevant dimensions so as to conform to the range of body sizes of workers.

5. Transmission of vibration to the hand and arm or to the spine may be reduced by engineering controls, such as mechanical isolation and damping in tool handles or vehicle seats. Alternatively, the duration of exposure may be reduced by administrative measures, such as rotation or rest breaks.

6. Exposing the hand to local cold may be avoided by insulating air-powered tools and directing exhaust air away from the tool handle. Also, parts stored in an unheated warehouse should be moved into the work area early enough so that they warm before use. Workers in cold environments need adequate insulating clothing and rest breaks in warm locations whenever possible.

Certain administrative controls have been proposed for primary prevention purposes, especially workplace exercise and stretching programs and medical screening of prospective employees to eliminate those with predisposing factors. Unfortunately, neither of these approaches has been shown to be effective for prevention. Exercise and stretching can logically be expected to increase an individual's strength, flexibility, and tolerance of the physical demands of work. However, only limited formal evaluations of these programs have been undertaken to date, and no benefits have been found for an entire working population either in prevention of new problems or in improvement of preexisting conditions. (In fact, in some cases, exercise might aggravate or worsen a problem once it has begun, because inflamed tendons, strained muscles, and compressed nerves require rest and elimination of the cause for recovery to occur.)

There is also no evidence that medical evaluation of prospective workers to eliminate those at high risk reduces future occurrences, because no individual factors have been shown conclusively to predict which workers will or will not develop musculoskeletal disorders after exposure to ergonomically stressful working conditions. (The only exception is that, for low back pain, it might be plausible to screen for a history of previous conditions, since a worker with prior low back pain is more likely to experience a recurrence. However, up to half of all people experience low back pain at some time, usually with complete recovery and without further disability or apparent risk of recurrence.)

Secondary preventive measures include reducing obstacles that (a) prevent or deter workers with symptoms from reporting them to their physicians or the workplace health service, (b) prevent affected workers from being allowed adequate recuperation time away from work, and (c) deter reassignment of affected workers to "light-duty" work. Initially, symptoms may disappear with cessation of work, only to recur after the employee resumes work. But if the injury gets worse, pain may continue even when work ceases. The longer a person works with pain on the same job, the slower the recovery and the greater the risk of permanent disability.

Health care providers should be trained in the appropriate interview and clinical examination procedures to diagnose occupational musculoskeletal disorders. Exposed workers and their supervisors should be informed of the symptoms and signs associated with these syndromes. Because not all workers with symptoms will be likely to seek medical attention promptly, active medical surveillance (interviews and physical examinations of all workers) should be carried out to identify both those in need of treatment and the jobs or areas where ergonomic redesign should be undertaken. Once reported, cases should be treated conservatively, and jobs should be analyzed for ergonomic features that may be modified.

The best treatment for cumulative trauma disorders is rest for the affected tendon, joint, or nerve; this includes rest from work, all household duties, and other manual activities. Rest is sometimes combined with local heat or cold compresses for temporary pain relief, anti-inflammatory medications, physical or occupational therapy, or use of splints. However, none of these treatments is likely to be effective if the individual continues to work without appropriate ergonomic modification of the workstation, tools, equipment, and/or work pace. Splint use on the job should be considered only if it does not interfere with work or require the worker to exert more force or strain on another joint to perform the task. Follow-up is important to ensure that job modifications have been effective, that light-duty jobs have been correctly selected to avoid continuing stress on the affected part, and that symptoms and signs do not progress.

Concerning tertiary measures, surgical operations are available for certain disorders, such as carpal tunnel syndrome and herniated spinal disks. However, surgery, cortisone injections, and use of other drugs are last resorts and should be avoided if at all possible. Surgery may be only temporarily effective if the worker is returned to an ergonomically stressful job that has not been modified; possible loss of strength and flexibility and increased vulnerability to technical insult following surgery make it essential that job assignments be selected carefully to avoid recurrence. Use of transitional workshops for affected workers and graded retraining under the supervision of an experienced physical therapist may also reduce the risk of recurrence.

LP

References

1. Putz-Anderson V, ed. *Cumulative Trauma Disorders: A Manual for Musculoskeletal Diseases of the Upper Limbs.* New York, NY: Taylor & Francis; 1988.

2. Silverstein BA, Fine LJ. *Evaluation of Upper Extremity and Low Back Cumulative Trauma Disorders: A Screening Manual.* Ann Arbor, Mich: University of Michigan; 1984. Technical Report, School of Public Health/Center for Ergonomics. (Available from Center for Ergonomics, 1205 Beal Ave, University of Michigan, Ann Arbor, MI 48109-2117.)

3. Bureau of Labor Statistics. *Annual Survey: 1987.* Washington, DC: US Government Printing Office; 1987.

4. Armstrong TJ. Ergonomics and cumulative trauma disorders. *Hand Clin.* 1986;2:553-565.

Further Reading

Armstrong TJ, Silverstein BA. Upper-extremity pain in the workplace: role of usage in causality. In: Hadler NM, ed. *Clinical Concepts in Regional Musculoskeletal Illness.* New York, NY: Grune & Stratton; 1987.

Cailliet R. *Soft Tissue Pain and Disability.* Philadelphia, Pa: FA Davis; 1983.

Wallace M, Buckle P. Ergonomic aspects of neck and upper limb disorders. *Intl Rev Ergonomics.* 1987;1:173-200.

SEE Carpal Tunnel Syndrome
Hand-Arm Vibration Syndrome
Low Back Pain Syndrome
Peripheral Nerve Entrapment Syndromes
Peripheral Neuropathy
Tendinitis, Tenosynovitis

An Overview of Occupational Cancer

Identification

Cancer includes a variety of diseases characterized by abnormal, unregulated cell growth. Nearly every organ of the body, and nearly every cell type, may be a cancer site. Collections of cancer cells are called tumors; these may be malignant, with potential for rapid growth and metastasis to distant sites, or benign, with low potential for spread. Each type of cancer has distinct epidemiological and clinical features.

Some cancers may be identified through screening before they cause noticeable signs or symptoms; examples include cervical cancer, which can be detected with Pap smears, and breast cancer, which can be detected with physical examination and/or mammography. Cancers may also be detected when they cause symptoms that prompt a patient to seek medical attention. Presenting symptoms vary with the type of cancer: lung cancer may present with cough, shortness of breath, or weight loss, whereas colon cancer may present with rectal bleeding, change in bowel habits, or weight loss. Laboratory diagnosis of cancer also varies with the type of cancer; most widely used are radiological techniques including plain x-rays, contrast studies, ultrasound studies, computed tomography (CT) scans, and magnetic resonance imaging (MRI). The most definitive means of detection is tissue diagnosis, by biopsy or at autopsy.

Occurrence

About 1 million Americans are diagnosed annually with some form of cancer, and more than 250 000 die of cancer each year. Cancer occurrence varies by age, sex, geographic location, and other factors. Each type of cancer has a distinct pattern of occurrence, but several general statements can be made:

1. *Incidence and mortality:* The overall cancer incidence in the United States is about 400 cases per 100 000 people per year. Incidence is highest for lung (62 per 100 000), colorectal (60 per 100 000), and breast cancer (55 per 100 000). Cancers that are rarer but of occupational interest include bladder cancer (19 per 100 000), leukemia (11 per 100 000), brain cancer (6 per 100 000), and mesothelioma (1 to 2 per 100 000). Many cancers are difficult to treat and prove fatal, so mortality rates for those cancers are approximately equal to incidence rates. For treatable cancers, such as skin cancer, lymphoma, and testicular cancer, mortality may be only a fraction of incidence.

2. *Gender:* Some cancers demonstrate a gender differential. For example, the incidence of pancreatic cancer is 50% higher in men than in women, and kidney cancer occurs about twice as often in men as in women.

3. *Age:* With the exception of childhood cancers, most types of cancer increase in incidence exponentially from the third decade to the eighth decade of life.

4. *Geography:* There are marked international and interregional variations in cancer rates. The incidence of stomach cancer, for example, is 91 per 100 000 in Japan, 43 per 100 000 in Iceland, 21 per 100 000 in Israel, and 7 per 100 000 in Nigeria. Although this variability may have a genetic component, other evidence, such as that from migrant studies, provides strong support for the concept of environmental causes of cancer (primarily dietary factors in the case of stomach cancer).

5. *Race:* Racial differences are noted in cancer incidence within nations. For example, the incidence of lung cancer in the U.S. is 110 per 100 000 black men, 75 per 100 000 white men,

27 per 100 000 Hispanic men, and 11 per 100 000 Native American men. These differences may also have a genetic component, but environmental differences, including occupational exposures, play a part.

6. *Social class:* Social class has a greater impact on survival than on incidence, but incidence does vary by class. For example, after adjustment for age and geographic location, the breast cancer incidence among white American women ranges from 69 per 100 000 in the lowest income category to 99 per 100 000 in the highest, and from 67 per 100 000 in the lowest education category to 103 per 100 000 in the highest.

Causes

A variety of factors has been implicated, some of which are noted below:

1. *Genetics:* Certain individuals have a propensity to develop cancer because of race (eg, whites are more susceptible to sunlight-induced skin cancer than blacks), genetic disease (eg, xeroderma pigmentosum), family tendency, or other genetic factors.

2. *Tobacco:* Tobacco use increases the incidence of cancers of the lung, pharynx, larynx, esophagus, bladder, and other organs.

3. *Diet:* Certain foods, food additives, and food contaminants, both natural and artificial, are carcinogenic.

4. *Occupation:* Various occupational exposures are established carcinogens: asbestos (lung cancer, mesothelioma, some gastrointestinal cancers), benzene (leukemia), and cadmium (prostate cancer). Certain industrial processes are associated with elevated cancer risks, some without identified specific carcinogenic agents; examples include bladder and lung cancer in the rubber industry. Major examples of occupational carcinogens appear in Table 1 at the end of this section.

5. *Ionizing and nonionizing radiation:* Examples include UV light exposure, which contributes to the incidence of basal cell carcinoma and melanoma; radon exposure, which causes lung cancer; and gamma radiation and neutrons, which increased leukemia, lymphoma, and thyroid and breast cancer among atom bomb survivors.

Pathophysiology

The mechanisms of carcinogenesis are incompletely understood. The process is thought to proceed through several sequential steps: initiation, promotion, and progression. Initiation involves unrepaired alterations of cellular DNA, or mutations, followed by "fixing" of those changes through cell replication. The DNA changes may involve alterations of normal genes, called proto-oncogenes, into genes that mediate malignant conversion, called oncogenes. One kind of carcinogen is therefore an initiator, which acts by altering DNA. Promotion, by which initiated cells develop into focal proliferations, may take years. It is stimulated by another kind of carcinogen, called a promoter. Promoters act through epigenetic mechanisms, such as increasing macromolecule synthesis. Progression may be a series of steps leading to the development of cancer. This part of the process is poorly understood.

An important consideration in carcinogenesis is latency, or the elapsed time between the start of a carcinogenic exposure and the appearance of a resultant cancer. The several steps mentioned above account for latency, which is in the range of several years for hematologic neoplasms and may reach several decades for solid tumors. Latency must be considered in epidemiological surveillance of workers who have been exposed to carcinogens.

Prevention

Primary prevention: Primary prevention is accomplished by eliminating exposure to carcinogens, changing a process to obviate the need for using a carcinogen, substituting noncarcinogenic chemicals for carcinogens, or enclosing carcinogens to avoid worker exposure, among other methods.

Secondary prevention: Early detection may be accomplished by cancer screening. Examples of screening in the occupational setting include urine cytology to screen for bladder cancer in benzidine dye workers and regular chest x-rays and sputum cytology to screen for lung cancer in workers at risk; both strategies are controversial. Early detection may also be achieved by educating people to recognize "warning signs" of cancer and to seek medical evaluation promptly.

Other Issues

Identification of carcinogens: Four kinds of complementary evidence have been used to identify carcinogens.

1. *Human Evidence:* The most definitive evidence comes from studies of human populations. Clinical observations may identify some causes; vinyl chloride as a cause of hepatic angiosarcoma and bis-chloromethyl ether (BCME) as a cause of lung cancer were identified this way. Cohort studies follow groups of exposed workers to determine cancer incidence and mortality, while case-control studies compare cancer patients with controls to identify high rates of exposure among the patients. Cohort studies have demonstrated high rates of lung cancer among asbestos insulators, bladder cancer among benzidine dye workers, and nasal cavity cancer among woodworkers. Case-control studies have demonstrated high rates of laryngeal cancer among asbestos workers, leukemia among solvent-exposed workers, and soft-tissue sarcomas among those exposed to phenoxyherbicides. Occupational cancer epidemiology is hampered by the long latency between exposure and disease onset, variations over time in work forces and work processes, the few registries of workers, the dearth of precise exposure information, the multicausal nature of many cancers, and the multiple exposures many workers have.

2. *Animal evidence:* Animal studies, usually with rodents, are often used to test for carcinogenicity. These are less expensive and more rapid than epidemiological studies, and they do not depend on human exposures. However, extrapolation between species, the need to use large doses, and differences in target organs make interpretation difficult. Nonetheless, an animal carcinogen is generally assumed to be a potential human carcinogen.

3. *In vitro evidence:* Cell systems, using both bacterial and mammalian cells, are used to test for mutagenicity. If exposure to a substance causes mutations, then the substance is assumed to have the same potential for human DNA and to be a potential carcinogen. In vitro results are less conclusive than human or animal evidence, but they are useful screening tests and help to identify mechanisms.

4. *Structure-activity relationships:* Chemicals that are structurally similar to known carcinogens may themselves have carcinogenic potential and may merit further attention and testing.

Categories of carcinogens: Based on available evidence, regulatory and advisory bodies have classified the carcinogenicity of various exposures. The major schemes are those of the International Agency for Research on Cancer (IARC), NIOSH, the American Conference of Governmental Industrial Hygienists (ACGIH), and the National Toxicology Program (NTP). Some of these results appear in Table 1.

Reported exposure to carcinogens: When individual or group exposure to a carcinogen is recognized, several responses are appropriate. The exposure should be abated through the primary prevention measures described elsewhere. The exposed individuals should be counseled to terminate their exposure, to use personal protective equipment if necessary, and to avoid concomitant carcinogenic exposures, such as cigarette smoke. Early medical evaluation is rarely helpful but may be reassuring to those exposed; follow-up medical screening may be appropriate.

The question of causation: Occupational causation may be difficult to prove in an individual case of cancer, even with a definite exposure history. This is because most cancers have many causes. Before causation can be asserted in an individual case, a history of exposure must be established, the exposure must be determined to have preceded disease onset by an adequate interval, and evidence must support the association of exposure and disease. If these conditions are met, some would argue that causation is established. On the other hand, a relative risk above a certain level might be required. For example, with a relative risk above 2.0, an individual case may be viewed as "more likely than not" to have been caused by the exposure. Another approach is to attempt to allocate causal "shares" among various antecedents, based on the risk associated with each.

Cancer clusters: Suspected cancer clusters should be approached and investigated openly, methodically, and expeditiously. Traditional epidemiological investigations of infectious disease outbreaks provide a suitable model: confirm the existence of a cluster, verify the tissue types and latency periods of cases, assemble occupational histories, identify common exposures through an industrial hygiene evaluation, and review the literature for evidence that an exposure may be carcinogenic. Unfortunately, such investigations are rarely rewarding.

No safe threshold: For carcinogens, there is no evidence for a safe threshold below which exposure of people represents no risk of resultant cancer. This situation and its implications for prevention of cancer are subjects of much discussion.

HF

Further Reading

Alderson M. *Occupational Cancer*. London, England: Butterworths; 1986.

American Conference of Governmental Industrial Hygienists. *Threshold Limit Values for Chemical Substances and Physical Agents in the Work Environment.* Cincinnati, Ohio: American Conference of Governmental Industrial Hygienists; 1986.

DeVita VT, Hellman S, Rosenberg S, eds. *Cancer: Principles and Practice of Oncology.* 2nd ed. Philadelphia, Pa: JB Lippincott; 1985.

Holland JS, Frei E, eds. *Cancer Medicine.* 2nd ed. Philadelphia, Pa: Lea & Febiger; 1982.

International Agency for Research on Cancer. *Overall Evaluations of Carcinogenicity: An Updating of IARC Monographs Volumes 1 to 42.* Lyons, France: International Agency for Research on Cancer; 1987. IARC Monographs on the Evaluation of Carcinogenic Risks to Humans Supplement 7.

International Labour Organization. *Occupational Cancer Prevention and Control.* 2nd ed. Geneva, Switzerland: International Labour Office; 1988. Occupational Safety and Health Series No. 39.

National Toxicology Program. *Fourth Annual Report on Carcinogens.* Washington, DC: Public Health Service, U.S. Department of Health and Human Services; 1985. Document NTP 85-002.

Rothman KJ. Clustering of disease. *Am J Pub Health.* 1987;77:14-15. Editorial.

Schottenfeld D, Fraumeni JF Jr, eds. *Cancer Epidemiology and Prevention.* Philadelphia, Pa: WB Saunders; 1982.

Schulte PA, Ehrenberg RL, Singal M. Investigation of occupational cancer clusters: theory and practice. *Am J Public Health.* 1987;77:52-56.

SEE Bladder Cancer
Brain Cancer
Colorectal Cancer
Esophageal Cancer
Hepatic Angiosarcoma
Laryngeal Cancer
Leukemia
Lung Cancer
Mesothelioma
Nasal Cancer
Pancreatic Cancer
Skin Cancer
Stomach Cancer
Thyroid Cancer

Table 1.—Some Substances and Industries that Pose Occupational Cancer Risk

Carcinogen	Human cancer site	Industry Setting	IARC designation	NTP designation	ACGIH designation
acrylonitrile	?lung, ?colon, ?brain, ?stomach, ?prostate	Acrylic fiber, chemical, and pesticide production	2A	2	A2
4-aminodiphenyl	bladder	Formerly used as rubber anti-oxidant and dye intermediate	1		1
arsenic and arsenic compounds	lung	Smelting, meta-llurgy pigment and glass pro-duction, and pesticide	1	1	—

Carcinogen	Human cancer site	Industry Setting	IARC designation	NTP designation	ACGIH designation
asbestos	lung, pleura, peritoneum, larynx, gastro-intestinal tract	Insulation: ships, buildings, pipes, brake shoes, etc.	1	1	A1
benzene	leukemia ?lymphona	Multiple uses in chemical products: adhesives, rubber, petro-chemicals, etc.	1	1	A2
benzidine & salts	bladder	Plastic, rubber, dye, chemical industries	1	1	A1
beryllium & certain beryllium compounds	?lung	Mining, electronics, chemical, electric, ceramics industries	2A	2	A2

Carcinogen	Human cancer site	Industry Setting	IARC designation	NTP designation	ACGIH designation
bis (chloro-methyl) ether (BCME)	lung	Contaminant of CMME	1	1	A1
chloromethyl methyl ether (CMME)	lung	Ion exchange resin, organic chemical manufacturing	1	1	—
chromium and chromium compounds	lung	Welding, etching, plating; steel and metal industries	1 (hexavalent) 3 (others)	1	A1
coal tars and coal tar pitches; soot	skin, lung ?esophagus ?liver ?leukemia	Petrochemical and steel industries; fossil fuel combustion byproduct	1	1 (**)	A1

Carcinogen	Human cancer site	Industry Setting	IARC designation	NTP designation	ACGIH designation
coke oven emissions	lung, bladder, skin	Steel industry	*	1	—
1,2-dibromo-3-chloropropane (DBCP)	A	Former nematocide	2B	2	—
1,2-dibromo-ethane (EDB)	A	Gasoline additive, solvent, pesticide	2A	2	A2
3,3'-dichloro-benzidine and salts	A	Polyurethane and pigment workers	2B	2	A2
1,4-dioxane	A	Solvent, degreaser	2A	2	—

Carcinogen	Human cancer site	Industry Setting	IARC designation	NTP designation	ACGIH designation
epichloroydrin	A	Chemical production	2A	2	—
ethylene oxide	A	Sterilizing agent	2A	2	A2
formaldehyde	A	Manufacture of woods, resins, leather, rubber, metals	2A	2	A2
4,4'-methylene-bis-2-chloro-aniline (MBOCA)	A	Polyurethane, epoxy resin, elastomer manufacturing	2A	2	A2
mineral oils (certain)	skin, scrotum	Lubricant in metal-working; solvents in printing, etc.	1 (***)	1 (**)	—

Carcinogen	Human cancer site	Industry Setting	IARC designation	NTP designation	ACGIH designation
beta-naphthylamine	bladder	Chemical, textile, dye, rubber industries; no longer in commercial use	1	1	A1
polychlorinated and polybrominated biphenyls (PCBs and PBBs)	A	Flame retardants (PBB's), transformer and capacitor fluids, solvents (PCBs)	2B (PBB's) 2A (PCB's)	2	-
beta-propiolactone	A	Production and use of plastics, resins, and viricides.	2B	2	A2
vinyl chloride	hepatic angiosarcoma	Polyvinyl plastic production	1	1	A1

INDUSTRIES

Industry	IARC designation	NTP designation
Aluminum production	1	—
Auramine manufacture	1	1
Boot and shoe manufacture and repair (certain exposures)	1	—
Coal gasification (older processes)	1	—
Coke production (certain exposures)	1	—
Furniture and cabinet making (wood dusts)	1	—
Iron and steel founding	1	—
Isopropyl alcohol manufacture (strong acid processes)	1	1
Nickel refining	—	1
Rubber industry	1	1
Underground hematite mining (with radon exposure)	1	1

KEY: "A" indicates that only animal evidence of carcinogenicity is currently satisfactory. When human evidence is satisfactory, the site is indicated.

The International Agency for Research on Cancer (IARC) designations include three groups: Group 1 (sufficient evidence of carcinogenicity to humans), Group 2 (probably carcinogenic; 2A represents high degree of evidence and 2B a lower degree of evidence), and Group 3 (cannot be classified as carcinogen).

The National Toxicology Program (NTP) designations include substances or processes "known to be carcinogenic" (here designated 1) and those "that may reasonably be anticipated to be carcinogenic (here designated 2).

The American Conference of Governmental Industrial Hygienists (ACGIH) designations include A1 (recognized to have carcinogenic or cocarcingenic potential for humans) and A2 (suspected of having carcinogenic potential for humans).

NOTES:
- * Coke oven emissions not directly evaluated as a substance,but coke production is Category 1.
- ** Category is "soots, tars, and mineral oils."
- *** Untreated and mildly treated mineral oils are Category 1, while highly treated mineral oils are Category 3.

Infectious Diseases

Table 1 lists selected work-related infectious diseases by occupation. Most high-risk jobs are in agriculture, human health care, and animal care.

Table 2 provides data on the U.S. incidence of selected infectious diseases that are sometimes work-related.

Table 1.—Selected Occupational Infectious Diseases, by Occupation

Occupation	Selected work-related infectious diseases
Bulldozer operator	Coccidioidomycosis, histoplasmosis
Butcher	Anthrax, erysipeloid, tularemia
Cat and dog handler	Cat-scratch disease, *Pasteurella multocida* cellulitis, rabies
Cave explorer	Rabies, histoplasmosis
Construction worker	Rocky Mountain spotted fever, coccidioidomycosis, histoplasmosis
Cook	Tularemia, salmonellosis, trichinosis
Cotton mill worker	Coccidioidomycosis
Dairy farmer	Milkers' nodules, Q fever, brucellosis
Delivery personnel	Rabies
Dentist	Hepatitis B
Ditch digger	Creeping eruption (cutaneous larva migrans), hookworm disease, ascariasis
Diver	*Mycobacterium marinum* (swimming pool granuloma)
Dock worker	Leptospirosis, swimmers' itch (Schistosoma species)

Occupation	Selected work-related infectious diseases
Farmer	Rabies, anthrax, brucellosis, Rocky Mountain spotted fever, tetanus, leptospirosis, plague, tularemia, coccidioidomycosis, histoplasmosis, sporotrichosis, hookworm disease, ascariasis
Fisherman, fish handler	Erysipeloid, swimming pool granuloma
Florist	Sporotrichosis
Food-processing worker	Salmonellosis
Forestry worker	California encephalitis, Lyme disease, Rocky Mountain spotted fever, tularemia
Fur handler	Tularemia
Gardener	Sporotrichosis, creeping eruption (cutaneous larva migrans)
Geologist	Plague, California encephalitis
Granary and warehouse worker	Murine typhus (endemic)
Hide, goat hair, and wool handler	Q Fever, anthrax, dermatophytoses
Hunter	Lyme disease, Rocky Mountain spotted fever, plague, tularemia, trichinosis
Laboratory worker	Hepatitis B
Livestock worker	Brucellosis, leptospirosis
Meat packer and slaughterhouse (abattoir) worker	Brucellosis, leptospirosis, Q fever, salmonellosis
Nurse	Hepatitis B, rubella, tuberculosis
Pet shop worker	*Pasteurella multocida* cellulitis, psittacosis, dermatophytoses
Physician	Hepatitis B, rubella, tuberculosis
Pigeon breeder	Psittacosis
Poultry handler	Newcastle disease, erysipeloid, psittacosis
Rancher	Lyme disease, rabies, Rocky Mountain spotted fever, Q fever, tetanus, plague, tularemia, trichinosis
Rendering plant worker	Brucellosis, Q fever
Sewer worker	Leptospirosis, hookworm disease, ascariasis
Shearer	Orf, tularemia

Occupation	Selected work-related infectious diseases
Shepherd	Orf, plague
Trapper	Leptospirosis, Lyme disease, tularemia, rabies, Rocky Mountain spotted fever
Veterinarian	Anthrax, brucellosis, erysipeloid, rabies, leptospirosis, *Pasteurella multocida* cellulitis, salmonellosis, cat-scratch disease, orf, tularemia, psittacosis
Wild animal handler	Rabies
Zoo worker	Psittacosis, tuberculosis

(Source: Levy BS, Wegman DH, eds. *Occupational Health: Recognizing and Preventing Work-related Disease.* 2nd ed. Boston: Little, Brown: 1988, p. 282.)

Table 2.—Annual U.S. incidence (1985) of infectious diseases that are sometimes work-related*

Diseases	Number of reported cases
Anthrax	0
Brucellosis	133
Hepatitis, viral, type B	25,808
Leptospirosis	34
Plague	16
Psittacosis	106
Rabies	1
Tetanus	70
Tuberculosis	21,106
Tularemia	163
Typhus fever, murine	25

*Actual number of cases that are work-related is unknown; data include both work- and non-work-related cases. Source: Centers for Disease Control. *MMWR.* 1986;34:774.

Permissible Exposure Limits and Recommended Exposure Limits

The following table shows Permissible Exposure Limits (PEL) promulgated by OSHA and Recommended Exposure Limits (REL) advised by NIOSH. PELs are published in 30 CFR 1910.1000; RELs have been published by NIOSH in several ways, including criteria documents, current intelligence bulletins, testimony before Congress, and comments on OSHA and MSHA rule making.

SUMMARY OF OSHA PELs AND NIOSH RELs

SUBSTANCE	FINAL OSHA PEL ppm[a]	mg/m3[b]	Type[d]	NIOSH RECOMMENDATION ppm[a]	mg/m3[b]	Type[c]
Acetaldehyde	100 150	180 270	TWA STEL	*Ca (LOQ 18)		
Acetic acid	10	25	TWA	10 15	25 37	TWA STEL
Acetic anhydride	5	20	CL	5	20	CL
Acetone	750 1000	1800 2400	TWA STEL	250	590	TWA
Acetone cyanohydrin		none		1	4	
Acetonitrile	40 60	70 105	TWA STEL	20	34	TWA
2-Acetylamino-fluorene		Ca - no PEL (see 1910.1014)			Ca (use 1910.1014)	
Acetylene		none		2500	2662	CL
Acetylene dichloride; see 1,2-Dichloro-ethylene						
Acetylene tetrabromide	1	14	TWA		—	
Acetylsalicylic acid (aspirin)		5	TWA		5	TWA
Acrolein	0.1 0.3	0.25 0.8	TWA TSTEL	0.1 0.3	0.25 0.8	TWA STEL
Acrylamide		0.03 sk	TWA		Ca 0.03 sk	TWA
Acrylic acid	10	30 sk	TWA	2	6 sk	TWA
Acrylonitrile	2 sk 10 sk (see 1910.1045)		TWA CL	Ca 1 sk 10 sk		TWA CL
Adiponitrile		none		4	18	TWA
Aldrin		0.25 sk	TWA		Ca 0.25 sk (LOQ 0.15)	TWA

SUMMARY OF OSHA PELs AND NIOSH RELs

	FINAL OSHA PEL			NIOSH RECOMMENDATION		
SUBSTANCE	ppm[a]	mg/m^3[b]	Type[d]	ppm[a]	mg/m^3[b]	Type[c]
Allyl alcohol	2	5 sk	TWA	2	5 sk	TWA
	4	10 sk	STEL	4	10 sk	STEL
Allyl chloride	1	3	TWA	1	3.1	TWA
	2	6	STEL	2	6TSTEL	
Allyl glycidyl ether	5	22	TWA	5	22 sk	TWA
(AGE)	10	44	STEL	10	44 sk	STEL
Allyl propyl disulfide	2	12	TWA	2	12	TWA
	3	18	STEL	3	18	STEL
alpha-Alumina						
Total dust		10	TWA		—	
Respirable fraction		5	TWA		—	
Aluminum (as Al)						
Metal						
Total dust		15	TWA		10	TWA
Respirable fraction		5	TWA		5	TWA
Pyro powders		5	TWA		5	TWA
Welding fumes		5	TWA		5	TWA
Soluble salts		2	TWA		2	TWA
Alkyls		**2	TWA		2	TWA
4-Aminodiphenyl		Ca - no PEL (see 1910.1011)			Ca (use 1910.1011)	
2-Aminoethanol; see ethanolamine						
2-Aminopyridine	0.5	2	TWA	0.5	2	TWA
Amitrol		0.2	TWA		Ca 0.2	TWA
Ammonia	35	27	STEL	25	18	TWA
				35	27	STEL
Ammonium chloride		10	TWA		10	TWA
fume		20	STEL	T20	STEL	
Ammonium sulfamate						
Total dust		10	TWA		10	TWA
Respirable fraction		5	TWA		5	TWA
n-Amyl acetate	100	525	TWA	100	525	TWA
sec-Amyl acetate	125	650	TWA	125	650	TWA

SUMMARY OF OSHA PELs AND NIOSH RELs

SUBSTANCE	FINAL OSHA PEL ppm[a]	mg/m[3b]	Type[d]	NIOSH RECOMMENDATION ppm[a]	mg/m[3b]	Type[c]
Aniline and homolo	2	8 sk	TWA	Ca—lowest feasible		
Anisidine (isomers)		0.5 sk	TWA		Ca 0.5 sk	TWA
Ortho-						
Para-						
Antimony and compounds (as Sb)		0.5	TWA		0.5	TWA
ANTU (alpha naphthyl-thiourea)		0.3	TWA		0.3	TWA
Arsenic, organic compounds (as As)		0.5	TWA			
Arsenic, inorganic compounds (as As)		0.01 (see 1910.1018)	TWA		Ca 0.002	CL
Arsine	0.05	0.2	TWA		Ca 0.002	CL
Asbestos		0.2 fbr/cc (see 1910.1001 & 1910.1101)	TWA	Ca 0.1 fbr/cc (see 1910.1001 & 1910.1101)		TWA
Asphalt		none (PEL delayed)			Ca 5	CL
Atrazine		5	TWA		5	TWA
Azinphos-methyl		0.2 sk	TWA		0.2 sk	TWA
Barium, soluble compounds (as Ba)		0.5	TWA		0.5	TWA
Barium sulfate						
Total dust		10	TWA		10	TWA
Respirable fraction		5	TWA		5	TWA
Benomyl						
Total dust		10	TWA		—	
Respirable fraction		5	TWA		—	
Benzene	1		TWA	Ca 0.1	0.32	TWA
	5		STEL	1	3.2	CL
	(see 1910.1028)					
Benzidine		Ca - no PEL (see 1910.1010)			Ca (use 1910.1010)	

SUMMARY OF OSHA PELs AND NIOSH RELs

SUBSTANCE	FINAL OSHA PEL ppm[a]	mg/m3[b]	Type[d]	NIOSH RECOMMENDATION ppm[a]	mg/m3[b]	Type[c]
Benzidine-based dyes		none		Ca—lowest feasible		
p-Benzoquinone; see Quinone						
Benzo(a)pyrene; see coal tar pitch volatiles						
Benzoyl peroxide		5	TWA		5	TWA
Benzyl chloride	1	5	TWA	1	5	CL
Beryllium and beryllium compounds (as Be)		0.002 0.005 0.025	TWA CL PK (30-min)	Ca–not to exceed 0.0005		
Bisphenyl; see diphenyl						
Bismuth telluride, undoped						
Total dust		15	TWA		10	TWA
Respirable fraction		5	TWA		5	TWA
Bismuth telluride, Se-doped		5	TWA		5	TWA
Borates, tetra sodium salts						
Anhydrous		10	TWA		1	TWA
Decahydrate		10	TWA		5	TWA
Pentahydrate		10	TWA		1	TWA
Boron oxide		10	TWA		10	TWA
Boron tribromide	1	10	CL	1	10	CL
Boron trifluoride	1	3	CL	1	3	CL
Bromacil	1	10	TWA	1	10	TWA
Bromine	0.1 0.3	0.7 2	TWA STEL	0.1 0.3	0.7 2	TWA STEL
Bromine pentafluoride	0.1	0.7	TWA	0.1	0.7	TWA

SUMMARY OF OSHA PELs AND NIOSH RELs

SUBSTANCE	FINAL OSHA PEL ppm[a]	mg/m^3[b]	Type[d]	NIOSH RECOMMENDATION ppm[a]	mg/m^3[b]	Type[c]
Bromoform	0.5	5 sk	TWA	0.5	5 sk	TWA
Butadiene (1,3-butadiene)	1000 In process of 6(b) rulemaking	2200	TWA	Ca—lowest feasible ,		
Butane	800	1900	TWA	800	1900	TWA
Butanethiol; see butyl mercaptan						
2-Butanone (methylethyl ketone)	200 300	590 885	TWA STEL	200 300	590 885	TWA, STEL
2-Butoxyethanol	25	120 sk	TWA	5	24 sk	TWA
2-Butoxyethanol acetate		none		5	31	
n-Butyl-acetate	150 200	710 950	TWA STEL	150 200	710 950	TWA STEL
sec-Butyl acetate	200	950	TWA	200	950	TWA
tert-Butyl acetate	200	950	TWA	200	950	TWA
Butyl acrylate	10	55	TWA	10	55	TWA
n-Butyl alcohol	50	150 sk	CL	50	150 sk	CL
sec-Butyl alcohol	100	305	TWA	100 150	305 455	TWA STEL
tert-Butyl alcohol	100 150	300 450	TWA STEL	100 150	300 450	TWA STEL
Butylamine	5	15 sk	CL	5	15 sk	CL
tert-Butyl chromate (as CrO_3)		0.1 sk	CL		Ca 0.001	TWA
n-Butyl glycidyl ether (BGE)	25	135	TWA	5.6	30	CL
n-Butyl lactate	5	25	TWA	5	25	TWA
Butyl mercaptan	0.5	1.5	TWA	0.5	1.8	CL
o-sec-Butylphenol	5	30 sk	TWA	5	30 sk	TWA

SUMMARY OF OSHA PELs AND NIOSH RELs

SUBSTANCE	FINAL OSHA PEL ppm[a]	mg/m^3[b]	Type[d]	NIOSH RECOMMENDATION ppm[a]	mg/m^3[b]	Type[c]
p-tert-Butyltoluene	10	60	TWA	10	60	TWA
	20	120	STEL	20	120	STEL
n-Butyronitrile		none		8	22	TWA
Cadmium fume		0.1	TWA	*Ca—lowest feasible		
(as Cd)		0.3	CL	(LOQ 0.01)		
	In process of 6(b) rulemaking					
Cadmium dust		0.2	TWA	*Ca—lowest feasible		
(as Cd)		0.6	CL	(LOQ 0.01)		
	In process of 6(b) rulemaking					
Calcium carbonate						
Total dust		15	TWA		10	TWA
Respirable fraction		5	TWA		5	TWA
Calcium cyanamide		0.5	TWA		0.5	TWA
Calcium hydroxide		5	TWA		5	TWA
Calcium oxide		5	TWA		2	TWA
Calcium silicate						
Total dust		15	TWA		10	TWA
Respirable fraction		5	TWA		5	TWA
Calcium sulfate						
Total dust		15	TWA		10	TWA
Respirable fraction		5	TWA		5	TWA
Camphor, synthetic		2	TWA		2	TWA
Caprolactam						
Dust		1	TWA		1	TWA
		3	STEL		3	STEL
Vapor	5	20	TWA	0.22	1	TWA
	10	40	STEL	0.66	3	STEL
Captafol (Difolatan[R])		0.1	TWA		Ca 0.1 sk	TWA
Captan		5	TWA		Ca 5	TWA
Carbaryl (Sevin[R])		5	TWA		5	TWA
				Repro. effect—minimize exposure during pregnancy		
Carbofuran (Furadan[R])		0.1	TWA		0.1	TWA

SUMMARY OF OSHA PELs AND NIOSH RELs

SUBSTANCE	FINAL OSHA PEL ppm[a]	mg/m³[b]	Type[d]	NIOSH RECOMMENDATION ppm[a]	mg/m³[b]	Type[c]
Carbon black		3.5	TWA		3.5 Ca 0.1 in presence of PAHs	TWA
Carbon dioxide	10000 30000	18000 54000	TWA STEL	5000 30000	9000 54000	TWA STEL
Carbon disulfide	4 12	12 sk 36 sk	TWA STEL	1 10	3 sk 30 sk	TWA STEL
Carbon monoxide	35 200	40 229	TWA CL	35 200	40 229	TWA CL
Carbon tetrabromide	0.1 0.3	1.4 4	TWA STEL	0.1 0.3	1.4 4	TWA STEL
Carbon tetrachloride	2	12.6	TWA	Ca 2	12.6	STEL (60-min)
Carbonyl fluoride	2 5	5 15	TWA STEL	2 5	5 15	TWA STEL
Catechol (pyrocatechol)	5	20 sk	TWA	5	20 sk	TWA
Cellulose						
Total dust		15	TWA		10	TWA
Respirable fraction		5	TWA		5	TWA
Cesium hydroxide		2	TWA		2	TWA
Cetylmercaptan		none		0.5	5.3	CL
Chlordane		0.5 sk	TWA		Ca 0.5 sk	TWA
Chlorinated camphene		0.5 sk 1 sk	TWA STEL		*Ca (LOQ 0.01) sk	
Chlorinated diphenyl oxide		0.5	TWA		0.5	TWA
Chlorine	0.5 1	1.5 3	TWA STEL	0.5 1	1.5 3	TWA STEL
Chlorine dioxide	0.1 0.3	0.3 0.9	TWA STEL	0.1 0.3	0.3 0.9	TWA STEL
Chlorine trifluoride	0.1	0.4	CL	0.1	0.4	CL

SUMMARY OF OSHA PELs AND NIOSH RELs

SUBSTANCE	FINAL OSHA PEL ppm[a]	mg/m[3b]	Type[d]	NIOSH RECOMMENDATION ppm[a]	mg/m[3b]	Type[c]
Chloroacetaldehyde	1	3	CL	1	3	CL
a-Chloroaceto-phenone (phenacyl chloride)	0.05	0.3	TWA	0.05	0.3	TWA
Chloroacetyl chloride	0.05	0.2	TWA	0.05	0.2	TWA
Chlorobenzene	75	350	TWA	—		
o-Chloro-benzylidene malononitrile	0.05	0.4 sk	CL	0.05	0.4 sk	CL
Chloro-bromomethane	200	1050	TWA	200	1050	TWA
2-Chloro-1,3 butadiene; see beta-chloroprene						
Chloro-difluoromethane	1000	3500	TWA	1000 1250	3500 4375	TWA STEL
Chlorodiphenyl (42% chlorine) (PCB)		1 sk			Ca 0.001	TWA
Chlorodiphenyl (54% chlorine) (PCB)		0.5 sk	TWA		Ca 0.001	TWA
1-Chloro,2,3-epoxy-propane; see Epichlorohydrin						
2-Chloroethanol; see ethylene chlorohydrin						
Chloroethylene; see vinyl chloride						
Chloroform (trichloromethane)	2	9.78	TWA	Ca 2	9.78	STEL (60-min)
bis(Chloromethyl) ether	Ca - no PEL (see 1910.1008)			Ca (use 1910.1008)		
Chloromethyl methyl ether	Ca - no PEL (see 1910.1006)			Ca (use 1910.1006)		

SUMMARY OF OSHA PELs AND NIOSH RELs

SUBSTANCE	FINAL OSHA PEL ppm[a]	mg/m3[b]	Type[d]	NIOSH RECOMMENDATION ppm[a]	mg/m3[b]	Type[c]
1-Chloro-1-nitropropane	2	10	TWA	2	10	TWA
Chloropenta-fluoroethane	1000	6320	TWA	1000	6320	TWA
Chloropicrin	0.1	0.7	TWA	0.1	0.7	TWA
beta-Chloroprene	10	35 sk	TWA	Ca 1	3.6	CL
o-Chlorostyrene	50 75	285 430	TWA STEL	50 75	285 430	TWA STEL
o-Chlorotoluene	50	250	TWA	50 75	250 375	TWA STEL
2-Chloro-6-trichloro-methyl pyridine						
Total dust		15	TWA		10 20	TWA STEL
Respirable fraction		5	TWA		5	TWA
Chlorpyrifos		0.2 sk	TWA		0.2 sk 0.6 sk	TWA STEL
Chromic acid and chromates (as CrO_3)		0.1	CL		Ca 0.001 (as Cr)	TWA
Chromium (II) compounds (as Cr)		0.5	TWA		0.5	TWA
Chromium (III) compounds (as Cr)		0.5	TWA		0.5	TWA
Chromium metal (as Cr)		1	TWA		0.5	TWA
Chromyl chloride		none			Ca .001 (Cr VI)	TWA
Chrysene; see coal tar pitch volatiles						
Clopidol						
Total dust		15	TWA		10 20	TWA STEL
Respirable fraction		5	TWA		5	TWA

SUMMARY OF OSHA PELs AND NIOSH RELs

	FINAL OSHA PEL			NIOSH RECOMMENDATION		
SUBSTANCE	ppm[a]	mg/m^3[b]	Type[d]	ppm[a]	mg/m^3[b]	Type[c]
Coal dust (less than 5% $Si0_2$), respirable quartz fraction		2	TWA		—	
Coal dust (greater than or equal to 5% $Si0_2$), respirable quartz fraction		0.1	TWA		—	
Coal tar pitch volatiles (benzene soluble fraction), anthracene, BaP, phenanthrene, acridine, chrysene, pyrene		0.2 (see 1910.1002 for definition)	TWA	Chrysene—control as a carcinogen	Ca 0.1 Cyclohexane extractable	TWA
Cobalt metal, dust, and fume (as Co)		0.05	TWA		0.05	TWA
Cobalt carbonyl (as Co)		0.1	TWA		0.1	TWA
Cobalt hydrocarbonyl (as Co)		0.1	TWA		0.1	TWA
Coke oven emissions		Ca 0.15 (see 1910.1029)	TWA		Ca 0.5—0.7 (Total particulates) as screening level	
Copper						
Fume (as Cu)		0.1	TWA		0.1	TWA
Dusts and mists (as Cu)		1	TWA		1	TWA
Cotton dust (raw)		1	TWA		0.2	(lint free) lowest feasible

This 8-hour TWA applies to respirable dust as measured by a vertical elutriator cotton dust sampler or equivalent instrument. The TWA applies to the cotton waste processing operations of waste recycling (sorting, blending, cleaning, and willowing) and garretting. See also 1910.1043 for cotton dust limits to other sectors.

SUBSTANCE	ppm[a]	mg/m^3[b]	Type[d]	ppm[a]	mg/m^3[b]	Type[c]
Crag herbicide (sesone)						
Total dust		10	TWA		10	TWA
Respirable fraction		5	TWA		5	TWA
Cresol, all isomers	5	22 sk	TWA	2.3	10	TWA

SUMMARY OF OSHA PELs AND NIOSH RELs

SUBSTANCE	FINAL OSHA PEL ppm[a]	mg/m3[b]	Type[d]	NIOSH RECOMMENDATION ppm[a]	mg/m3[b]	Type[c]
Crotonaldehyde	2	6	TWA	2	6	TWA
Crufomate		5	TWA		5 20	TWA STEL
Cumene	950	245 sk	TWA	50	245 sk	TWA
Cyanamide		2	TWA		2	TWA
Cyanides (as CN)		5	TWA	4.7	5	CL (10-min)
Cyanogen	10	20	TWA	10	20	TWA
Cyanogen chloride	0.3	0.6	CL	0.3	0.6	CL
Cyclohexane	300	1050	TWA	300	1050	TWA
Cyclohexanol	50	200 sk	TWA	50	200 sk	TWA
Cyclohexanone	25	100 sk	TWA	25	100 sk	TWA
Cyclohexene	300	1015	TWA	300	1015	TWA
Cyclohexylamine	10	40	TWA	10	40	TWA
Cyclohexylmercaptan		none		0.5	2.4	CL
Cyclonite		1.5 sk	TWA		1.5 sk 3 sk	TWA STEL
Cyclopentadiene	75	200	TWA	75	200	TWA
Cyclopentane	600	1720	TWA	600	1720	TWA
Cyhexatin		5	TWA		5	TWA
2,4-D (dichlorophenoxyacetic acid)		10	TWA		10	TWA
Decaborane	0.05 0.15	0.3 sk 0.9 sk	TWA STEL	0.05 0.15	0.3 sk 0.9 sk	TWA STEL
Decylmercaptan		none		0.5	3.6	CL
Demeton (Systox[R])		0.1 sk	TWA		0.1 sk	TWA

SUMMARY OF OSHA PELs AND NIOSH RELs

SUBSTANCE	FINAL OSHA PEL ppm[a]	mg/m3[b]	Type[d]	NIOSH RECOMMENDATION ppm[a]	mg/m3[b]	Type[c]
Diacetone alcohol (4-hydroxy-4-methyl-2-pentanone)	50	240	TWA	50	240	TWA
2,4-Diaminoanisole		none		Ca—lowest feasible		
1,2-Diaminoethane; see ethylenediamine						
o-Dianisidine based dyes		none		Ca—minimize exposure		
Diazinon		0.1 sk	TWA		0.1 sk	TWA
Diazomethane	0.2	0.4	TWA	0.2	0.4	TWA
Diborane	0.1	0.1	TWA	0.1	0.1	TWA
1,2-Dibromo-3-chloropropane	Ca 0.001 (see 1910.1044)		TWA	Ca (use 1910.1044)		
2-N-Dibutylamino ethanol	2	14	TWA	2	14 sk	TWA
Dibutyl phosphate	1 2	5 10	TWA STEL	1 2	5 10	TWA STEL
Dibutyl phthalate		5	TWA		5	TWA
Dichloroacetylene	0.1	0.4	CL	Ca 0.1	0.4	CL
o-Dichlorobenzene	50	300	CL	50	300	CL
p-Dichlorobenzene	75 110	450 675	TWA STEL	*Ca (LOQ 1.7)		
3,3'-Dichloro-benzidine	Ca - no PEL (see 1910.1007)			Ca (use 1910.1007)		
Dichlorodifluoro-methane	1000	4950	TWA	1000	4950	TWA
1,3-Dichloro-5,5-dimethyl hydantoin		0.2 0.4	TWA STEL		0.2 0.4	TWA STEL
Dichlorodiphenyltri-chloroethane (DDT)		1 sk	TWA		*Ca 0.5 (LOQ 0.1)	TWA
1,1-Dichloroethane	100	400	TWA	100	400	TWA

SUMMARY OF OSHA PELs AND NIOSH RELs

SUBSTANCE	FINAL OSHA PEL ppm[a]	mg/m3[b]	Type[d]	NIOSH RECOMMENDATION ppm[a]	mg/m3[b]	Type[c]
1,2-Dichloro-ethylene	200	790	TWA	200	790	TWA
Dichloroethyl ether	5 10	30 sk 60 sk	TWA STEL	Ca 5 10	30 sk 60 sk	TWA STEL
Dichloromethane; see methylene chloride						
Dichloromono-fluoromethane	10	40	TWA	10	40	TWA
1,1-Dichloro-1-nitroethane	2	10	TWA	2	10	TWA
1,2-Dichloropropane; see propylene dichloride						
1,3-Dichloropropene	1	5 sk	TWA	Ca 1	5 sk	TWA
2,2-Dichloropro-pionic acid	1	6	TWA	1	6	TWA
Dichlorotetra-fluoroethane	1000	7000	TWA	1000	7000	TWA
Dichlorvos (DDVP)		1 sk	TWA		1 sk	TWA
Dicrotophos		0.25 sk	TWA		0.25 sk	TWA
Dicyclopentadiene	5	30	TWA	5	30	TWA
Dicyclopentadienyl iron						
Total dust		10	TWA		10	TWA
Respirable fraction		5	TWA		5	TWA
Dieldrin		0.25 sk	TWA		Ca 0.25 sk	TWA
Diesel Exhaust		none		Ca—lowest feasible		
Diethanolamine	3	15	TWA	3	15	TWA
Diethylamine	10 25	30 75	TWA STEL	10 25	30 75	TWA STEL
2-Diethylamino-ethanol	10	50 sk	TWA	10	50 sk	TWA
Diethylene triamine	1	4	TWA	1	4 sk	TWA

SUMMARY OF OSHA PELs AND NIOSH RELs

SUBSTANCE	FINAL OSHA PEL ppm[a]	mg/m^{3b}	Type[d]	NIOSH RECOMMENDATION ppm[a]	mg/m^{3b}	Type[c]
Diethyl ether; see ethyl ether						
Diethyl ketone	200	705	TWA	200	705	TWA
Diethyl phthalate		5	TWA		5	TWA
Difluorodibromo-methane	100	860	TWA	100	860	TWA
Diglycidyl ether (DGE)	0.1	0.5	TWA	Ca 0.1	0.5	TWA
Dihydroxybenzene; see hydroquinore						
Diisobutyl ketone	25	150	TWA	25	150	TWA
Diisopropylamine	5	20 sk	TWA	5	20 sk	TWA
4-Dimethylamino-azobenzene	Ca - no PEL (see 1910.1015)			Ca (use 1910.1015)		
Dimethoxymethane; see methylal						
Dimethyl acetamide	10	35 sk	TWA	10	35 sk	TWA
Dimethylamine	10	18	TWA	10	18	TWA
Dimethylaminobenzene; see xylidine						
bis (2-Dimethylamino-ethyl) ether	none			Minimize exposure to NIAX catalyst ESN		
Dimethylamino-propionitrile	none			Minimize exposure to NIAX catalyst ESN		
Dimethylaniline (N,N-dimethyl-aniline)	5 10	25 sk 50 sk	TWA STEL	5 10	25 sk 50 sk	TWA STEL
Dimethylbenzene; see xylene						
Dimethyl carbamoyl chloride	none			Ca—lowest feasible		

SUMMARY OF OSHA PELs AND NIOSH RELs

SUBSTANCE	FINAL OSHA PEL ppm[a]	mg/m^{3b}	Type[d]	NIOSH RECOMMENDATION ppm[a]	mg/m^{3b}	Type[c]
Dimethyl-1,2-dibromo-2,2-dichloroethyl phosphate (naled)		3 sk	TWA		3 sk	TWA
Dimethylformamide	10	30 sk	TWA	10	30 sk	TWA
2,6-Dimethyl-4-heptanone see diisobutyl ketone						
1,1-Dimethyl-hydrazine	0.5	1 sk	TWA	Ca 0.06	0.15	CL (120-min)
Dimethylphthalate		5	TWA		5	TWA
Dimethyl sulfate	0.1	0.5 sk	TWA	Ca 0.1	0.5 sk	TWA
Dinitrolmide (3,5-dinitro-o-toluamide)		5	TWA		5	TWA
Dinitrobenzene (all isomers)	(alpha) (beta) (meta)	1 sk	TWA		1 sk	TWA
Dinitro-o-cresol		0.2 sk	TWA		0.2 sk	TWA
Dinitrotoluene		1.5 sk	TWA		Ca 1.5 sk	TWA
Dioxane (diethylene dioxide)	25	90 sk	TWA	Ca 1	3.6	CL, (30-min)
Dioxathion (Delnav)		0.2 sk	TWA		0.2 sk	TWA
Diphenyl (biphenyl)	0.2	1	TWA	0.2	1	TWA
Diphenylamine		10	TWA		10	TWA
Diphenylamine diisocyanate; see methylene bisphenyl isocyanate						
Dipropylene glycol methyl ether	100 150	600 sk 900 sk	TWA STEL	100 150	600 sk 900 sk	TWA STEL
Dipropyl ketone	50	235	TWA	50	235	TWA
Diquat		0.5	TWA		0.5	TWA

SUMMARY OF OSHA PELs AND NIOSH RELs

	FINAL OSHA PEL			NIOSH RECOMMENDATION		
SUBSTANCE	ppm[a]	mg/m[3b]	Type[d]	ppm[a]	mg/m[3b]	Type[c]
Di-sec octyl phthalate (di-2-ethylhexyl-phthalate)		5 10	TWA STEL		Ca 5 10	TWA STEL
Disulfiram		2	TWA		2	TWA
Disulfoton		0.1 sk	TWA		0.1 sk	TWA
2,6-Di-tert-butyl-p-cresol		10	TWA		10	TWA
Diuron		10	TWA		10	TWA
Divinyl benzene	10	50	TWA	10	50	TWA
Dodecylmercaptan		none		0.5	4.1	CL
Emery						
Total dust		10	TWA		—	
Respirable fraction		5	TWA		—	
Endosulfan		0.1 sk	TWA		0.1 sk	TWA
Endrin		0.1 sk	TWA		0.1 sk	TWA
Enflurane		none		2	15.1	CL (60-min)
Epichlorohydrin	2	8 sk	TWA		*Ca (LOQ 2.5) Occupational exposure to be minimized	
EPN		0.5 sk	TWA		0.5 sk	TWA
1,2-Epoxypropane; see propylene oxide						
2,3-Epoxy-1-propanol; see glycidol						
Ethanethiol; see ethyl mercaptan						
Ethanolamine	3 6	8 15	TWA STEL	3 6	8 15	TWA STEL
Ethion		0.4 sk	TWA		0.4 sk	TWA

SUMMARY OF OSHA PELs AND NIOSH RELs

SUBSTANCE	FINAL OSHA PEL ppm[a]	mg/m3[b]	Type[d]	NIOSH RECOMMENDATION ppm[a]	mg/m3[b]	Type[c]
2-Ethoxyethanol	200	740 sk	TWA	Reduce exposure to lowest feasible concentration		
	In process of 6(b) rulemaking					
2-Ethoxyethyl acetate (cellosolve acetate)	100	540 sk	TWA	Reduce exposure to lowest feasible concentration		
	In process of 6(b) rulemaking					
Ethyl acetate	400	1400	TWA	400	1400	TWA
Ethyl acrylate	5	20 sk	TWA		*Ca	
	25	100 sk	STEL		(LOQ 4)	
Ethyl alcohol (ethanol)	1000	1900	TWA	1000	1900	TWA
Ethylamine	10	18	TWA	10	18	TWA
Ethyl amyl ketone (5-methyl-3-heptanone)	25	130	TWA	25	130	TWA
Ethyl benzene	100	435	TWA	100	435	TWA
	125	545	STEL	125	545	STEL
Ethyl bromide	200	890	TWA		—	
	250	1110	STEL			
Ethyl butyl ketone (3-heptanone)	50	230	TWA	50	230	TWA
Ethyl chloride	1000	2600	TWA	Ca—handle with caution		
Ethyl ether	400	1200	TWA		—	
	500	1500	STEL			
Ethyl formate	100	300	TWA	100	300	TWA
Ethyl mercaptan	0.5	1	TWA	0.5	1.3	CL
Ethyl silicate	10	85	TWA	10	85	TWA
Ethylene chlorohydrin	1	3 sk	CL	1	3 sk	CL
Ethylenediamine	10	25	TWA	10	25	TWA
Ethylene dibromide	20		TWA	Ca 0.045		TWA
	30		CL	0.13		CL
	50 (5-min)		PK			
	In process of 6(b) rulemaking					

SUMMARY OF OSHA PELs AND NIOSH RELs

	FINAL OSHA PEL			NIOSH RECOMMENDATION		
SUBSTANCE	ppm[a]	mg/m^3[b]	Type[d]	ppm[a]	mg/m^3[b]	Type[c]
Ethylene dichloride	1 2	4 8	TWA STEL	Ca 1 2	4 8	TWA STEL
Ethylene glycol	50	125	CL		—	
Ethylene glycol dinitrate		0.1 sk	STEL		0.1 sk	STEL
Ethylene glycol methyl acetate; see methyl cellosolve acetate						
Ethylene glycol mono-butyl ether; see 2-Butoxyethanol						
Ethylene glycol monobutyl ether acetate; see 2-Butoxyethanol acetate						
Ethyleneimine		Ca - no PEL (see 1910.1012)			Ca (use 1910.1012)	
Ethylene oxide	1 5	(see 1910.1047(a)(2) for operations excluded)	TWA CL	Ca<0.1 5	0.18 9	TWA CL (10-min)
Ethylene thriourea		none			Ca—minimize exposure	
Ethylidene chloride; see 1,1-dichloroethane						
Ethylidene norbornene	**5	25	CL	5	25	CL
N-Ethylmorpholine	5	23 sk	TWA	5	23 sk	TWA
Fenamiphos		0.1 sk	TWA		0.1 sk	TWA
Fensulfothion (Dasanit)		0.1	TWA		0.1	TWA
Fenthion		0.2 sk	TWA		—	
Ferbam		10	TWA		10	TWA
Ferrovanadium dust		1 3	TWA STEL		1 3	TWA STEL

SUMMARY OF OSHA PELs AND NIOSH RELs

	FINAL OSHA PEL			NIOSH RECOMMENDATION		
SUBSTANCE	ppm[a]	mg/m^3[b]	Type[d]	ppm[a]	mg/m^3[b]	Type[c]
Fibrous glass dust	none (PEL delayed)			3 million fibers/m^3 (fibers ≤3.5 ≤m dia and ≥10 ≤m length) 5 mg/m^3 TWA (total fibrous glass)		TWA
Fluorides (as F)		2.5	TWA		2.5	TWA
Fluorine	0.1	0.2	TWA	0.1	0.2	TWA
Fluorotrichloro methane (trichlorofluoromethane)	1000	5600	CL	1000	5600	CL
Fluroxene		none		2	10.3	CL (60-min)
Fonofos		0.1 sk			0.1 sk	TWA
Formaldehyde	1 2 See Table Z-2 for operations or sectors excluded from 1910.1048 for which limit(s) is(are) stated		TWA STEL	Ca 0.016 0.1		TWA CL
Formamide	20 30	30 45	TWA STEL	10	15 sk	TWA
Formic acid	5	9	TWA	5	9	TWA
Furfural	2	8 sk	TWA		—	
Furfuryl alcohol	10 15	40 sk 60 sk	TWA STEL	10 15	40 sk 60 sk	TWA STEL
Gallium Arsenide		none			Ca 0.002	CL
Gasoline	300 500	900 1500	TWA STEL	*Ca (LOQ 15)		
Germanium tetrahydride	0.2	0.6	TWA	0.2	0.6	TWA
Glutaraldehyde	0.2	0.8	CL	0.2	0.8	CL
Glycerin (mist)						
Total particulate		10	TWA		—	
Respirable fraction		5	TWA		—	

SUMMARY OF OSHA PELs AND NIOSH RELs

SUBSTANCE	FINAL OSHA PEL ppm[a]	mg/m3[b]	Type[d]	NIOSH RECOMMENDATION ppm[a]	mg/m3[b]	Type[c]
Glycidol	25	75	TWA	25	75	TWA
Glycol monoethyl ether; see 2-ethoxyethanol						
Glycolonitrile		none		2	5	CL
Grain dust (oat, wheat, barley)		10	TWA		4	TWA
Graphite, natural respirable dust		2.5	TWA		2.5	TWA
Graphite, synthetic						
Total dust		10	TWA		—	
Respirable fraction		5	TWA		—	
Guthion[R]; see Azinphos methyl						
Gypsum						
Total dust		15	TWA		10	TWA
Respirable fraction		5	TWA		5	TWA
Hafnium		0.5	TWA		0.5	TWA
Halothane		none		2	16.2	CL (60-min)
Heptachlor		0.5 sk	TWA		Ca 0.5 sk	TWA
Heptane (n-heptane)	400 500	1600 2000	TWA STEL	85 440	350 1800	TWA CL
n-Heptylmercaptan		none		0.5	2.7	CL
Hexachloro-butadiene	0.02	0.24	TWA	Ca 0.02	0.24 sk	TWA
Hexachlorocyclo-pentadiene	0.01	0.1	TWA	0.01	0.1	TWA
Hexachloroethane	1	10 sk	TWA	Ca 1	10 sk	TWA
Hexachloro-naphthalene		0.2 sk	TWA		0.2 sk	TWA (10-min)
Hexafluoro-acetone	**0.1	0.7 sk	TWA	0.1	0.7 sk	TWA

SUMMARY OF OSHA PELs AND NIOSH RELs

	FINAL OSHA PEL			NIOSH RECOMMENDATION		
SUBSTANCE	ppm[a]	mg/m³[b]	Type[d]	ppm[a]	mg/m³[b]	Type[c]
Hexamethylene diisocyanate (HDI)		none			0.035 0.14	TWA CL
Hexamethyle phosphoramide		none		Ca—lowest feasible		
n-Hexane	50	180	TWA	50	180	TWA
Hexane isomers	500 1000	1800 3600	TWA STEL	100 510	350 1800	TWA CL
2-Hexanone (methyl n-butyl ketone)	5	20	TWA	1	4	TWA
Hexone (methyl isobutyl ketone)	50 75	205 300	TWA STEL	50 75	205 300	TWA STEL
sec-Hexyl acetate	50	300	TWA	50	300	TWA
Hexylene glycol	25	125	CL	25	125	CL
n-Hexylmercaptan		none		0.5	2.7	CL
Hydrazine	0.1	0.1 sk	TWA	Ca 0.03	0.04	CL (120-min)
Hydrogenated terphenyls	0.5	5	TWA	0.5	5	TWA
Hydrogen bromide	3	10	CL	3	10	CL
Hydrogen chloride	5	7	CL	5	7	CL
Hydrogen cyanide	4.7	5 sk	STEL	4.7	5 sk	STEL
Hydrogen fluoride (as F)	3 6		TWA STEL	3 6	2.5 5.0	TWA STEL
Hydrogen peroxide	1	1.4	TWA	1	1.4	TWA
Hydrogen selenide (as Se)	0.05	0.2	TWA	0.05	0.2	TWA
Hydrogen sulfide	10 15	14 21	TWA STEL	10	15	CL
Hydroquinone		2	TWA		2	CL
2-Hydroxypropyl acrylate Indene	0.5 10	3 sk 45	TWA TWA	0.5 10	3 sk 45	TWA TWA

SUMMARY OF OSHA PELs AND NIOSH RELs

	FINAL OSHA PEL			NIOSH RECOMMENDATION		
SUBSTANCE	ppm[a]	mg/m^{3b}	Type[d]	ppm[a]	mg/m^{3b}	Type[c]
Indium and compounds (as In)		0.1	TWA		0.1	TWA
Iodine	0.1	1	CL	0.1	1	CL
Iodoform	0.6	10	TWA	0.6	10	TWA
Iron oxide dust and fume (as Fe) total particulate		10	TWA		5	TWA
Iron pentacarbonyl (as Fe)	0.1 0.2	0.8 1.6	TWA STEL	0.1 0.2	0.8 1.6	TWA STEL
Iron salts (soluble) (as Fe)		1	TWA		1	TWA
Isoamyl acetate	100	525	TWA	100	525	TWA
Isoamyl alcohol (primary and secondary)	100 125	360 450	TWA STEL	100 125	360 450	TWA STEL
Isobutyl acetate	150	700	TWA	150	700	TWA
Isobutyl alcohol	50	150	TWA	50	150	TWA
Isobutyronitrile		none		8	22	TWA
Isooctyl alcohol	50	270 sk	TWA	50	270 sk	TWA
Isophorone	4	23	TWA	4	23	TWA
Isophorone diisocyanate	0.005 sk 0.02 sk		TWA STEL	0.005 0.02	0.045 sk 0.180 sk	TWA STEL
2-Isopropoxy-ethanol	25	105	TWA		—	
Isopropyl acetate	250 310	950 1185	TWA STEL		—	
Isopropyl alcohol	400 500	980 1225	TWA STEL	400 500	980 1225	TWA STEL
Isopropylamine	5 10	12 24	TWA STEL		—	

SUMMARY OF OSHA PELs AND NIOSH RELs

SUBSTANCE	FINAL OSHA PEL ppm[a]	mg/m^3[b]	Type[d]	NIOSH RECOMMENDATION ppm[a]	mg/m^3[b]	Type[c]
N-Isopropylaniline	2	10 sk	TWA	2	10 sk	TWA
Isopropyl ether	500	2100	TWA	500	2100	TWA
Isopropyl glycidyl ether (IGE)	50 75	240 360	TWA STEL	50	240	CL
Kaolin						
Total dust		10	TWA		10	TWA
Respirable fraction		5	TWA		5	TWA
Kepone		none			Ca 0.001	TWA
Kerosene		none			100	TWA
Ketene	0.5 1.5	0.9 3	TWA STEL	0.5 1.5	0.9 3	TWA STEL
Lead inorganic (as Pb)		0.05	TWA		<0.1	
	(see 1910.1025) For independent battery breaking, non-ferrous foundries, secondary copper, lead pigments, lead chemical, ship building, stevedoring, and brass and bronze ingot manufacturing, paragraph (e)(1) is under court remand			Air level to be maintained so that worker blood level remain ≤0.060 mg/100g of whole blood		
Limestone						
Total dust		15	TWA	10		TWA
Respirable fraction		5	TWA	5		TWA
Lindane		0.5 sk	TWA	0.5 sk		TWA
Lithium hydride		0.025	TWA	0.025		TWA
L.P.G. (liquefied petroleum gas)	1000	1800	TWA	1000	1800	TWA
Magnesite						
Total dust		15	TWA		10	TWA
Respirable fraction		5	TWA		5	TWA
Magnesium oxide fume						
Total particulate		10	TWA		—	
Malathion		10 sk	TWA		10 sk	TWA

SUMMARY OF OSHA PELs AND NIOSH RELs

SUBSTANCE	FINAL OSHA PEL ppm[a]	mg/m³[b]	Type[d]	NIOSH RECOMMENDATION ppm[a]	mg/m³[b]	Type[c]
Maleic anhydride	0.25	1	TWA	0.25	1	TWA
Malononitrile		none		3	8	TWA
Manganese compounds (as Mn)		5	CL		1 3	TWA STEL
Manganese fume (as Mn)		1 3	TWA STEL		1 3	TWA STEL
Manganese cyclo-pentadienyl tricarbonyl (as Mn)		0.1 sk	TWA		0.1 sk	TWA
Manganese tetroxide (as Mn)		1	TWA		—	
Marble						
Total dust		15	TWA		10	TWA
Respirable fraction		5	TWA		5	TWA
Mercury (aryl and inorganic) (as Hg)		0.1 sk	CL		0.1 sk	CL
Mercury (organo) alkyl compounds (as Hg)		**0.01 sk 0.03 sk	TWA STEL		0.01 sk 0.03 sk	TWA STEL
Mercury (vapor) (as Hg)		0.05 sk	TWA		0.05 sk	TWA
Mesityl oxide	15 25	60 100	TWA STEL	10	40	TWA
Methacrylic acid	20	70 sk	TWA	20	70 sk	TWA
Methanethiol; see methyl mercaptan						
Methomyl (Lannate)		2.5	TWA		2.5	TWA
Methoxychlor		10	TWA		*Ca (LOQ 0.07)	
2-Methoxyethanol; see methyl cellosolve						
Methoxyflurane		none		2	13.5	CL
4-Methoxyphenol		5	TWA		5	TWA

SUMMARY OF OSHA PELs AND NIOSH RELs

SUBSTANCE	FINAL OSHA PEL ppm[a]	mg/m[3b]	Type[d]	NIOSH RECOMMENDATION ppm[a]	mg/m[3b]	Type[c]
Methyl acetate	200 250	610 760	TWA STEL	200 250	610 760	TWA STEL
Methyl acetylene (propyne)	1000	1650	TWA	1000	1650	TWA
Methyl acetylene-propadiene mixture (MAPP)	1000 1250	1800 2250	TWA STEL	1000 1250	1800 2250	TWA STEL
Methyl acrylate	10	35 sk	TWA	10	35 sk	TWA
Methylacrylonitrile	1	3 sk	TWA	1	3 sk	TWA
Methylal (dimethoxymethane)	1000	3100	TWA	1000	3100	TWA
Methyl alcohol	200 250	260 sk 325 sk	TWA STEL	200 250	260 sk 325 sk	TWA STEL
Methylamine	10	12	TWA	10	12	TWA
Methyl amyl alcohol; see methyl isobutyl carbinol						
Methyl n-amyl ketone	100	465	TWA	100	465	TWA
Methyl bromide	5	20 sk	TWA	*Ca (LOQ 4.7)		
Methyl butyl ketone; see 2 hexanone						
Methyl cellosolve (2-methoxy-ethanol)	25 In process of 6(b) rulemaking	80 sk	TWA	Reduce exposure to lowest feasible concentration		
Methyl cellosolve acetate (2-methoxyethyl acetate)	25 In process of 6(b) rulemaking	120 sk	TWA	Reduce exposure to lowest feasible concentration		
Methyl chloride	50 100	105 210	TWA STEL	*Ca—lowest feasible (LOQ 1.6)		
Methyl chloroform (1,1,1-trichloro-ethane)	350 450	1900 2450	TWA STEL	350	1900	CL

SUMMARY OF OSHA PELs AND NIOSH RELs

	FINAL OSHA PEL			NIOSH RECOMMENDATION		
SUBSTANCE	ppm[a]	mg/m[3b]	Type[d]	ppm[a]	mg/m[3b]	Type[c]
Methyl 2-cyanoacrylate	2 4	8 16	TWA STEL	2 4	8 16	TWA STEL
Methylcyclohexane	400	1600	TWA	400	1600	TWA
Methylcyclohexanol	50	235	TWA	50	235	TWA
o-Methylcyclo hexanone	50 75	230 sk 345 sk	TWA STEL	50 75	230 sk 345 sk	TWA STEL
Methylcyclopentadienyl manganese tricarbonyl (as Mn)		0.2 sk	TWA		0.2 sk	TWA
Methyl demeton		0.5 sk	TWA		0.5 sk	TWA
4,4'-Methylene bis (2-chloroaniline) (MBOCA)	0.02	0.22 sk	TWA		Ca 0.003 sk	TWA
Methylene bis (4-cyclo-hexylisocyanate)	0.01	0.11	CL	0.01	0.11	CL
Methylene bispheny isocyanate (MDI)	0.02	0.2	CL	0.005 0.020	0.050 0.200	TWA CL (10-min)
Methylene chloride	500 1000 2000		TWA CL PK (5-min in 2 hrs)	Ca—lowest feasible		
	In process of 6(b) rulemaking					
4,4'-Methylenedi-aniline		0.01 .1	TWA STEL	CA—lowest feasible (LOQ 0.03) 1990		
	In process of 6(b) rulemaking					
Methyl ethyl ketone (MEK); see 2-Butanone						
Methyl ethyl ketone peroxide (MEKP)	0.7	5	CL	0.2	1.5	CL
Methyl formate	100 150	250 375	TWA STEL	100 150	250 375	TWA STEL

SUMMARY OF OSHA PELs AND NIOSH RELs

SUBSTANCE	FINAL OSHA PEL ppm[a]	mg/m^3[b]	Type[d]	NIOSH RECOMMENDATION ppm[a]	mg/m^3[b]	Type[c]
Methyl hydrazine (mono-methyl hydrazine)	0.2	0.35 sk	CL	Ca 0.04	0.08	CL (120-min)
Methyl iodide	2	10 sk	TWA	Ca 2	10 sk	TWA
Methyl isoamyl ketone	50	240	TWA	50	240	TWA
Methyl isobutyl carbinol	25 40	100 sk 165 sk	TWA STEL	25 40	100 sk 165 sk	TWA STEL
Methyl isobutyl ketone; see Hexone						
Methyl isocyanate	0.02	0.05 sk	TWA	0.02	0.05 sk	TWA
Methyl isopropyl ketone	200	705	TWA	200	705	TWA
Methyl mercaptan	0.5	1	TWA	0.5	1	CL
Methyl methacrylate	100	410	TWA	100	410	TWA
Methyl parathion		0.2 sk	TWA		0.2 sk	TWA
Methyl propyl ketone; see 2-Pentanone						
Methyl silicate	1	6	TWA	1	6	TWA
alpha-Methyl styrene	50 100	240 485	TWA STEL	50 100	240 485	TWA STEL
Metribuzin		5	TWA		5	TWA
Mica; see silicates						
Mineral wool fiber		none (PEL delayed)		3 million fibers/m^3 (fibers ≤3.5 ≤g dia and ≥ 10 ≤m length) 5 mg/m^3 (total mineral wool)		TWA TWA
Molybdenum (as Mo)						
Soluble compounds		5	TWA		—	
Insoluble compounds						
Total dust		10	TWA		—	

SUMMARY OF OSHA PELs AND NIOSH RELs

SUBSTANCE	FINAL OSHA PEL ppm[a]	mg/m3[b]	Type[d]	NIOSH RECOMMENDATION ppm[a]	mg/m3[b]	Type[c]
Monocrotophos (Azodrin[R])		0.25	TWA		0.25	TWA
Monomethyl aniline	0.5	2 sk	TWA	0.5	2 sk	TWA
Morpholine	20 30	70 sk 105 sk	TWA STEL	20 30	70 sk 105 sk	TWA STEL
Naphtha (coal tar)	100	400	TWA	100	400	TWA
Naphthalene	10 15	50 75	TWA STEL	10 15	50 75	TWA STEL
Naphthalene diiso- cyanate (NDI)		none			0.04 0.17	TWA CL
alpha-Naphthylamine		Ca - no PEL (see 1910.1004)			Ca (use 1910.1004)	
beta-Naphthylamine		Ca - no PEL (see 1910.1009)			Ca (use 1910.1009)	
Niax Catalyst ESN		none			minimize exposure	
Nickel carbonyl (as Ni)	0.001	0.007	TWA	Ca 0.001	0.007	TWA
Nickel, metal and insoluble compounds as (Ni)		1	TWA		Ca 0.015	TWA
Nickel, soluble compounds (as Ni)		0.1	TWA		Ca 0.015 (inorganic nickel)	TWA
Nickel sulfide—roasting (as Ni)		none			Ca 0.015	TWA
Nicotine		0.5 sk	TWA		0.5 sk	TWA
Nitric acid	2 4	5 10	TWA STEL	2 4	5 10	TWA STEL
Nitric oxide	25	30	TWA	25	30	TWA
p-Nitroaniline		3 sk	TWA		3 sk	TWA
Nitrobenzene	1	5 sk	TWA	1	5 sk	TWA

SUMMARY OF OSHA PELs AND NIOSH RELs

SUBSTANCE	FINAL OSHA PEL ppm[a]	mg/m^3[b]	Type[d]	NIOSH RECOMMENDATION ppm[a]	mg/m^3[b]	Type[c]
4-Nitrobiphenyl		Ca - no PEL (see 1910.1003)			Ca (1910.1003)	
p-Nitrochlorobenzene		1 sk	TWA		*Ca sk (LOQ 0.25)	
Nitroethane	100	310	TWA	100	310	TWA
Nitrogen dioxide	1	1.8	STEL	1	1.8	STEL
Nitrogen trifluoride	10	29	TWA	10	29	TWA
Nitroglycerin		0.1 sk	STEL		0.1 sk	STEL
Nitromethane	100	250	TWA		—	
2-Nitro-naphthalene		none			Ca—lowest feasible	
1-Nitropropane	25	90	TWA	25	90	TWA
2-Nitropropane	10	35	TWA		*Ca (LOQ 1.4)	
N-Nitrosodimethylamine		Ca - no PEL (see 1910.1016)			Ca (1910.1016)	
Nitrotoluene o-isomer m-isomer p-isomer	2	11 sk	TWA	2	11 sk	TWA
Nitrotrichloromethane; see Chloropicrin						
Nitrous oxide		none	25	30	TWA	
Nonane	200	1050	TWA	200	1050	TWA
n-Nonylmercaptan		none	0.5	3.3	CL	
Octachloronaphthalene		0.1 sk 0.3 sk	TWA STEL		0.1 sk 0.3 sk	TWA STEL
Octadecylmercaptan		none		0.5	5.9	CL
Octane	300	1450 375	TWA 1800	75 STEL	350 385	TWA 1800CL

SUMMARY OF OSHA PELs AND NIOSH RELs

SUBSTANCE	FINAL OSHA PEL ppm[a]	mg/m[3b]	Type[d]	NIOSH RECOMMENDATION ppm[a]	mg/m[3b]	Type[c]
n-Octylmercaptan		none		0.5	3.0	CL
Oil mist, mineral		5	TWA		5 10	TWA STEL
Osmium tetroxide (as Os)	0.0002 0.0006	0.002 0.006	TWA STEL	0.0002 0.0006	0.002 0.006	TWA STEL
Oxalic acid		1 2	TWA STEL		1 2	TWA STEL
Oxygen difluoride	**0.05	0.1	CL	0.05	0.1	CL
Ozone	0.1 0.3	0.2 0.6	TWA STEL	0.1	0.2	CL
Paraffin wax fume		2	TWA		2	TWA
Paraquat, respirable dust		0.1 sk	TWA		0.1 sk	TWA
Parathion		0.1 sk	TWA		.05 sk	TWA
Particulates not otherwise regulated						
Total dust		15	TWA		—	
Respirable fraction		5	TWA		—	
Pentaborane	0.005 0.015	0.01 0.03	TWA STEL	0.005 0.015	0.01 0.03	TWA STEL
Pentachloroethane		none		Ca—minimize exposure		
Pentachloronaphthalene		0.5 sk	TWA		0.5 sk	TWA
Pentachlorophenol		0.5 sk	TWA		0.5 sk	TWA
Pentaerythritol						
Total dust		10	TWA		10	TWA
Respirable fraction		5	TWA		5	TWA
Pentane	1600 750	1800 2250	TWA STEL	120 610	350 1800	TWA CL
2-Pentanone (Methyl propyl ketone)	200 250	700 875	TWA STEL	150	530	TWA
Pentylmercaptan		none		0.5	2.1	CL

SUMMARY OF OSHA PELs AND NIOSH RELs

SUBSTANCE	FINAL OSHA PEL ppm[a]	mg/m³[b]	Type[d]	NIOSH RECOMMENDATION ppm[a]	mg/m³[b]	Type[c]
Perchloroethylene (Tetrachloroethylene)	25	170	TWA	*Ca—minimize exposure (LOQ 0.4)		
Perchloromethyl mercaptan	0.1	0.8	TWA	0.1	0.8	TWA
Perchloryl fluoride	3 6	14 28	TWA STEL	3 6	14 28	TWA STEL
Perlite						
Total dust		15			10	TWA
Respirable fraction		5	TWA		5	TWA
Petroleum distillates (Naphtha)	400	1600	TWA		350 1800	TWA CL
Phenol	5	19 sk	TWA	5 15.6	19 sk 60 sk	TWA CL
Phenothiazine		5 sk	TWA		5 sk	TWA
N-Phenyl-beta-naphthylamine		none		Ca—lowest feasible		
p-Phenylene diamine		0.1 sk	TWA		0.1 sk	TWA
Phenyl ether, vapor	1	7	TWA	1	7	TWA
Phenyl ether-biphenyl mixture, vapor	1	7	TWA	1	7	TWA
Phenylethylene; see Styrene						
Phenyl glycidyl ether (PGE)	1	6	TWA	Ca 1	6	CL
Phenylhydrazine	5 10	20 sk 45 sk	TWA STEL	Ca 0.14	0.6 sk	CL (120-min)
Phenyl mercaptan	0.5	2	TWA	0.1	0.5	CL
Phenylphosphine	**0.05	0.25	CL	0.05	0.25	CL
Phorate		0.05 sk 0.2 sk	TWA STEL		0.05 sk 0.2 sk	TWA STEL

SUMMARY OF OSHA PELs AND NIOSH RELs

SUBSTANCE	FINAL OSHA PEL ppm[a]	mg/m^{3b}	Type[d]	NIOSH RECOMMENDATION ppm[a]	mg/m^{3b}	Type[c]
Phosdrin (Mevinphos[R])	0.01 0.03	0.1 sk 0.3 sk	TWA STEL	0.01 0.03	0.1 sk 0.3 sk	TWA STEL
Phosgene (Carbonyl chloride)	0.1	0.4	TWA	0.1 0.2	0.4 0.8	TWA CL
Phosphine	0.3 1	0.4 1	TWA STEL	0.3 1	0.4 1	TWA STEL
Phosphoric acid		1 3	TWA STEL		1 3	TWA STEL
Phosphorus (yellow)		0.1	TWA		0.1	TWA
Phosphorus oxychloride	0.1	0.6	TWA	0.1 0.5	0.6 3	TWA STEL
Phosphorus pentachloride		1	TWA		1	TWA
Phosphorus pentasulfide		1 3	TWA STEL		1 3	TWA STEL
Phosphorus trichloride	0.2 0.5	1.5 3	TWA STEL	0.2 0.5	1.5 3	TWA STEL
Phthalic anhydride	1	6	TWA	1	6	TWA
m-Phthalodinitrile		5	TWA		5	TWA
Picloram						
Total dust		10	TWA		—	
Respirable fraction		5	TWA		—	
Picric acid		0.1 sk	TWA		0.1 sk 0.3 sk	TWA STEL
Pindone (2-Pivalyl-1,3-indandione)		0.1	TWA		0.1	TWA
Piperazine dihydrochloride		5	TWA		5	TWA
Plaster of Paris						
Total dust		15	TWA		10	TWA
Respirable fraction		5	TWA		5	TWA

SUMMARY OF OSHA PELs AND NIOSH RELs

SUBSTANCE	FINAL OSHA PEL ppm[a]	mg/m^{3b}	Type[d]	NIOSH RECOMMENDATION ppm[a]	mg/m^{3b}	Type[c]
Platinum (as Pt)						
Metal		1	TWA		1	TWA
Soluble salts		0.002	TWA		0.002	TWA
Portland cement						
Total dust		10	TWA		10	TWA
Respirable fraction		5	TWA		5	TWA
Potassium hydroxide		2	CL		2	CL
Propane	1000	1800	TWA	1000	1800	TWA
Propane sultone		none		Ca—lowest feasible		
Propargyl alcohol	1	2 sk	TWA	1	2 sk	TWA
beta-Propriolactone	Ca - no PEL (see 1910.1013)			Ca (use 1910.1013)		
Propionic acid	10	30	TWA	10	30	TWA
				15	45	STEL
Propionitrile		none		6	14	TWA
Propoxur (Baygon)		0.5	TWA		0.5	TWA
n-Propyl acetate	200	840	TWA	200	840	TWA
	250	1050	STEL	250	1050	STEL
n-Propyl alcohol	200	500	TWA	200	500 sk	TWA
	250	625	STEL	250	625 sk	STEL
Propylene dichloride	75	350	TWA	*Ca (LOQ 0.03)		
	110	510	STEL			
Propylene glycol dinitrate	0.05	0.3	TWA	0.05	0.3 sk	TWA
Propylene glycol mono-methyl ether	100	360	TWA	100	360	TWA
	150	540	STEL	150	540	STEL
Propylene imine	2	5 sk	TWA	Ca 2	5 sk	TWA
Propylene oxide	20	50	TWA	*Ca (LOQ 8.4)		
n-Propylmercaptan		none		0.5	1.6	CL

SUMMARY OF OSHA PELs AND NIOSH RELs

SUBSTANCE	FINAL OSHA PEL ppm[a]	mg/m^{3b}	Type[d]	NIOSH RECOMMENDATION ppm[a]	mg/m^{3b}	Type[c]
n-Propyl nitrate	25	105	TWA	25	105	TWA
	40	170	STEL	40	170	STEL
Propyne; see Methyl acetylene						
Pyrethrum		5	TWA		5	TWA
Pyridine	5	15	TWA	5	15	TWA
Quinone	0.1	0.4	TWA	0.1	0.4	TWA
Resorcinol	10	45	TWA	10	45	TWA
	20	90	STEL	20	90	STEL
Rhodium (as Rh), metal fume and insoluble compounds		0.1	TWA		0.1	TWA
Rhodium (as Rh), soluble compounds		0.001	TWA		0.001	TWA
Ronnel		10	TWA		10	TWA
Rosin core solder pyrolysis products as formaldehyde		0.1	TWA		Ca 0.1	CL
Rotenone		5	TWA		5	TWA
Rouge						
Total dust		10	TWA		—	
Respirable fraction		5	TWA		—	
Selenium compounds (as Se)		0.2	TWA		0.2	TWA
Selenium hexafluoride (as Se)	0.05	0.4	TWA	0.05	0.4	TWA
Silica, amorphous, precipitated and gel		6	TWA		6	TWA
Silica, amorphous, diatomaceous earth, containing less than 1% crystalline silica		6	TWA		—	

SUMMARY OF OSHA PELs AND NIOSH RELs

SUBSTANCE	FINAL OSHA PEL ppm[a]	mg/m^3[b]	Type[d]	NIOSH RECOMMENDATION ppm[a]	mg/m^3[b]	Type[c]
Silica, crystalline cristobalite (as quartz), respirable dust		0.05	TWA		Ca 0.05	TWA
Silica, crystalline quartz (as quartz), respirable dust		0.1	TWA		Ca 0.05	TWA
Silica, crystalline tripoli (as quartz), respirable dust		0.1	TWA		Ca 0.05	TWA
Silica, crystalline tridymite (as quartz), respirable dust		0.05	TWA		Ca 0.05	TWA
Silica, fused, respirable dust		0.1	TWA		Ca 0.05	TWA
Silicates (less than 1% crystalline silica)						
Mica (respirable dust)		3	TWA		3	TWA
Soapstone, total dust		6	TWA		6	TWA
Soapstone, respirable dust		3	TWA		3	TWA
Talc (containing asbestos): use asbestos limit	(see 29 CFR 1910.1001)			Ca 100,000/m^3 (fibers >5 ≤m long) 400 liter sample		
Talc (containing no asbestos), respirable dust		2	TWA		2	TWA
Tremolite	(see 29 CFR 1910.1101)	2 fbrs/cc	TWA	Ca 0.1 fbrs/cc (see 1910.1101)		TWA
Silicon						
Total dust		10	TWA		10	TWA
Respirable fraction		5	TWA		5	TWA
Silicon carbide						
Total dust		10	TWA		10	TWA
Respirable fraction		5	TWA		5	TWA
Silicon tetrahydride	5	7	TWA	5	7	TWA

SUMMARY OF OSHA PELs AND NIOSH RELs

	FINAL OSHA PEL			NIOSH RECOMMENDATION		
SUBSTANCE	ppm[a]	mg/m^3[b]	Type[d]	ppm[a]	mg/m^3[b]	Type[c]
Silver, metal and soluble compounds (as Ag)		0.01	TWA		0.01	TWA
Soapstone; see Silicates						
Sodium azide						
(as HN_3)	0.1 sk		CL	0.1 sk		CL
(as NaN_3)		0.3 sk	CL		0.3 sk	CL
Sodium bisulfite		5	TWA		5	TWA
Sodium fluoroacetate		0.05 sk 0.15 sk	TWA STEL		0.05 sk 0.15 sk	TWA STEL
Sodium hydroxide		2	CL		2	CL
Sodium metabisulfite		5	TWA		5	TWA
Starch						
Total dust		15	TWA		10	TWA
Respirable fraction		5	TWA		5	TWA
Stibine	0.1	0.5	TWA	0.1	0.5	TWA
Stoddard solvent	100	525	TWA		350 1800	TWA CL
Strychnine		0.15	TWA		0.15	TWA
Styrene	50 100	215 425	TWA STEL	50 100	215 425	TWA STEL
Subtilisins (Proteolytic enzymes)		0.00006	STEL (60-min)		0.00006	STEL (60-min)
Succinonitrile		none	6	20	TWA	
Sucrose						
Total dust			15	TWA		10TWA
Respirable fraction			5	TWA		5TWA
Sulfur dioxide	2 5	5 10	TWA STEL	2 5	5 10	TWA STEL
Sulfur hexa-fluoride	1000	6000	TWA	1000	6000	TWA
Sulfuric acid		1	TWA		1	TWA

SUMMARY OF OSHA PELs AND NIOSH RELs

SUBSTANCE	FINAL OSHA PEL ppm[a]	mg/m[3b]	Type[d]	NIOSH RECOMMENDATION ppm[a]	mg/m[3b]	Type[c]
Sulfur monochloride	1	6	CL	1	6	CL
Sulfur pentafluoride	**0.01	0.1	CL	0.01	0.1	CL
Sulfur tetrafluoride	0.1	0.4	CL	0.1	0.4	CL
Sulfuryl fluoride	5 10	20 40	TWA STEL	5 10	20 40	TWA STEL
Sulprofos		1	TWA		1	TWA
Systox[R], see Demetron						
2,4,5-T		10	TWA		10	TWA
Talc; see Silicates						
Tantalum, metal and oxide dust		5	TWA		5 10	TWA STEL
TEDP (sulfotep)		0.2 sk	TWA		0.2 sk	TWA
Tellurium and compounds (as Te)		0.1	TWA		0.1	TWA
Tellurium hexafluoride (as Te)	0.02	0.2	TWA	0.02	0.2	TWA
Temephos Total dust Respirable fraction		 10 5	 TWA TWA		 10 5	 TWA TWA
TEPP		0.05 sk	TWA		0.05 sk	TWA
2,3,7,8-Tetrachlorodibenzo-p-dioxin (TCDD)		none		Ca—lowest feasible		
Terphenyls	0.5	5	CL	0.5	5	CL
1,1,2,2-Tetrachloroethane	1	7 sk	TWA	Ca 1	7 sk	TWA
Tetrachloroethylene; see perchloroethylene						
Tetrachloromethane; see carbon tetrachloride						
Tetrachloronaphthalene		2 sk	TWA		2 sk	TWA

SUMMARY OF OSHA PELs AND NIOSH RELs

SUBSTANCE	FINAL OSHA PEL ppm[a]	mg/m[3b]	Type[d]	NIOSH RECOMMENDATION ppm[a]	mg/m[3b]	Type[c]
Tetraethyl lead (as Pb)		0.075 sk	TWA		0.075 sk	TWA
Tetrahydrofuran	200 250	590 735	TWA STEL	200 250	590 735	TWA STEL
Tetramethyl lead (as Pb)		0.075 sk	TWA		0.075 sk	TWA
Tetramethyl succinonitrile	0.5	3 sk	TWA	0.5	3 sk	TWA
Tetranitromethane	1	8	TWA	1	8	TWA
Tetrasodium pyrophosphate		5	TWA		5	TWA
Tetryl (2,4,6-Trinitro phenyl-methyl-nitramine)		1.5 sk	TWA		1.5 sk	TWA
Thallium, soluble compounds (as Tl)		0.1 sk	TWA		0.1 sk	TWA
4,4'-Thiobis(6-tert, butyl-m-cresol)						
Total dust		10	TWA		10	TWA
Respirable fraction		5	TWA		5	TWA
Thioglycolic acid	1	4 sk	TWA	1	4 sk	TWA
Thionyl chloride	1	5	CL	1	5	CL
Thiram		5	TWA		5	TWA
Tin, inorganic compounds (except oxides) (as Sn)		2	TWA		2	TWA
Tin, organic compounds (as Sn)		0.1 sk	TWA		0.1 sk	TWA
Tin oxide (as Sn)		2	TWA		2	TWA
Titanium dioxide		10	TWA		*Ca (LOQ 0.2)	
o-Tolidine		none			Ca 0.02	CL (60-min)
o-Tolidine-based dyes		none		Ca—minimize exposure		

SUMMARY OF OSHA PELs AND NIOSH RELs

	FINAL OSHA PEL			NIOSH RECOMMENDATION		
SUBSTANCE	ppm[a]	mg/m^3[b]	Type[d]	ppm[a]	mg/m^3[b]	Type[c]
Toluene	100	375	TWA	100	375	TWA
	150	560	STEL	150	560	STEL
Toluenediamine (TDA)		none		Ca—lowest feasible		
Toluene-2,4-diiso-	0.005	0.04	TWA	Ca—reduce to lowest		
cyanate (TDI)	0.02	0.15	STEL	feasible cancer		
m-Toluidine	2	9 sk	TWA		—	
					(LOQ 0.15) sk	
o-Toluidine	5	22 sk	TWA	Ca—lowest feasible		
p-Toluidine	2	9 sk	TWA	*Ca sk		
				(LOQ 0.15)		
Toxaphene; see chlorinated camphene						
Tremolite; see silicates						
Tributyl phosphate	0.2	2.5	TWA	0.2	2.5	TWA
Trichloroacetic acid	1	7	TWA	1	7	TWA
1,2,4-Trichloro- benzene	5	40	CL	5	40	CL
1,1,1-Trichloroethane; see methyl chloroform						
1,1,2-Trichloro- ethane	10	45 sk	TWA	Ca 10	45 sk	TWA
Trichloroethylene	50	270	TWA	Ca 25		TWA
	200	1080	STEL			
Trichloromethane; see chloroform						
Trichloronaphthalene		5 sk	TWA		5 sk	TWA
1,2,3-Trichloro- propane	10	60	TWA	Ca 10	60 sk	TWA
1,1,2-Trichloro-	1000	7600	TWA	1000	7600	TWA
1,2,2 trifluoroethane	1250	9500	STEL	1250	9500	STEL

SUMMARY OF OSHA PELs AND NIOSH RELs

	FINAL OSHA PEL			NIOSH RECOMMENDATION		
SUBSTANCE	ppm[a]	mg/m^{3}[b]	Type[d]	ppm[a]	mg/m^{3}[b]	Type[c]
Triethylamine	10 15	40 60	TWA STEL		—	
Trifluoro-bromomethane	1000	6100	TWA	1000	6100	TWA
Trimellitic anhydride	0.005	0.04	TWA	0.005	0.04	TWA
				Should be handled in workplace as an extremely toxic subtance		
Trimethylamine	10 15	24 36	TWA STEL	10 15	24 36	TWA STEL
Trimethyl benzene	25	125	TWA	25	125	TWA
Trimethyl phosphite	2	10	TWA	2	10	TWA
2,4,6-Trinitrophenyl; see picric acid						
2,4,6-Trinitrophenyl-methyl nitramine; see Tetryl						
2,4,6-Trinitrotoluene (TNT)		0.5 sk	TWA		0.5 sk	TWA
Triorthocresyl phosphate		0.1 sk	TWA		0.1 sk	TWA
Triphenyl amine		5	TWA		5	TWA
Triphenyl phosphate		3	TWA		3	TWA
Tungsten (as W)						
Insoluble compounds		5 10	TWA STEL		5 10	TWA STEL
Soluble compounds		1 3	TWA STEL		1 3	TWA STEL
Tungsten						
(containing >2% cobalt)		none			0.01 (Co)	TWA
(containing >3% nickel)					0.015 (Ni)	TWA
Turpentine	100	560	TWA	100	560	TWA
1-Undecanethiol		none		0.5	3.9	CL

SUMMARY OF OSHA PELs AND NIOSH RELs

SUBSTANCE	FINAL OSHA PEL ppm[a]	mg/m[3b]	Type[d]	NIOSH RECOMMENDATION ppm[a]	mg/m[3b]	Type[c]
Uranium (as U)						
Soluble compounds		0.05	TWA		Ca 0.05	TWA
Insoluble compounds		0.2	TWA		Ca 0.2	TWA
		0.6	STEL		Ca 0.6	STEL
n-Valeraldehyde	50	175	TWA	50	175	TWA
Vanadium						
Respirable dust (as V_2O_5)		0.05	TWA		0.05	
Fume (as V_2O_5)		0.05	TWA		0.05	CL
Vegetable oil mist						
Total dust		15	TWA		10	TWA
Respirable fraction		5	TWA		5	TWA
Vinyl acetate	10	30	TWA	4	15	CL
	20	60	STEL			
Vinyl benzene; see styrene						
Vinyl bromide	5	20	TWA	*Ca (LOQ 0.2) (use 1910.1017)		
Vinyl chloride	1		TWA	Ca		
	5 (use 1910.1017)		CL	(use 1910.1017)		
Vinylcyanide; see acrylonitrile						
Vinyl cyclohexene dioxide	10	60 sk	TWA	Ca 10	60 sk	TWA
Vinyl fluoride		none		1		TWA
				5 (use 1910.1017)		CL
Vinylidene chloride (1,1-dichloroethylene)	1	4	TWA	*Ca (LOQ 0.4) (use 1910.1017) vinyl chloride		

SUMMARY OF OSHA PELs AND NIOSH RELs

SUBSTANCE	FINAL OSHA PEL ppm[a]	 mg/m^3[b]	 Type[d]	NIOSH RECOMMENDATION ppm[a]	 mg/m^3[b]	 Type[c]
Vinylidene fluoride		none		1 5 (use 1910.1017)		TWA CL
Vinyl toluene	100	480	TWA	100	480	TWA
VM&P Naphtha	300 400	1350 1800	TWA STEL		350 1800	TWA CL
Warfarin		0.1	TWA		0.1	TWA
Welding fumes (total particuate)		5	TWA	Ca—lowest feasible		
Wood dust, all soft and hard woods, except western red cedar		5 10	TWA STEL		1	TWA
Wood dust, western red cedar		2.5	TWA		1	TWA
Xylenes (o-, m-, p- isomers)	100 150	435 655	TWA STEL	100 150	435 655	TWA STEL
m-Xylene alpha, alpha'-diamine		0.1 sk	CL		0.1 sk	CL
Xylidine	2	10 sk	TWA	2	10 sk	TWA
Yttrium		1	TWA		1	TWA
Zinc chloride fume		1 2	TWA STEL		1 2	TWA STEL
Zinc chromate (as $Cr0_3$)		0.1	CL		Ca 0.001	TWA
Zinc oxide fume		5 10	TWA STEL		5 10	TWA STEL
Zinc oxide Total dust		 10	 TWA		 5	 TWA 15CL
Respirable fraction		5	TWA			

SUMMARY OF OSHA PELs AND NIOSH RELs

SUBSTANCE	FINAL OSHA PEL ppm[a]	mg/m^{3} [b]	Type[d]	NIOSH RECOMMENDATION ppm[a]	mg/m^{3} [b]	Type[c]
Zinc stearate						
Total dust		10	TWA		10	TWA
Respirable fraction		5	TWA		5	TWA
Zirconium compounds		5	TWA		5	TWA
(as Zr)		10	STEL		10	TWA
(except Zirconium Tetrachloride)					—	

Health effects noted at PEL.
*Because this substance is a potential occupational carcinogen, the NIOSH policy for exposure is lowest feasible limit. The limit of quantification is noted.
**Enforcement of limit indefinitely stayed by OSHA.
Denotes dusts that NIOSH suggests has health effects beyond those expected from nuisance dusts.

Environmental Health Law

Toxic substances and hazardous materials used in the workplace frequently escape into the air, soil, surface water, and groundwater. Releases to the environment occur as the result of industrial processes, domestic waste generation, accidents, fires, and improper disposal of toxic substances and hazardous materials. Toxic substances also leave the workplace on the contaminated clothing of workers.

Historically, environmental regulations have focused on end-of-the-pipe control technology to reduce emissions from the workplace and regulate the disposal of hazardous wastes generated in the production process. Beginning in the mid-1980s, policymakers, citizen groups, and employers began recognizing the failure of this approach and looking for ways to prevent pollution by reducing the use of toxic substances at the front end of production.

This trend has been referred to as "source reduction," "toxics use reduction," and "pollution prevention." The approach is quite simple. If the use of toxic substances is reduced at the beginning of the production process, workers are not exposed and many of the traditional costs related to pollution control and hazardous waste disposal are reduced or eliminated. The following hierarchy of pollution prevention strategies has been adopted by some states:

- Input substitution
- Product reformulation
- Production process redesign or modification
- Production process modernization
- Improved operation and maintenance of production process equipment and methods

By the end of 1990, the EPA had established the Office of Pollution Prevention, and 21 states had adopted or considered some form of pollution prevention, toxics use reduction, or hazardous waste reduction legislation. Successes by some states in several industries have led to "Pollution Prevention Pays" programs. Until pollution prevention is widely implemented, public health practitioners will have to rely on the existing environmental regulatory framework for assistance in investigating the relationship between environmental exposures and diseases.

Nonworkplace environmental exposures often differ from workplace exposures in that concentrations may be lower, duration of exposure may be longer, and route of body entry may be different. The population exposed outside the workplace is also more diverse; it includes children, elderly people, disabled persons, and those with existing illnesses. The health status of an exposed person and the characteristics of the exposure may affect presentation courses and outcome of any associated illness.

Unlike the statutes and agencies directed at controlling occupational disease and injury, environmental health statutes and agencies that regulate toxic substances and hazardous materials must address the complexity of the environment and the diversity of the general population. As a result, a patchwork of federal laws has been enacted to cover air and water quality (Clean Air Act, Clean Water Act, and Safe Drinking Water Act); toxic substances (Toxic Substances Control Act [TSCA]); hazardous waste transportation and disposal (Hazardous Materials Transportation Act [HMTA], Federal Railway Safety Act, and Resource Conservation and Recovery Act [RCRA]); pesticides (Federal Insecticide, Fungicide, and Rodenticide Act [FIFRA]); foods and drugs (Federal Food, Drug, and Cosmetic Act [FFDCA]); hazardous waste site cleanup (Comprehensive Environmental Response, Compensation, and Liability Act [CERCLA, Superfund] and Superfund Amendments and Reauthorization Act [SARA]); and consumer products (Consumer Product Safety Act). Some of these environmental laws, most notably FIFRA, SARA, and RCRA, also have provisions aimed at worker protection.

Each of these statutes confers responsibility on different federal agencies. The EPA has authority to implement the Clean Air and Clean Water acts, TSCA, FIFRA, CERCLA/SARA (Superfund), and RCRA; the Department of Transportation has responsibility for the Hazardous Materials Transportation Act and the Federal Railway Safety Act; the Food and Drug Administration for the FFDCA; and the Consumer Product Safety Commission for the Consumer Product Safety Act.

These federal laws can be broadly grouped into four general categories: (a) environmental protection statutes; (b) laws covering chemical manufacture, use, and toxicity assessment; (c) laws on transporting chemicals and hazardous substances; and (d) laws regulating the cleanup of hazardous waste sites. Table 1 presents an overview of the federal laws and their general areas of coverage.

Table 1.—Federal Toxic Substances Statutes by Categories of Coverage

Category of Coverage	Federal Statute
Environmental Protection	Clean Air Act
	Clean Water Act
	Safe Drinking Water Act
	Resource Conservation and Recovery Act
Chemical Manufacture, Use, and Assessment	Federal Food, Drug, and Cosmetic Act
	Federal Insecticide, Fungicide and Rodenticide Act
	Toxic Substance Control Act
	Superfund Amendments and Reauthorization Act
	Consumer Product Safety Act

Table 1.—Federal Toxic Substances Statutes by Categories of Coverage (continued)

Category of Coverage	Federal Statute
Transportation	Hazardous Materials Transportation Act Resource Conservation and Recovery Act Federal Railway Safety Act
Hazardous Waste Site Clean-up	Comprehensive Environmental Response, Compensation, and Liability Act (as amended by SARA)

Environmental Protection

In 1969, the National Environmental Policy Act was passed, setting the public policy direction for the many environmental statutes that were enacted during the 1970s. It declared a national policy that encouraged "productive and enjoyable harmony between man and his environment" and required preparation of environmental impact statements for "major Federal actions significantly affecting the quality of the human environment."

The first two of these new environmental protection statutes were the Clean Air Act of 1970 and the Clean Water Act of 1972, both of which amended earlier statutes. The focus of these laws is primarily on cleaning up common pollutants such as smoke and sulfur and nitrogen oxides in the air, oxygen-depleting discharges into surface waters, and solid wastes disposed on the land. By the end of the 1970s, environmental protection laws had begun to focus on the threat to human health that toxic substances could pose, even at low concentrations.

The Clean Air Act, as amended in 1977, focused on toxic air emissions and required the EPA to set three kinds of standards:

1. *National Primary and Secondary Ambient Air Quality Standards* define the maximum allowable concentration for air pollutants. (a) *Primary standards,* which establish a concentration level necessary to protect the public health, must be adequate to protect even those citizens who are particularly sensitive. (b) *Secondary standards,* which are designed to protect the public welfare from adverse effects of air emissions on soils, water, vegetation, wildlife, weather, visibility, property, economic values, and personal comfort and well-being, have been criticized as being cost-oblivious, because they have no provisions for balancing health and welfare benefits with the costs of achieving them.
2. *Standards of Performance for New Stationary Sources* are based on the premise that extraordinary flexibility in choice of location, technology, and economic circumstances exists for a new pollution source; therefore, the generator should be held to the highest performance standard. These standards require that the best available control technology (BACT) be used to comply with the standards.
3. *National Emissions Standards for Hazardous Air Pollutants* were intended to regulate at the source of emission those air pollutants that pose the greatest threat to public health because of their hazardous properties. Unfortunately, EPA has been extremely reluctant to issue these standards, despite numerous citizen suits aimed at forcing it to act. The Clean Air Act has become the prototype for federal citizen suit legislation, stating that "any person" may commence a civil action on his own behalf against "any person," including the United States and any other governmental instrumentality or agency.

The Clean Water Act of 1972 (as amended in 1977) synthesizes previous water pollution control legislation, sets limits on the discharge of toxic substances into surface water, and establishes water quality standards by setting maximum concentration levels of pollutants in surface water.

The Safe Drinking Water Act of 1975 (amended in 1986) is directed at protecting groundwater and drinking water sources. To accomplish this goal, EPA has established maximum contaminant levels (MCLs) for drinking water, a program to control the underground injection of toxic substances, and a program to protect underground wellheads.

The Resource Conservation and Recovery Act (RCRA) of 1976 (as amended in 1984) focuses on the recycling and disposal of solid wastes. It established a system of cradle-to-grave management of hazardous wastes, which included requirements for the treatment, storage, and disposal of hazardous waste and a manifest reporting system to monitor the shipment of hazardous waste offsite. It also created a program for controlling leaking underground storage tanks. Most significant are its public policy objectives, included in the 1984 amendments, that direct the EPA to phase out land disposal and deep-well injection of hazardous waste by minimizing the generation of hazardous waste through encouragement of process substitution, materials recovery, recycling, reuse, and treatment.

Chemical Manufacture, Use, and Assessment

In addition to those laws that focus on workplace exposures and the release of toxic substances into the environment, some federal laws cover exposure to toxic substances by consumers, users, and applicators.

The Federal Food, Drug, and Cosmetic Act (FFDCA) of 1938 is one of the early federal laws that regulated the presence of chemicals in food, drugs, and cosmetics. The act was aimed at protecting the public from adulterated or misbranded consumer products by setting premanufacturing standards and requiring premarketing clearances and assessment of potential adverse effects on human health. Substances known to cause cancer in animals were banned as food additives (by the

"Delaney Clause"), and pesticide residues on raw agricultural products were prohibited.

Control of pesticides also initially focused on consumer protection by regulating the adulterating and mislabeling of the product. In 1972, the Federal Insecticide, Fungicide, and Rodenticide Act (FIFRA) was passed, which requires pesticides to be registered and tested for their toxic effects, and establishes protections for pesticide applicators. Under FIFRA, pesticides must be tested prior to their registration and use. All health or environmental data submitted to EPA in the registration process are available to the public. However, out of 40 000 pesticides on the market, only 70 had been registered by May 1984.

The Toxics Substances Control Act (TSCA) was enacted in 1976 to require chemical producers to provide EPA with premanufacture notification of their intent to manufacture, process, or distribute a chemical substance that might present an unreasonable risk of injury to health or the environment. Chemical producers are also required to provide information on any significant risks posed by the chemicals they market. EPA may also require premanufacturing testing of chemical substances before they can enter the market for distribution.

Transportation

During the transport of toxic substances and hazardous wastes, workers and the general public are at great risk of exposure. The Department of Transportation (DOT), pursuant to the Hazardous Materials Transportation Act (HMTA) of 1975, has the authority to regulate the shipment of substances that may pose a threat to health, safety, property, or the environment when transported by air, water, rail, or highway. Depending on the mode of transportation, the DOT branch with jurisdiction might be the Federal Aviation Administration, the Federal Highway Administration, the Federal Railroad Administration, or the United States Coast Guard. In addition to the labeling, packaging, and placarding requirements of DOT, transporters of hazardous waste are required under the Resource Conservation and Recovery Act (RCRA) to register with EPA and carry hazardous waste manifests.

Hazardous Waste Site Cleanup

Despite all the attempts at pollution prevention through federal regulation, there are over 50 000 hazardous waste sites that pose significant health and environmental risks. In response to this situation, the Comprehensive Environmental Response, Compensation, and Liability Act (CERCLA) was enacted in 1980 and established a $1.6 billion "Superfund" to clean up abandoned hazardous waste sites. Superfund monies are generated by a tax on crude oil, petroleum products, and feedstock chemicals. CERCLA included provisions for the National Response Center, which was created by the Clean Water Act, to be notified of releases of hazardous substances. Additionally, it authorized the revision of the National Contingency Plan, also created by the Clean Water Act, to develop responses to releases of hazardous substances. As part of this plan, EPA was to create a national priorities list of hazardous waste sites presenting the greatest threat to public health or to the welfare of the environment.

Using Federal Environmental Laws To Get Information About Toxic Substances

Superfund Amendments and Reauthorization Act (SARA) of 1986

In 1986, CERCLA was amended and the Superfund was increased to $8.5 billion. As part of the Superfund Amendments and Reauthorization Act (SARA) of 1986, new sections were added to provide for emergency planning, public knowledge about hazardous substances in the community, and training for workers who either are exposed to hazardous waste or must respond to hazardous waste emergencies. Additionally, SARA requires the creation of state emergency response commissions and local emergency planning committees (LEPCs).

SARA also established a nationwide, public right to know about local chemical usage and hazardous substance releases. The act requires state and local communities to develop preparedness plans for chemical accident emergencies. A facility that uses hazardous substances must provide its LEPC with either (a) a list of each chemical used and its hazardous components, organized by category of health or physical hazard; or (b) a material safety data sheet (MSDS) for each chemical. Each facility is also required to report annually on hazardous chemical usage and on releases of toxic chemicals. This information must be available to the public through the LEPCs.

In the event of an emergency, SARA also provides for any health professional to gain immediate access to information on any specific chemical, even if this information is allegedly a trade secret. In nonemergency situations, information may be temporarily withheld if a trade secret claim has been made at the time of initial reporting to LEPC. However, trade secrecy is a difficult claim to make, and inaccurate or frivolous claims may result in a $25,000 penalty from EPA.

SARA also provides for community assistance grants and health studies. Technical assistance grants are available to local communities seeking to participate in cleaning up hazardous waste sites on the national priorities list. These grants are generally used to hire technical experts to interpret studies and evaluate remedial proposals.

In 1980, CERCLA created the Agency for Toxic Substances and Disease Registry (ATSDR) within the Public Health Service. The agency is required to assess potential health impacts of exposure to toxic substances; one way it does this is by medical monitoring of community residents for health effects due to toxic exposures. Both SARA and RCRA enable ATSDR to conduct health assessments. Additionally, ATSDR is required to create national registries of (a) serious diseases and illnesses and

(b) exposed persons. It is also charged with creating an inventory of medical and scientific reports and research studies on the health effects of toxic substances.

Citizen Suits

Despite this complex regulatory matrix, there are some common themes. In general, federal laws provide opportunities for concerned citizens to gain information about toxic substances, to be involved in planning for and evaluating situations concerning toxic substances, and to bring legal actions against polluters and against agencies that fail to enforce environmental regulations.

Although many environmental laws contain enforcement provisions, most of these provisions are rarely used. The citizen suit provisions of the Clean Air Act of 1970 set in motion the inclusion of similar provisions in more than 15 subsequent environmental legislative actions. Agencies do act on their own, but, with limited resources, they often must be prodded by others to act. A 1984 report by the Environmental Law Institute, which analyzed citizen enforcement actions under EPA-administered statutes, revealed a sharp rise in citizen suits under the Clean Water Act since 1982; of the 349 enforcement actions brought in 1983 and 1984, 64% were initiated by citizen suits. The report concluded that this increase in citizen suits was the result of a decrease in EPA enforcement since 1981. In addition to citizen suits, organized activity by affected communities is frequently required to compel agencies to respond to specific problems.

State and Local Statutes

Many states and localities have laws or regulations that may help citizens gain access to information about chemical substances. These laws may include expanded rights to know what is used and discharged into the environment, as well as more stringent requirements for permits or waste disposal. State public health departments, departments of environmental quality, and poison control centers may also be asked to assist in acquiring, interpreting, or using information.

Pollution Prevention: A Solution

Recognition, diagnosis, and treatment of occupationally and environmentally related diseases are important, but without a mechanism for reducing exposures, these efforts are incomplete. Both occupational and environmental laws have recognized the importance of reducing exposure through the substitution of toxic substances or the reduction of hazardous wastes generated.

OSHA and MSHA advocate the replacement of toxic substances in the workplace with substances that are less toxic. This is the first step to be taken, even before engineering controls or personal protective equipment are considered as means to reduce exposure.

The National Environmental Policy Act (NEPA), SARA (Superfund), TSCA, and RCRA have clearly stated policy objectives of reducing the production of hazardous waste through recycling, waste reduction, and substitution.

SE

Further Reading

The Citizens' Toxics Protection Manual. Washington, DC: National Campaign Against Toxic Hazards; 1988.

Environment Reporter. Washington, DC: Bureau of National Affairs.

Environmental Law Reporter. Washington, DC: Environmental Law Institute.

Foecke T. A new mandate for pollution prevention. *Pollution Prevention Rev.* 1990;1(1):91-97.

Geiser K. Toxics use reduction and pollution prevention. *New Solutions.* 1990;1(1):43-50.

Ginsberg R. What's In a Name? Pollution Prevention Implementation. *New Solutions.* 1990;1(2):54-65.

Huisingh D, et al. *Proven Profits from Pollution Prevention: Case Studies in Resource Conservation and Waste Reduction.* Washington, DC: Institute for Local Self-Reliance; 1986.

Moore AO. *Making Polluters Pay: A Citizens' Guide to Legal Action and Organizing.* Washington, DC: Environmental Action Foundation; 1987.

Rodgers WH. *Environmental Law: Air and Water.* Vols 1 and 2. St. Paul, Minn: West Publishing Co.; 1986.

Royston MG. *Pollution Prevention Pays.* New York, NY: Pergamon Press; 1979.

Selected Environmental Law Statutes: 1986 Edition. St. Paul, Minn: West Publishing Co.; 1989.

Statutes

National Environmental Policy Act of 1969, 42 USCA §§4321-4370a.

Clean Air Act, as amended, 42 USCA §§7401-7626.

Federal Water Pollution Control Act, as amended by the Clean Water Act of 1977, 33 USCA §§1251-1387.

Safe Drinking Water Act, 42 USCA §§300F-300J-26.

Toxic Substances Control Act, 15 USCA §§2601-2671.

Hazardous Materials Transportation Act, 49 USCA §§1801-1813.

Federal Railway Safety Act, 42 USCA §§42 et seq.

Resource Conservation and Recovery Act of 1976, 42 USCA §§6901-6992k.

Federal Insecticide, Fungicide, and Rodenticide Act, 7 USCA §§136-136y.

Federal Food, Drug, and Cosmetic Act, 21 USCA §§301-392.

Comprehensive Environmental Response, Compensation, and Liability Act of 1980, 42 USCA §§9601-9675.

Emergency Planning and Community Right-to-Know Act (Superfund Amendments and Reauthorization Act of 1986—Title III), 42 USCA §§11001-11050.

Consumer Product Safety Act, 15 USCA §§2051 et seq, 5 USCA §§5314-5315.

Index